SPACEFLIGHT DYNAMICS

McGraw-Hill Series in Aeronautical and Aerospace Engineering

John D. Anderson, Jr., University of Maryland
Consulting Editor

Anderson: *Fundamentals of Aerodynamics*
Anderson: *Hypersonic and High Temperature Gas Dynamics*
Anderson: *Introduction to Flight*
Anderson: *Modern Compressible Flow: With Historical Perspective*
D'Azzo and Houpis: *Linear Control System Analysis and Design*
Kane, Likins and Levinson: *Spacecraft Dynamics*
Nelson: *Flight Stability and Automatic Control*
Peery and Azar: *Aircraft Structures*
Rivello: *Theory and Analysis of Flight Structures*
Schlichting: *Boundary Layer Theory*
White: *Viscous Fluid Flow*
Wiesel: *Spaceflight Dynamics*

SPACEFLIGHT DYNAMICS

William E. Wiesel

Professor of Astronautical Engineering
Air Force Institute of Technology
Wright-Patterson Air Force Base, Ohio

McGraw-Hill Publishing Company

New York   St. Louis   San Francisco   Auckland   Bogotá   Caracas
Hamburg   Lisbon   London   Madrid   Mexico   Milan
Montreal   New Delhi   Oklahoma City   Paris   San Juan
São Paulo   Singapore   Sydney   Tokyo   Toronto

To Karen

TL 1050
.W54
1989

018463901

This book was set in Times Roman.
The editors were Anne T. Brown, Lyn Beamesderfer, and Scott Amerman;
the production supervisor was Louise Karam.
The cover was designed by Amy E. Becker.
R. R. Donnelley & Sons Company was printer and binder

SPACEFLIGHT DYNAMICS

34567890 DOCDOC 943210

ISBN 0-07-070106-7

Library of Congress Cataloging-in-Publication Data

Wiesel, William E.
  Spaceflight dynamics/William E. Wiesel.
    p.   cm.
  Includes bibliographies and index.
  ISBN 0-07-070106-7
  1. Astrodynamics. I. Title
  TL1050.W54 1989
  629.4'11—dc19   88-28242

ABOUT THE AUTHOR

Dr. William E. Wiesel received his B.S. degree in astronomy from the University of Massachusetts in 1970. He was awarded the M.S. (1971) and Ph.D. (1974) by Harvard University, also in astronomy. His doctoral research involved applying statistical mechanics to resonance problems in the asteroid belt.

Entering the U.S. Air Force in 1974, he was assigned to the First Aerospace Control Squadron, located in Colorado Springs, Colorado. This organization is tasked with tracking all orbiting objects. During four years in Colorado, he directed the development of considerable portions of the current Spacetrack software system.

He joined the faculty of the Air Force Institute of Technology (AFIT) in 1977 as an Assistant Professor, was promoted to Associate Professor of Astronautics in 1980, and to Professor in 1988. He is the author of over 20 technical articles on solar system astronomy, optimal orbit transfer, control of periodic systems, and helicopter control systems.

CONTENTS

PREFACE

This book grew out of two courses required of all students in astronautical engineering at the Air Force Institute of Technology. The intent of the book is admittedly ambitious: to cover, at the introductory level, the entire spectrum of dynamically related areas in astronautics. Although students may then go on to specialize in one subdiscipline or another, they will be conversant with the basics in a broad group of areas. Wherever possible, I have made an attempt to indicate how the text material interfaces with related disciplines, especially control theory, propulsion, and aerodynamics. Students will find problems requiring them to cross these disciplinary boundaries at the end of several of the chapters.

The book is arranged to cover the basics of three-dimensional particle dynamics (Chapter 1) and three-dimensional rigid-body dynamics (Chapter 4) as early as possible. Each of these fundamental chapters is immediately followed by one or two chapters on the major application of that area to astronautics. These are basic orbital theory for particle dynamics (Chapters 3 and 4) and satellite attitude dynamics for rigid-body theory (Chapter 5). This should furnish more than sufficient material for the first course of a two-semester sequence, with the remainder of the text covered in the second course.

Alternate arrangements are possible, of course. If the first course is to concentrate on particle dynamics, the chapter order would be 1, 2, 3, 10, and 11 in the first course and 4, 5, and 6 in the second course, with the remainder of the material at the discretion of the instructor.

This book is deliberately not a textbook in orbital mechanics. When the space age began, orbital mechanics was in the domain of astronomy, and textbooks written from this viewpoint have often been employed in the astronautical engineering curriculum. Orbital mechanics has enjoyed a prominence in the astronautical engineering curriculum which is not mirrored, for example, by the importance of aircraft navigation in aeronautical engineering programs. While many of the characteristics of orbital systems are dictated by the realities of orbital mechanics, and while I have the greatest appreciation for the richness of celestial mechanics as a discipline (it is my own speciality), I have chosen not to emphasize this material. Rather, the "classical" solution of the two-body problem is the only one (of many) presented, and the remaining "orbits" chapters emphasize applications. The only acknowledgment of perturbation theory is Chapter 10 on the restricted problem.

ACKNOWLEDGMENTS

McGraw-Hill and the author would like to thank the following reviewers for their very helpful comments and suggestions: John D. Anderson, Jr., University of Maryland; Russell M. Cummings, California Polytechnic State University; Robert G. Melton, Pennsylvania State University; Harold S. Morton, Jr., University of Virginia; Larry Silverberg, North Carolina State University, and Steven B. Skaar, Iowa State University.

William E. Wiesel

CHAPTER
1

PARTICLE
DYNAMICS

1.1 INTRODUCTION

The science of dynamics was the first of the so-called exact sciences. With the publication of Isaac Newton's *Principia* in 1687, a new era began. For the first time in any area of the natural world a series of postulates, the three laws of motion, were stated, and as a consequence predictions could be made for any dynamical system which were *as accurate as the available observations*. This was a revolutionary concept, and it defines, even today, the expectations we have for science. Before the *Principia,* the study of the physical world was termed *natural philosophy,* and it had far more in common with what we now think of as philosophy than does modern science. After Newton, science aspired to the exactness achieved in pure mathematics.

However, mathematicians can state any postulates they wish, and then study their consequences. This is not true in science, where the object is to describe the real world, or in engineering, where the ultimate aim is to manipulate the real world. For centuries after Newton, experiments were conducted to verify the correctness of his laws of motion and law of gravity. Observations of the motion of objects in the solar system played a leading role in this, since orbital motion is essentially free of frictional forces. And indeed, the first discrepancy between nature and newtonian predictions was found in the motion of the planet Mercury. This discrepancy was explained by general relativity.

In this chapter we will become familiar with Newton's three laws of motion and their application to particles and systems of particles. We will pay particular attention to motion in three dimensions and to the use of nonrectangular coordinates. Along the way, we will lay the groundwork for the rest of the book.

1.2 NEWTON'S LAWS

In his *Principia,* Newton announced his three laws of motion. They are:

1. A particle at rest remains at rest and a particle in motion remains in uniform, straight-line motion if the applied force is zero.
2. The force on a particle equals the mass of the particle times its inertial acceleration.
3. For every applied force, there is an equal and opposite reaction force.

These three laws, along with Newton's law of gravity, constitute a complete description of the laws of mechanics. In modern vector notation, the second law takes the form

$$\mathbf{F} = m\mathbf{a} \tag{1.1}$$

where $\mathbf{F}$ is the total force on the particle and $\mathbf{a}$ is its acceleration with respect to an inertial frame. Notice that if the applied force is zero, we have

$$\frac{d^2\mathbf{r}}{dt^2} = 0 \tag{1.2}$$

This equation immediately integrates to give Newton's first law. The first law is thus a simple consequence of the second law, and was probably included by Newton to emphasize his differences with aristotelian mechanics, where the natural state of motion of a particle was considered to be rest. The second law of motion is the normal working tool of the dynamicist. The third law is used when dealing with systems of particles. It is responsible for the working of a rocket and may be termed the *anti-warp-drive law,* since it forbids any method of propulsion which does not involve paired equal and opposite forces.

However, some care must be exercised in applying Newton's second law. The second derivative $\mathbf{a} = d^2\mathbf{r}/dt^2$ embodied in the acceleration vector *must be calculated with respect to an inertial frame of reference.* To Newton, this meant a frame of reference at absolute rest, although absolute rest is now known to be impossible to define. It is the task of the dynamicist to find a frame of reference which is "inertial enough" for the task at hand. For example, a frame of reference tied to the earth's surface is inertial enough to treat the dynamics of a baseball and would also suffice for a study of projectile motion, so long as the range is "short enough." In fact, the earth does rotate. Rotational effects in projectile motion were not included until the advent of long-range naval gunnery at the end of the nineteenth century. Over ranges of 20 mi or so, these effects amount to several hundred yards deviation from the trajectory calculated assuming the earth's surface is an inertial frame. Since this figure is larger than the size of the targets of long-range naval guns, it was necessary to admit that the earth did rotate. A "more inertial" frame of reference can be defined with its origin at the center of the earth and not participating in the rotation of the earth. So, what is to be considered inertial enough depends on the problem at hand and the accuracy requirements for the system being considered. It is not necessary to work every dynamics problem in a frame of reference with its origin at the center of our galaxy and referenced to distant quasars at the edge of the observable universe.

Once a frame of reference has been selected, the forces acting on the particle must be identified before the second law can be applied. Consider the baseball game shown in Figure 1.1. The forces acting on the baseball are gravity $\mathbf{F}_g = -mg\mathbf{k}$ downward and air drag $\mathbf{F}_d = -C_d A \rho v^2 \mathbf{v}/2v$ backward along the velocity vector, or in the $-\mathbf{v}/v$ direction. In this expression, C_d is the drag coefficient, A is the area of the baseball, and ρ is the air density. Figure 1.1, showing the forces acting on the projectile, is called a *free-body diagram* for the mass. We can now write the second law in vector form as

$$m\frac{d^2\mathbf{r}}{dt^2} = -mg\mathbf{k} - \tfrac{1}{2}C_d A \rho v \mathbf{v} \tag{1.3}$$

or, since the position vector $\mathbf{r} = x\mathbf{i} + y\mathbf{j} + z\mathbf{k}$, the velocity vector $\mathbf{v} = \dot{x}\mathbf{i} + \dot{y}\mathbf{j} + \dot{z}\mathbf{k}$, and the acceleration $\mathbf{a} = \ddot{x}\mathbf{i} + \ddot{y}\mathbf{j} + \ddot{z}\mathbf{k}$, we can rewrite (1.3) in component form as

$$m\ddot{x} = -\tfrac{1}{2}C_d A \rho \dot{x} \sqrt{\dot{x}^2 + \dot{y}^2 + \dot{z}^2} \tag{1.4}$$

$$m\ddot{y} = -\tfrac{1}{2}C_d A \rho \dot{y} \sqrt{\dot{x}^2 + \dot{y}^2 + \dot{z}^2} \tag{1.5}$$

$$m\ddot{z} = -\tfrac{1}{2}C_d A \rho \dot{z} \sqrt{\dot{x}^2 + \dot{y}^2 + \dot{z}^2} - mg \tag{1.6}$$

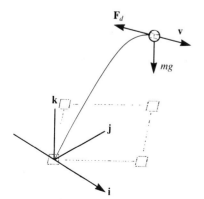

FIGURE 1.1
Projectile motion near the earth's surface.

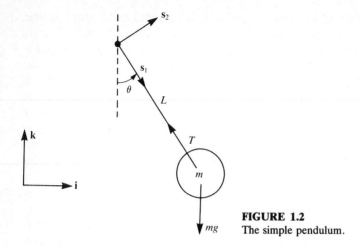

FIGURE 1.2
The simple pendulum.

Equations (1.4) to (1.6) are termed *equations of motion* for the system. They are three coupled, nonlinear, second-order ordinary differential equations. Once the equations of motion for a system have been obtained, the first task of the dynamicist is finished. The second task, of course, is obtaining the solution to the equations of motion. This is usually much harder. With patience, practice, and skill, the equations of motion can be obtained for *any* dynamical system. Solutions are available for relatively few systems and often are not available for realistic systems. Equations (1.4) to (1.6) only admit of a simple solution when $C_d = 0$, that is, when air drag is neglected.

However, our bag of tools is not yet sufficient to handle all systems. Let us try the same technique on the simple pendulum shown in Figure 1.2. Write the acceleration vector $\mathbf{a} = \ddot{x}\mathbf{i} + \ddot{z}\mathbf{k}$ as before. Gravity acts on the mass downward, and in addition there is a tension force T in the pendulum rod. Applying Newton's second law, we have

$$m\ddot{x} = -T \sin \theta \tag{1.7}$$

$$m\ddot{z} = -mg + T \cos \theta \tag{1.8}$$

Equations (1.7) and (1.8) do *not* constitute valid equations of motion for the pendulum. First, θ is not one of the variables x or z for this system. We could insert the fact that $\sin \theta = x/L$ and $\cos \theta = \pm(L^2 - x^2)^{1/2}/L$ to eliminate the extra variable θ. We are then faced with the fact that we do not know the force T. Unlike gravity or air drag, we do not have an expression for the force T in terms of the variables of the system. The tension in the pendulum rod is a *force of constraint*. Its purpose is to enforce the fact that $L = (x^2 + z^2)^{1/2}$ should be a constant. This force will take on any magnitude necessary, within reason, to ensure that this remains true. If the constraint is violated because the pendulum rod breaks, then the dynamics revert to the projectile case considered earlier. As yet we have not included the constraint in the formulation of the equations of motion.

The problem arises because the variables x and z are very poorly adapted to this system. The angle θ alone is sufficient to describe where the mass is at any moment, given the fact that L is constant. Since only one coordinate is needed to describe the position of the mass, the pendulum is termed a *one-degree-of-freedom system.* In general, when a constraint force forbids motion in a given direction, the number of degrees of freedom is reduced from its maximum of three. We need only find one second-order equation of motion for each degree of freedom.

Newton's second law, expressed in vector notation, essentially forces us into the use of rectangular coordinates. Also, we must always reference our second derivative for the acceleration vector back to an inertial frame for Newton's second law to be valid. However, the rotating frame of reference $\mathbf{s}_1$, $\mathbf{s}_2$ has several advantages for use in *breaking* $\mathbf{F} = m\mathbf{a}$ *into its components.* The tension force will only appear along the $\mathbf{s}_1$ direction, and the component of $\mathbf{F} = m\mathbf{a}$ in this direction will yield an *equation of constraint,* not an equation of motion. The $\mathbf{s}_2$ direction should isolate the θ dynamics of the particle. Since we, as dynamicists, have little interest in the magnitude of the force T (so long as it does its job), this will allow us to throw away the equation of constraint and to work only with the one equation of motion for this problem.

So, before we can go much further, we will need to study how to calculate derivatives of vectors *with respect to an inertial frame but expressed in the unit vectors of a noninertial frame.*

1.3 THE VELOCITY

Consider the space station shown in Figure 1.3. It has radius R and rotates about its center at angular speed ω rad/s. An astronaut stands on the inside rim and walks around the inner edge. Now, we can write the position vector of the astronaut as

$$\mathbf{r} = R \cos \theta \mathbf{s}_1 + R \sin \theta \mathbf{s}_2 \qquad (1.9)$$

This form is convenient, since the astronaut will remain at a distance R from the center of the station while moving about, and only the value of θ will change. Now, let us attempt to calculate the velocity of the astronaut with respect to the station. If we treat the unit vectors $\mathbf{s}_1$ and $\mathbf{s}_2$ *as if they were constants,* we obtain

$$\frac{{}^s d}{dt}\mathbf{r} = -R \sin \theta \dot{\theta}\mathbf{s}_1 + R \cos \theta \dot{\theta}\mathbf{s}_2 \qquad (1.10)$$

This is the velocity vector of the astronaut *with respect to the frame of reference attached to the station.* It is not the *inertial* velocity of the astronaut, since the station itself is rotating and the unit vectors $\mathbf{s}_1$ and $\mathbf{s}_2$ are also changing with respect to inertial space. If we return to (1.9) and recalculate the derivative,

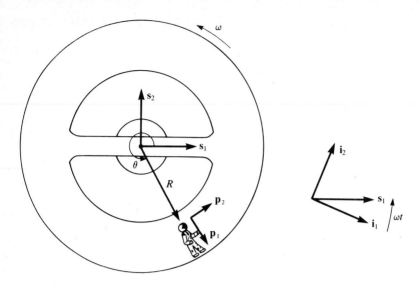

FIGURE 1.3
A strolling astronaut in a rotating space station.

including the rate of change of the unit vectors of the **s** frame, we have

$$\frac{{}^i d}{dt}\mathbf{r} = -R \sin \theta \dot{\theta} \mathbf{s}_1 + R \cos \theta \dot{\theta} \mathbf{s}_2$$

$$+ R \cos \theta \frac{{}^i d}{dt} \mathbf{s}_1 + R \sin \theta \frac{{}^i d}{dt} \mathbf{s}_2 \qquad (1.11)$$

In view of (1.10), this becomes

$$\frac{{}^i d}{dt}\mathbf{r} = \frac{{}^s d}{dt}\mathbf{r} + R \cos \theta \frac{{}^i d}{dt} \mathbf{s}_1 + R \sin \theta \frac{{}^i d}{dt} \mathbf{s}_2 \qquad (1.12)$$

So, the velocity vector seen from different frames **i** and **s** will differ by the extra terms in (1.12) due to the rate of change of the unit vectors of one frame seen from the other. Equation (1.12) justifies the **s** superscript in (1.10) and the claim that (1.10) is the **s**-*frame* derivative of **r**. The extra terms in (1.12) would vanish if this were an **s**-frame derivative, since the **s**-unit vectors are treated as constants in this case.

Now, to calculate the derivatives in (1.12), consider Figure 1.4. In a small instant of time dt the tips of the **s**-unit vectors move to new positions, and since their lengths remain unity, we have

$$d\mathbf{s}_1 = \omega \, dt \mathbf{s}_2 \qquad d\mathbf{s}_2 = -\omega \, dt \mathbf{s}_1 \qquad (1.13)$$

This will yield the necessary derivatives quite easily by division by dt and the

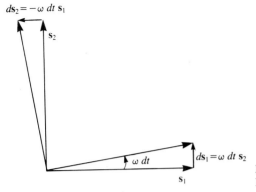

FIGURE 1.4
Derivatives of the s-frame unit vectors.

taking of limits. We obtain the result

$$\frac{{}^i d}{dt}\mathbf{r} = \frac{{}^s d}{dt}\mathbf{r} - \omega R \sin\theta \mathbf{s}_1 + \omega R \cos\theta \mathbf{s}_2 \tag{1.14}$$

This can be simplified still further if we define the *angular-velocity vector* $\boldsymbol{\omega}^{si}$ as

$$\boldsymbol{\omega}^{si} = \omega \mathbf{s}_3 \tag{1.15}$$

that is, a vector along the instantaneous rotation axis and whose direction is in a right-handed sense. The symbol $\boldsymbol{\omega}^{si}$ is to be read as "the angular velocity of the **s** frame with respect to the **i** frame," so it is necessary to have two superscripts. Then, (1.14) can be rewritten in the form

$$\frac{{}^i d}{dt}\mathbf{r} = \frac{{}^s d}{dt}\mathbf{r} + \boldsymbol{\omega}^{si} \times \mathbf{r} \tag{1.16}$$

Although we have found (1.16) for the special case shown in Figure 1.3, this result is far more general. Since it is expressed as operations on *vectors*, not their components, it is true for any other problem. For example, should the astronaut walk into the page or climb a ladder into the hub of the station, only the s-frame velocity term would change. Equation (1.16) would remain valid. Also, since (1.16) is a vector relation, *it is true independent of the choice of basis vectors in which it is expressed.* If you did not notice that there was something strange about (1.14), let us rewrite it here as

$$\frac{{}^i d}{dt}\mathbf{r} = \mathbf{s}_1(-R \sin\theta\, \dot{\theta} - \omega R \sin\theta)$$

$$+ \mathbf{s}_2(R \cos\theta\, \dot{\theta} + \omega R \cos\theta) \tag{1.17}$$

The unusual thing about (1.17) is that it is the inertial velocity of the astronaut, *expressed in* the *s-frame unit vectors.* If we write the inertial position vector directly in the inertial unit vectors as $\mathbf{r} = R \cos(\theta + \omega t)\mathbf{i}_1 + R \sin(\theta + \omega t)\mathbf{i}_2$, we can obtain the inertial velocity directly as

$$\frac{^{\mathbf{i}}d}{dt}\mathbf{r} = -R\sin(\theta + \omega t)(\dot{\theta} + \omega)\mathbf{i}_1$$

$$+ R\cos(\theta + \omega t)(\dot{\theta} + \omega)\mathbf{i}_2 \qquad (1.18)$$

We will see in the next section that (1.18) is the same as (1.17). They represent the same inertial-velocity vector but expressed in components of different reference frames.

In fact, neither the inertial $\mathbf{i}$ nor space-station $\mathbf{s}$ frames yield the simplest expression for the velocity. Return to Figure 1.3, and imagine that the astronaut carries along a personal reference frame $\mathbf{p}$ while walking around the rim. During the walk the position vector always has the simple expression $\mathbf{r} = R\mathbf{p}_1$. The angular velocity of the $\mathbf{p}$ frame will now include not only the rotation rate of the station but also the angular velocity of the astronaut, $d\theta/dt$. So, the angular-velocity vector of the astronaut's frame $\boldsymbol{\omega}^{\mathbf{pi}}$ is simply $(d\theta/dt + \omega)\mathbf{p}_3$. Applying (1.16) in this case gives

$$\frac{^{\mathbf{i}}d}{dt}\mathbf{r} = \frac{^{\mathbf{p}}d}{dt}\mathbf{r} + \boldsymbol{\omega}^{\mathbf{pi}} \times \mathbf{r}$$

$$= (\dot{\theta} + \omega)\mathbf{p}_3 \times R\mathbf{p}_1$$

$$= (\dot{\theta} + \omega)R\mathbf{p}_2 \qquad (1.19)$$

since in the $\mathbf{p}$ frame the astronaut never moves. This result says that the astronauts inertial velocity must always be in the tangential direction $\mathbf{p}_2$ so long as the astronaut's wanderings are constrained to the rim of the station. Expression (1.19) is also completely equivalent to the inertial velocity *expressed in the station frame,* (1.17), and also equivalent to the inertial velocity *expressed in the inertial frame itself,* (1.18). However, this last form is the simplest of all.

1.4   COORDINATES AND ROTATIONS

A vector is usually represented by its coordinates with respect to a given frame of reference, as in

$$\mathbf{A}^{\mathbf{i}} = A_{i1}\mathbf{i}_1 + A_{i2}\mathbf{i}_2 + A_{i3}\mathbf{i}_3 \qquad (1.20)$$

which expresses the vector $\mathbf{A}$ in terms of its components along the unit basis vectors $\mathbf{i}_1$, $\mathbf{i}_2$, and $\mathbf{i}_3$. Now, in another frame $\mathbf{s}$ the same vector will have components

$$\mathbf{A}^{\mathbf{s}} = A_{s1}\mathbf{s}_1 + A_{s2}\mathbf{s}_2 + A_{s3}\mathbf{s}_3 \qquad (1.21)$$

Now, neither of these expressions *is* the vector $\mathbf{A}$. A vector is a geometric object, possessing direction and a length. So, a vector $\mathbf{A}$ has an infinity of representations in terms of the basis vectors of different reference frames. The coordinates A_i and A_s are different, but they represent the same vector.

To calculate the components of a vector $\mathbf{A}$, we take its dot product with the basis vectors of a reference frame. By calculating the dot product of $\mathbf{A}$ with $\mathbf{i}_1$, $\mathbf{i}_2$, and $\mathbf{i}_3$ in succession using (1.20), we obtain the advertised components of $\mathbf{A}$. However, if we take exactly the same dot products, only using the expression (1.21) for the vector $\mathbf{A}$, we find

$$
\begin{aligned}
A_{i1} &= \mathbf{A} \cdot \mathbf{i}_1 \\
&= A_{s1}\mathbf{i}_1 \cdot \mathbf{s}_1 + A_{s2}\mathbf{i}_1 \cdot \mathbf{s}_2 + A_{s3}\mathbf{i}_1 \cdot \mathbf{s}_3 \\
A_{i2} &= \mathbf{A} \cdot \mathbf{i}_2 \\
&= A_{s1}\mathbf{i}_2 \cdot \mathbf{s}_1 + A_{s2}\mathbf{i}_2 \cdot \mathbf{s}_2 + A_{s3}\mathbf{i}_2 \cdot \mathbf{s}_3 \\
A_{i3} &= \mathbf{A} \cdot \mathbf{i}_3 \\
&= A_{s1}\mathbf{i}_3 \cdot \mathbf{s}_1 + A_{s2}\mathbf{i}_3 \cdot \mathbf{s}_2 + A_{s3}\mathbf{i}_3 \cdot \mathbf{s}_3
\end{aligned}
\tag{1.22}
$$

This can be written more compactly if we write the vector as a column matrix of its components. Equations (1.22) then become

$$
\begin{bmatrix} A_{i1} \\ A_{i2} \\ A_{i3} \end{bmatrix} = \begin{bmatrix} \mathbf{i}_1 \cdot \mathbf{s}_1 & \mathbf{i}_1 \cdot \mathbf{s}_2 & \mathbf{i}_1 \cdot \mathbf{s}_3 \\ \mathbf{i}_2 \cdot \mathbf{s}_1 & \mathbf{i}_2 \cdot \mathbf{s}_2 & \mathbf{i}_2 \cdot \mathbf{s}_3 \\ \mathbf{i}_3 \cdot \mathbf{s}_1 & \mathbf{i}_3 \cdot \mathbf{s}_2 & \mathbf{i}_3 \cdot \mathbf{s}_3 \end{bmatrix} \begin{bmatrix} A_{s1} \\ A_{s2} \\ A_{s3} \end{bmatrix}
\tag{1.23}
$$

Again, it should be emphasized that (1.23) does not transform the vector $\mathbf{A}$ itself. It is the rule by which the *components* of the vector $\mathbf{A}$ are transformed from the $\mathbf{s}$ frame to the $\mathbf{i}$ frame. Both sides of (1.23) represent the same vector. The matrix in between the two column vectors is termed a *rotation matrix*. Since the dot product of two unit vectors is the cosine of the angle between them, it is referred to in older works as a *direction cosine matrix*. Note also that a rotation matrix is itself simply the components of one set of basis vectors with respect to the other set of basis vectors. The columns of the matrix above are the $\mathbf{s}$-unit vectors written in their $\mathbf{i}$-frame components, while the rows of the matrix are the $\mathbf{i}$-unit vectors written in terms of their $\mathbf{s}$-frame components.

Let us abbreviate the matrix above, which rotates from the $\mathbf{s}$ frame to the $\mathbf{i}$ frame, as R^{is}, the superscripts indicating the order of the transformation. It should be clear that the reverse process can also be carried through: If (1.20) is dotted with each of the $\mathbf{s}$-unit vectors, we will obtain the reverse transformation matrix R^{is}. This will be the transpose of R^{is}. Also, since (1.23) constitutes three linear equations in three unknowns, it may be solved by multiplying by the inverse of R^{is}. Hence, the inverse of a rotation matrix is its transpose, the only type of matrix for which this is true.

Figure 1.5 shows simple rotations about the three coordinate axes of the $\mathbf{i}$ frame. It is useful to work out the *elementary rotation matrices* for these single-rotation cases. By inspection of the figure and the use of (1.23) for the

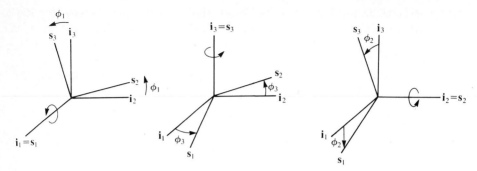

FIGURE 1.5
Simple rotations about each of the **i**-frame coordinate axes.

order of the dot products, we have

$$R_1^{is}(\phi_1) = \begin{bmatrix} 1 & 0 & 0 \\ 0 & \cos \phi_1 & -\sin \phi_1 \\ 0 & \sin \phi_1 & \cos \phi_1 \end{bmatrix} \qquad (1.24)$$

$$R_2^{is}(\phi_2) = \begin{bmatrix} \cos \phi_2 & 0 & \sin \phi_2 \\ 0 & 1 & 0 \\ -\sin \phi_2 & 0 & \cos \phi_2 \end{bmatrix} \qquad (1.25)$$

$$R_3^{is}(\phi_3) = \begin{bmatrix} \cos \phi_3 & -\sin \phi_3 & 0 \\ \sin \phi_3 & \cos \phi_3 & 0 \\ 0 & 0 & 1 \end{bmatrix} \qquad (1.26)$$

More complex rotation matrices can be constructed by combinations of these elementary rotation matrices.

As an example, the inertial velocity of our strolling astronaut in the last section appeared simplest when expressed in the unit vectors of the **p** frame. The definition of this frame shows that the rotation from the **p** frame to the **i** frame is an R_3 rotation, and the angle $\phi_3 = \theta + \omega t$. Recalling that the inertial velocity was $^i d\mathbf{r}/dt = R(\dot{\theta} + \omega)\mathbf{p}_2$, we may transform this back to its inertial-frame components as

$$\frac{^i d}{dt}\mathbf{r} = \begin{bmatrix} \cos (\theta + \omega t) & -\sin (\theta + \omega t) & 0 \\ \sin (\theta + \omega t) & \cos (\theta + \omega t) & 0 \\ 0 & 0 & 1 \end{bmatrix} \begin{bmatrix} 0 \\ R(\dot{\theta} + \omega) \\ 0 \end{bmatrix}$$

$$= \begin{bmatrix} -R \sin (\theta + \omega t)(\dot{\theta} + \omega) \\ R \cos (\theta + \omega t)(\dot{\theta} + \omega) \\ 0 \end{bmatrix} \qquad (1.27)$$

This is exactly the result we obtained for the inertial velocity expressed in components in the inertial frame, (1.18). A similar rotation will show that the result of the last section in **s**-frame components is also identical to the inertial-frame result. All that has been changed is the "language" of the unit vectors.

We are now in a position to derive the velocity-transformation law by somewhat more formal techniques. To take a derivative with respect to the **i** frame of a vector expressed in its **s**-frame components, $\mathbf{A^s}$, we can proceed as follows. Transform the vector back to its **i**-frame components

$$\mathbf{A^i} = R^{\mathbf{is}}\mathbf{A^s} \tag{1.28}$$

where the superscripts on the vectors simply indicate the coordinate frame used to calculate the coordinates. Now, the right side of (1.28) is a square matrix multiplying a column vector. We can therefore apply the rules for differentiation of matrices and column vectors to obtain

$$\frac{d}{dt}\mathbf{A^i} = \frac{d}{dt}(R^{\mathbf{is}}\mathbf{A^s}) = \dot{R}^{\mathbf{is}}\mathbf{A^s} + R^{\mathbf{is}}\dot{\mathbf{A}}^{\mathbf{s}} \tag{1.29}$$

We need not distinguish frames of reference on either side of (1.29), since this equation is a shorthand for the three component equations (1.23). The left side, however, is the derivative of the three **i**-frame components of $\mathbf{A}$, so this is the inertial-frame derivative. On the right side, the quantity $\dot{\mathbf{A}}^{\mathbf{s}}$ is the derivative of the three **s**-frame components of $\mathbf{A}$, so this is the **s** derivative. If we now transform this result back into the **s**-unit vectors, we have

$$\frac{^{\mathbf{i}}d}{dt}\mathbf{A^s} = R^{\mathbf{is}^T}\dot{\mathbf{A}}^{\mathbf{i}} = R^{\mathbf{is}^T}R^{\mathbf{is}}\frac{^{\mathbf{s}}d}{dt}\mathbf{A^s} + R^{\mathbf{is}^T}\dot{R}^{\mathbf{is}}\mathbf{A^s} \tag{1.30}$$

A rotation matrix multiplying its inverse becomes the identity matrix, of course, and disappears, leaving

$$\frac{^{\mathbf{i}}d}{dt}\mathbf{A} = \frac{^{\mathbf{s}}d}{dt}\mathbf{A} + R^{\mathbf{is}^T}\dot{R}^{\mathbf{is}}\mathbf{A^s} \tag{1.31}$$

Now, the 3-by-3 square matrix in the last term is *skew-symmetric*; that is, its negative is equal to its transpose. To see this, differentiate the statement $R^T R = 1$, where 1 is the identity matrix. This gives

$$\dot{R}^T R + R^T \dot{R} = 0 \tag{1.32}$$

Using the matrix identity $(PQ)^T = Q^T P^T$, this becomes

$$R^T \dot{R} = -\dot{R}^T R = -(R^T \dot{R})^T \tag{1.33}$$

which shows that the matrix product is indeed skew-symmetric.

Now, if the transpose of this matrix $R^T\dot{R}$ equals the negative of the matrix itself, its diagonal elements must be zero, and the off-diagonal elements must reverse sign across the diagonal. That is, it must have the form

$$R^T \dot{R} = \begin{bmatrix} 0 & -\omega_3 & \omega_2 \\ \omega_3 & 0 & -\omega_1 \\ -\omega_2 & \omega_1 & 0 \end{bmatrix} \tag{1.34}$$

The choice of the sign conventions for off-diagonal elements above will shortly become apparent. Calculating the matrix-vector product in the last term of (1.31), we have

$$R^{is^T} \dot{R}^{is} \mathbf{A}^s = \begin{bmatrix} \omega_2 A_{s3} - \omega_3 A_{s2} \\ \omega_3 A_{s1} - \omega_1 A_{s3} \\ \omega_1 A_{s2} - \omega_2 A_{s1} \end{bmatrix} = \boldsymbol{\omega}^{si} \times \mathbf{A} \tag{1.35}$$

where $\boldsymbol{\omega}^{si}$ must be the angular-velocity vector we met earlier. All three terms of (1.31) can now be rewritten in forms which are independent of coordinate-frame basis vectors. This gives the result

$$\frac{{}^i d}{dt} \mathbf{A} = \frac{{}^s d}{dt} \mathbf{A} + \boldsymbol{\omega}^{si} \times \mathbf{A} \tag{1.36}$$

This is the rule for transforming derivatives with respect to one frame into derivatives with respect to another frame. However, since there is nothing in the above calculations which require that $\mathbf{A}$ be a position vector, (1.36) is much more general. It is an operator relation, which can be written as

$$\frac{{}^i d}{dt} (\ ) = \frac{{}^s d}{dt} (\ ) + \boldsymbol{\omega}^{si} \times (\ ) \tag{1.37}$$

The open slot ( ) above is the place where we can insert any vector. Equation (1.37) then allows us to calculate an inertial derivative while working in the unit vectors of another, noninertial frame of reference.

1.5 THE ACCELERATION

We can now calculate the inertial velocity when it is expressed in the unit vectors of a noninertial frame. However, Newton's second law requires us to calculate the inertial acceleration, not the inertial velocity. Again, there may be benefits to expressing the inertial acceleration in its components in a non-inertial frame. To derive the equivalent expansion for the acceleration, begin with the expansion for the velocity and take a second derivative with respect to the inertial frame:

$$\frac{{}^i d^2}{dt^2} (\mathbf{R} + \mathbf{r}) = \frac{{}^i d^2}{dt^2} \mathbf{R} + \frac{{}^i d}{dt} \left[\frac{{}^s d}{dt} \mathbf{r} + \boldsymbol{\omega}^{si} \times \mathbf{r} \right] \tag{1.38}$$

We have included the vector $\mathbf{R}$ which locates the origin of the s frame with respect to the i frame, shown in Figure 1.6. To evaluate the derivative of the

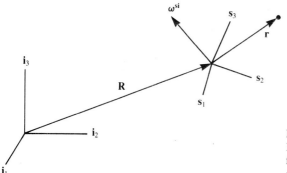

FIGURE 1.6
Inertial acceleration of a point referenced to translating, rotating coordinate axes.

bracketed term above, replace the inertial derivative with its equivalent

$$\frac{^{i}d}{dt}(\ ) = \frac{^{s}d}{dt}(\ ) + \boldsymbol{\omega}^{si} \times (\ )$$

(1.39)

in terms of **s**-frame derivatives. Combining (1.38) and (1.39), we obtain

$$\frac{^{i}d^{2}}{dt^{2}}(\mathbf{R} + \mathbf{r}) = \frac{^{i}d^{2}}{dt^{2}}\mathbf{R} + \frac{^{s}d^{2}}{dt^{2}}\mathbf{r} + 2\boldsymbol{\omega}^{si} \times \frac{^{s}d}{dt}\mathbf{r}$$

$$+ \left[\frac{^{s}d}{dt}\boldsymbol{\omega}^{si}\right] \times \mathbf{r} + \boldsymbol{\omega}^{si} \times (\boldsymbol{\omega}^{si} \times \mathbf{r})$$

(1.40)

The first term on the right-hand side above is the inertial acceleration of the origin of the **s** frame, while the second term is the acceleration of the particle seen from the **s** frame itself. The remaining three terms are corrections for the fact that the **s** frame may be rotating, and possibly rotating at a nonconstant rate. The third term is named the *Coriolis acceleration,* and the last term is known as the *centripetal* (or *center-seeking*) *acceleration.* The term involving the derivative of $\boldsymbol{\omega}^{si}$ has not received its own name.

Equation (1.40) may be applied directly to calculate the acceleration of a particle, or the process used in deriving (1.40) (two applications of the velocity law) may also be used to obtain the same result. For example, consider the case of the astronaut standing in the rotating space station considered previously. As shown in Figure 1.7, the position vector of the astronaut is $\mathbf{r} = R\mathbf{p}_{1}$ in the astronaut's personal reference frame **p**. The angular-velocity vector of his frame is $\boldsymbol{\omega}^{pi} = \omega\mathbf{p}_{3}$, and this is a constant. Also, in his own reference frame, which of course moves to follow the astronaut, he never has a velocity ($^{p}d\mathbf{r}/dt = 0$) or a nonzero acceleration ($^{p}d^{2}\mathbf{r}/dt^{2} = 0$). The expression (1.40) then becomes

$$\frac{^{i}d^{2}}{dt^{2}}\mathbf{r} = \boldsymbol{\omega}^{pi} \times (\boldsymbol{\omega}^{pi} \times \mathbf{r}) = -\omega^{2}R\mathbf{p}_{1}$$

(1.41)

This result states that the astronaut is accelerating inward, toward the axis of

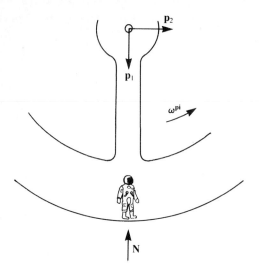

FIGURE 1.7
Acceleration calculation for a "stationary" astronaut in a rotating space station.

rotation. This is to be expected from the name of the remaining term, the centripetal acceleration. Using Newton's second law, this acceleration must be produced by the normal force $N\mathbf{p}_1$ which stops the astronaut from falling through the floor of the space station. The unbalanced force N can be calculated by writing $\mathbf{F} = m\mathbf{a}$ as

$$N\mathbf{p}_1 = -m\omega^2 R\mathbf{p}_1 \qquad (1.42)$$

If the rotation of the space station is to simulate earth-normal gravity, then it must be spun at a rate ω such that $\omega^2 R = g$.

But what about the astronaut, who has been told that it is "centrifugal force" which is being used to produce artificial gravity? Where is this force in (1.42)? The astronaut is in the interior of the space station, perhaps removed from any clues that he is in a noninertial reference frame. Subjectively, any of us would be fooled into interpreting *our personal reference frame as an inertial frame*. The astronaut thus tends to rewrite (1.42) as

$$\mathbf{F} = N\mathbf{p}_1 + m\omega^2 R\mathbf{p}_1 = m''\mathbf{a}'' = 0 \qquad (1.43)$$

Because he is at rest in what seems to be an inertial frame, the astronaut thinks that his *inertial* acceleration is zero. The real inertial $m\mathbf{a}$ term is taken over to the other side of the equation, where it becomes a "force." In fact, it is the force which holds the astronaut to the floor in the statics problem of (1.43). However, this is dangerous. If the forces on the astronaut really summed to zero, then his motion through space would be a straight line. In fact, his motion through inertial space is circular, and centrifugal force is really an acceleration term taken over onto the wrong side of $\mathbf{F} = m\mathbf{a}$. The danger, of course, is that centrifugal force can be added to the free-body diagram as if it were a real force. Then, if the acceleration is calculated correctly, the term has been included *twice*.

As another example of how dangerous this can be, you may have read that satellites orbit the earth by balancing the attraction of gravity against the repulsion of centrifugal force. By Newton's first law, then, all orbits are straight lines, since the forces sum to zero. This absurd conclusion again highlights the pitfalls of taking acceleration terms over to the force side of $\mathbf{F} = m\mathbf{a}$. In fact, the above argument is the flat-earth theory of satellite motion.

However, there are other effects that could eventually convince the astronaut that his frame **p** is not inertial. Suppose that the requirement to jog several miles around the station has been imposed to ensure that the crew stays healthy. If the astronaut jogs at constant angular rate $\dot{\theta}$ about the station, then the angular velocity of his personal frame is $\boldsymbol{\omega}^{\mathbf{pi}} = (\dot{\theta} + \omega)\mathbf{p}_3$. The inertial acceleration of the astronaut will then be

$$\mathbf{a} = -(\dot{\theta} + \omega)^2 R\mathbf{p}_1 \tag{1.44}$$

By jogging in the direction opposite the station's rotation, a lazy astronaut can considerably lessen the effective gravity, and if he can achieve a running velocity of $\dot{\theta} = -\omega$, he achieves takeoff speed. Masochists, on the other hand, will run in the same direction as the rotation and will be able to considerably increase their effective weight.

Figure 1.8 shows the simple pendulum considered in section 1.2. We now have the tools to tackle this problem effectively. The **s** frame sketched in the figure will isolate the constraint force T along one coordinate axis, as well as making it possible to introduce the angle θ as the variable for the problem. The position vector of the mass is $\mathbf{r} = L\mathbf{s}_1$. Since this is constant seen from the **s**

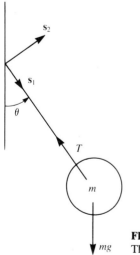

FIGURE 1.8
The simple pendulum using rotating axes.

frame, we have

$$\frac{^{s}d}{dt}\mathbf{r} = 0 \qquad \frac{^{s}d^2}{dt^2}\mathbf{r} = 0 \qquad\qquad (1.45)$$

The angular velocity of the **s** frame is $\boldsymbol{\omega}^{si} = \dot{\theta}\mathbf{s}_3$. Since this changes with time, the angular-acceleration vector is

$$\frac{^{s}d}{dt}\boldsymbol{\omega}^{si} = \ddot{\theta}\mathbf{s}_3 \qquad\qquad (1.46)$$

(Notice that the derivative of the angular-velocity vector is the same with respect to either of the frames **s** or **i**. This can be easily shown with the velocity-derivative rule.) The nonzero terms in (1.40) then give

$$\frac{^{i}d^2}{dt^2}\mathbf{r} = \dot{\boldsymbol{\omega}}^{si} \times \mathbf{r} + \boldsymbol{\omega}^{si} \times (\boldsymbol{\omega}^{si} \times \mathbf{r})$$

$$= -\dot{\theta}^2 L\mathbf{s}_1 + \ddot{\theta}L\mathbf{s}_2 \qquad\qquad (1.47)$$

The total force, also resolved into its **s**-frame components, is

$$\mathbf{F} = -T\mathbf{s}_1 + mg\cos\theta\mathbf{s}_1 - mg\sin\theta\mathbf{s}_2 \qquad\qquad (1.48)$$

Then, writing Newton's second law resolved along the **s**-unit vectors, we obtain

$\mathbf{s}_1$: $\qquad\qquad -mL\dot{\theta}^2 = -T + mg\cos\theta \qquad\qquad (1.49a)$

$\mathbf{s}_2$: $\qquad\qquad mL\ddot{\theta} = -mg\sin\theta \qquad\qquad (1.49b)$

The first of these equations is an *equation of constraint*. It gives information on the magnitude of the constraint force T, once the remainder of the problem is solved. However, to a dynamicist, it is far less interesting than the second equation, which is the equation of motion for a pendulum. It does not involve the constraint force T. Also, the use of the **s** frame has allowed the introduction of θ as the variable for the problem. This means that we are no longer tied to the use of rectangular coordinates in an inertial frame as variables for the problem.

We will spend several sections becoming familiar with the calculation of inertial accelerations using the techniques of the last several sections. The reader is very strongly encouraged to work most of the problems at the end of this chapter to develop facility with these techniques. Without correct *kinematics*, the description of motion, you cannot do dynamics.

1.6 THE SPHERICAL PENDULUM

As a three-dimensional example of an acceleration calculation, consider the spherical pendulum shown in Figure 1.9. As with the simple pendulum, the length of the pendulum rod will be the constant L, and the constraint force T

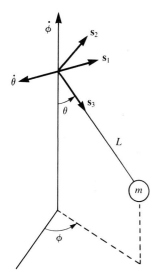

FIGURE 1.9
The spherical pendulum.

in the rod will simply enforce this requirement. The variables θ and ϕ are therefore much more natural coordinates for this problem than the rectangular coordinates x, y, and z, since θ and ϕ already include the statement of the constraint within their definition. Although the spherical pendulum moves through three dimensions, it only has two *degrees of freedom.*

We could attempt to write the inertial position vector $\mathbf{r}$ in terms of θ and ϕ as

$$\mathbf{r} = L \sin \theta \cos \phi \mathbf{i}_1 + L \sin \theta \sin \phi \mathbf{i}_2 - L \cos \phi \mathbf{i}_3 \qquad (1.50)$$

and then differentiate twice. This would indeed produce the inertial acceleration of the mass m in terms of the variables θ and ϕ, but another problem would develop. The total force on the mass can be written

$$\mathbf{F} = -mg \sin \theta \mathbf{s}_2 + (mg \cos \theta - T)\mathbf{s}_3 \qquad (1.51)$$

in the **s** frame. This isolates the constraint force T along one coordinate axis and prevents it from corrupting the two equations of motion. If we were to work in the unit vectors of the inertial frame, however, the constraint force T would appear in all three component equations of $\mathbf{F} = m\mathbf{a}$, and we would be forced to perform some algebraic manipulations to eliminate it from two equations. Working in components of the **s** frame eliminates this problem.

So, we need to express the inertial acceleration in its components in the **s** frame. In this frame the position vector is simply $\mathbf{r} = L\mathbf{s}_3$. Since this is constant, the first and second derivatives of this with respect to the **s** frame are zero:

$$\frac{^s d}{dt}\mathbf{r} = 0 \qquad \frac{^s d^2}{dt^2}\mathbf{r} = 0 \qquad (1.52)$$

The angular-velocity vector of the s frame with respect to the inertial frame is found by using the right-hand rule and by the fact that angular velocities add as vectors. Figure 1.9 shows the directions of positive angular rates for the individual coordinates θ and ϕ. Resolving these along the s-unit vectors, we have

$$\boldsymbol{\omega}^{si} = -\dot{\theta}\mathbf{s}_1 + \dot{\phi}\sin\theta\,\mathbf{s}_2 - \dot{\phi}\cos\theta\,\mathbf{s}_3 \qquad (1.53)$$

The angular-acceleration vector is the s-frame derivative of this, or

$$\frac{{}^{s}d}{dt}\boldsymbol{\omega}^{si} = -\ddot{\theta}\mathbf{s}_1 + (\ddot{\phi}\sin\theta + \dot{\phi}\dot{\theta}\cos\theta)\mathbf{s}_2$$
$$+ (-\ddot{\phi}\cos\theta + \dot{\phi}\dot{\theta}\sin\theta)\mathbf{s}_3 \qquad (1.54)$$

The inertial acceleration is given by

$$\frac{{}^{i}d^2}{dt^2}\mathbf{r} = \frac{{}^{s}d}{dt}\boldsymbol{\omega}^{si}\times\mathbf{r} + \boldsymbol{\omega}^{si}\times(\boldsymbol{\omega}^{si}\times\mathbf{r}) \qquad (1.55)$$

since all other terms in the acceleration expansion (1.40) are zero.

The first cross product is

$$\boldsymbol{\omega}^{si}\times\mathbf{r} = \begin{vmatrix} \mathbf{s}_1 & \mathbf{s}_2 & \mathbf{s}_3 \\ -\dot{\theta} & \dot{\phi}\sin\theta & -\dot{\phi}\cos\theta \\ 0 & 0 & L \end{vmatrix}$$
$$= L\dot{\phi}\sin\theta\,\mathbf{s}_1 + L\dot{\theta}\mathbf{s}_2 \qquad (1.56)$$

Notice, for future reference, that this quantity is also the inertial velocity of the mass m expressed in its s-frame components. The second cross product yields the centripetal-acceleration term

$$\boldsymbol{\omega}^{si}\times(\boldsymbol{\omega}^{si}\times\mathbf{r}) = \begin{vmatrix} \mathbf{s}_1 & \mathbf{s}_2 & \mathbf{s}_3 \\ -\dot{\theta} & \dot{\phi}\sin\theta & -\dot{\phi}\cos\theta \\ L\dot{\phi}\sin\theta & L\dot{\theta} & 0 \end{vmatrix}$$
$$= L\dot{\theta}\dot{\phi}\cos\theta\,\mathbf{s}_1 - L\dot{\phi}^2\sin\theta\cos\theta\,\mathbf{s}_2$$
$$- (L\dot{\theta}^2 + L\dot{\phi}^2\sin^2\theta)\mathbf{s}_3 \qquad (1.57)$$

Similarly, the angular-acceleration term produces a contribution to the inertial acceleration of

$$\frac{{}^{s}d}{dt}\boldsymbol{\omega}^{si}\times\mathbf{r} = (L\ddot{\phi}\sin\theta + L\dot{\phi}\dot{\theta}\cos\theta)\mathbf{s}_1 + L\ddot{\theta}\mathbf{s}_2 \qquad (1.58)$$

Combining (1.57) and (1.58), the inertial acceleration is

$$\frac{{}^{i}d^2}{dt^2}\mathbf{r} = \mathbf{s}_1(L\ddot{\phi}\sin\theta + 2L\dot{\phi}\dot{\theta}\cos\theta)$$
$$+ \mathbf{s}_2(L\ddot{\theta} - L\dot{\phi}^2\cos\theta\sin\theta)$$
$$+ \mathbf{s}_3(-L\dot{\theta}^2 - L\dot{\phi}^2\sin^2\theta) \qquad (1.59)$$

expressed in the s frame.

As an alternate method to the same end, begin with the inertial velocity, (1.56), and use the velocity expansion to take a second inertial derivative. Then

$$\frac{^id^2}{dt^2}\mathbf{r} = \left(\frac{^sd}{dt} + \boldsymbol{\omega}^{si}\times\right)\frac{^id}{dt}\mathbf{r} \tag{1.60}$$

The cross product above is

$$
\boldsymbol{\omega}^{si}\times\frac{^id}{dt}\mathbf{r} = \begin{vmatrix} \mathbf{s}_1 & \mathbf{s}_2 & \mathbf{s}_3 \\ -\dot{\theta} & \dot{\phi}\sin\theta & -\dot{\phi}\cos\theta \\ L\dot{\phi}\sin\theta & L\dot{\theta} & 0 \end{vmatrix}
$$
$$
= L\dot{\theta}\dot{\phi}\cos\theta\,\mathbf{s}_1 - L\dot{\phi}^2\sin\theta\cos\theta\,\mathbf{s}_2
$$
$$
- (L\dot{\theta}^2 + L\dot{\phi}^2\sin^2\theta)\mathbf{s}_3 \tag{1.61}
$$

Combining this result with the easily calculated

$$\frac{^sd}{dt}\mathbf{v}^i = (L\ddot{\phi}\sin\theta + L\dot{\phi}\dot{\theta}\cos\theta)\mathbf{s}_1 + L\ddot{\theta}\mathbf{s}_2 \tag{1.62}$$

we obtain the same result as in (1.59) above.

Now, with the inertial acceleration (1.59) and the force (1.51) both expressed in the **s** frame, $\mathbf{F} = m\mathbf{a}$ becomes

$$m(L\ddot{\phi}\sin\theta + 2L\dot{\phi}\dot{\theta}\cos\theta) = 0$$
$$m(L\ddot{\theta} - L\dot{\phi}^2\sin\theta\cos\theta) = -mg\sin\theta \tag{1.63}$$
$$-m(L\dot{\theta}^2 + L\dot{\phi}^2\sin^2\theta) = -T + mg\cos\theta$$

The first two of these equations are the desired equations of motion for the spherical pendulum. The tension force is isolated in the third equation, and this is the equation of constraint. It simply gives the tension force as a function of the other variables in the system. It is not necessary, however, to obtain the tension force, since the first two equations contain all the information needed to solve for the motion of the system. As one check, notice that if ϕ is constant, the first equation reduces to $0 = 0$, showing that this is a possible solution. The second equation of motion becomes

$$mL\ddot{\theta} = -mg\sin\theta \tag{1.64}$$

the equation of motion for a simple pendulum.

1.7 BASEBALL ON A SPACE COLONY

Now, consider moving the baseball game we considered earlier (Figure 1.1) to a large space colony of the type studied by Gerard O'Neill. This is a large rotating cylinder, large enough to make a baseball game possible, and rotating sufficiently rapidly that earth-normal artificial gravity obtains on the inner surface. As before, place the origin of the coordinate frame at home plate and

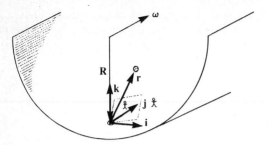

FIGURE 1.10
A baseball game on a space colony.

the y axis out through second base. The problem is shown in Figure 1.10. We wish to write equations of motion for the baseball in the coordinate frame which the ballplayers will think of as being inertial. However, in this situation it is not inertial. The reference frame rotates with the rotation of the colony, and its origin accelerates to stay a distance R from the axis of the cylinder.

Begin this problem by writing the acceleration of the baseball as

$$\mathbf{a}^i = \mathbf{a}^r + \mathbf{A} + 2\boldsymbol{\omega} \times \mathbf{v}^r + \dot{\boldsymbol{\omega}} \times \mathbf{r} + \boldsymbol{\omega} \times (\boldsymbol{\omega} \times \mathbf{r}) \tag{1.65}$$

The angular velocity of the reference frame is the rotational angular velocity of the colony, $\boldsymbol{\omega} = \omega_o \mathbf{j}$. The acceleration of the origin of the coordinate frame is the centripetal acceleration of its origin, which is $\mathbf{A} = \boldsymbol{\omega} \times (\boldsymbol{\omega} \times \mathbf{R}) = +\omega^2 R\mathbf{k}$ after calculating the cross products. The baseball is located by the position vector $\mathbf{r}$:

$$\mathbf{r} = x\mathbf{i} + y\mathbf{j} + z\mathbf{k} \tag{1.66}$$

and has velocity and acceleration with *respect to the rotating frame* of

$$\mathbf{v}^r = \dot{x}\mathbf{i} + \dot{y}\mathbf{j} + \dot{z}\mathbf{k} \tag{1.67}$$

$$\mathbf{a}^r = \ddot{x}\mathbf{i} + \ddot{y}\mathbf{j} + \ddot{z}\mathbf{k} \tag{1.68}$$

The fact that these are the rotating-frame derivatives is shown by the fact that (1.67) and (1.68) are found from (1.66) by taking derivatives treating the unit vectors $\mathbf{i}$, $\mathbf{j}$, and $\mathbf{k}$ as *constants*. We are now in a position to evaluate the cross products in (1.65). The first cross product in the centripetal-acceleration term is

$$\boldsymbol{\omega} \times \mathbf{r} = \begin{vmatrix} \mathbf{i} & \mathbf{j} & \mathbf{k} \\ 0 & \omega & 0 \\ x & y & z \end{vmatrix} = \omega z \mathbf{i} - \omega x \mathbf{k} \tag{1.69}$$

Since the position vector $\mathbf{r}$ and the rotating-frame velocity vector $\mathbf{v}^r$ differ only by dots over the latter, $\boldsymbol{\omega} \times \mathbf{v}^r = \omega \dot{z} \mathbf{i} - \omega \dot{x} \mathbf{k}$. The second cross product in the centripetal-acceleration term is

$$\boldsymbol{\omega} \times (\boldsymbol{\omega} \times \mathbf{r}) = \begin{vmatrix} \mathbf{i} & \mathbf{j} & \mathbf{k} \\ 0 & \omega & 0 \\ \omega z & 0 & -\omega x \end{vmatrix} = -\omega^2 x \mathbf{i} - \omega^2 z \mathbf{k} \tag{1.70}$$

We now have all the pieces needed to evaluate the inertial acceleration (1.65) expressed in the rotating-frame unit vectors.

The second part of setting up this problem is writing the forces acting on the baseball. It would be a mistake, however, to include "artificial gravity" as a force acting on the baseball. In fact, other than air drag, *no forces at all are acting on the baseball.* Artificial gravity is really an acceleration term taken over onto the "wrong" side of $\mathbf{F} = m\mathbf{a}$, as we will see. The fact that the baseball is within a rotating shell is irrelevant: The baseball is in free space, not connected to the space colony in any way. If we ignore air drag, then $\mathbf{F} = 0$.

We can now put together Newton's second law and write $\mathbf{F} = m\mathbf{a}$ in component form as

i: $$0 = m(\ddot{x} + 2\omega\dot{z} - \omega^2 z) \tag{1.71}$$

j: $$0 = m(\ddot{y}) \tag{1.72}$$

k: $$0 = m(\ddot{z} - 2\omega\dot{x} - \omega^2 z + \omega^2 R) \tag{1.73}$$

Now, this problem does contain the usual projectile problem as a special case. If $z \ll R$ (for example, the ball stays near the outer edge of the colony) and ωx and $\omega z \ll \omega R$ (that is, the speed of the ball is much less than the rotational speed of the station), then (1.71) to (1.73) simplify to

$$\ddot{x} = 0 \qquad \ddot{y} = 0 \qquad \ddot{z} = -\omega^2 R \tag{1.74}$$

Here, it is the acceleration of the origin of the frame, now on the "force side" of $\mathbf{F} = m\mathbf{a}$, that is the artificial-gravity term. That is, in the space station $g = \omega^2 R$. However, it is quite possible that either a fastball approaching the plate or a line drive leaving it could approach the rotational velocity of the colony itself. Then, the dynamics become more complicated, since the extra terms in (1.71) and (1.73) cannot be neglected.

The y equation of motion is simple, and its solution is simply $y = \dot{y}_o t + y_o$, a straight line. The x and z equations of motion form a coupled pair of constant-coefficient linear differential equations. In later sections we will often solve this type of system by defining a state vector, say, $\mathbf{X}^T = (x, z)$, and writing the coupled equations (1.71) and (1.73) in matrix form:

$$\begin{bmatrix} 1 & 0 \\ 0 & 1 \end{bmatrix}\begin{bmatrix} \ddot{x} \\ \ddot{z} \end{bmatrix} + \begin{bmatrix} 0 & 2\omega \\ -2\omega & 0 \end{bmatrix}\begin{bmatrix} \dot{x} \\ \dot{z} \end{bmatrix} + \begin{bmatrix} -\omega^2 & 0 \\ 0 & -\omega^2 \end{bmatrix}\begin{bmatrix} x \\ z \end{bmatrix} = \begin{bmatrix} 0 \\ -\omega^2 R \end{bmatrix} \tag{1.75}$$

However, that is not really necessary in the present case.

For, while the dynamics of the baseball may be quite complicated when seen from the rotating frame, and while the players (programmed as they are into believing their surroundings are inertial) will operate in this frame of reference, the situation is much simpler seen from the inertial frame. For in the inertial frame, since there is no force acting on the ball, *all trajectories of the baseball are straight lines at constant speed.* Consider the case of a simple vertical pop fly, shown in Figure 1.11. Seen from the rotating frame (on the left in the figure), the ball goes straight up, and then the artificial-gravity term

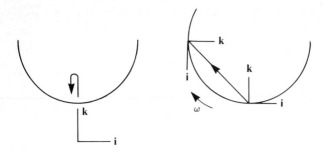

FIGURE 1.11
A pop fly as seen by an observer in the rotating frame (left) and from the inertial frame (right).

in (1.74) causes it to fall back down. Seen from the inertial frame, however, the initial speed of the ball is mostly due to the rotation of the space station. The ball proceeds across the open space of the colony, as shown on the right of Figure 1.11, and again meets the shell of the station at the place to which home plate has rotated.

1.8 ENERGY FOR ONE PARTICLE

In the early part of the eighteenth century the realization spread that while Newton's second law provided a complete basis to describe the motion of dynamical systems, certain other quantities were often exceedingly useful. The foremost among these is what we now term *energy,* but which was first referred to as *vis viva.* The latin term translates literally as "living force." The faith in its importance was vindicated when Lagrange and William R. Hamilton showed that Newton's laws could be completely replaced with an alternate description of the laws of motion based on energy principles.

However, in this section we will encounter energy in the context of newtonian mechanics. We can begin with Newton's second law

$$\mathbf{F} = m\mathbf{a} \tag{1.76}$$

and perform any legal mathematical operation to both sides. Most possibilities do not lead to any generally interesting result, but if we take the dot product of both sides of (1.76) with the inertial velocity $\mathbf{v}$, something magic happens. We have

$$\mathbf{F} \cdot \mathbf{v} = m\mathbf{a} \cdot \mathbf{v} = \frac{d}{dt}\left(\frac{1}{2}m\mathbf{v} \cdot \mathbf{v}\right) \tag{1.77}$$

The quantity $m\mathbf{a} \cdot \mathbf{v}$ is always the time rate of change of the *kinetic energy,* $T = (m\mathbf{v} \cdot \mathbf{v})/2$. Since this is a scalar, we do not need to be careful about specifying a coordinate frame for the derivative operation on the right side of (1.77). The quantity on the other side of (1.77) is termed the *power,* or the rate at which the force $\mathbf{F}$ does *work.* Note that this definition of work requires that the object have a component of velocity along the direction of the force, or no work is done. Thus on an aircraft, the drag force $\mathbf{D}$ does work, since

$\mathbf{D} \cdot \mathbf{v} \neq 0$. But the lift vector $\mathbf{L}$ is by definition perpendicular to the velocity $\mathbf{v}$, so the lift force does no work on the aircraft, in spite of the fact that it *does* hold it up.

But, this is not all there is to the concept of energy. The force $\mathbf{F}$ is very often a function of the position of the particle alone [that is, $\mathbf{F} = \mathbf{F}(\mathbf{r})$], and not a function of the time t or the particle velocity $\mathbf{v}$. If the force is a function of the position vector alone, then we can hope to find a scalar function $V(\mathbf{r})$ such that

$$\mathbf{F}(\mathbf{r}) \cdot \mathbf{v} = -\frac{dV}{dt} \tag{1.78}$$

where V, if it exists, is termed the *potential energy*. Expand both sides of (1.78) by the chain rule, assuming that we are using rectangular coordinates. We have

$$\mathbf{F} \cdot \mathbf{v} = F_x \dot{x} + F_y \dot{y} + F_z \dot{z}$$

$$= -\frac{\partial V}{\partial x} \dot{x} - \frac{\partial V}{\partial y} \dot{y} - \frac{\partial V}{\partial z} \dot{z} \tag{1.79}$$

If we equate the coefficients of the velocity components on both sides, we obtain three component equations relating the partial derivatives of V to the force components. These can be summarized by the one vector equation

$$\mathbf{F} = -\nabla V \tag{1.80}$$

where ∇ is the vector-gradient operator. In this form, (1.80) is independent of the coordinate frame used to express the force and the potential.

Not all forces possess potential-energy functions. A potential which is a function of either time or the velocity components would have a time derivative of

$$\frac{d}{dt} V(\mathbf{r}, \mathbf{v}, t) = \frac{\partial V}{\partial x} \dot{x} + \frac{\partial V}{\partial y} \dot{y} + \frac{\partial V}{\partial z} \dot{z}$$

$$+ \frac{\partial V}{\partial \dot{x}} \ddot{x} + \frac{\partial V}{\partial \dot{y}} \ddot{y} + \frac{\partial V}{\partial \dot{z}} \ddot{z} + \frac{\partial V}{\partial t} \tag{1.81}$$

The extra terms on the right side of (1.81) cannot be reconciled with the left side of (1.79), where they are missing. However, even if the force is *not* an explicit function of time or velocity, a potential function may still not exist. For a potential function to exist, we must have

$$-\nabla \times \nabla V = -\nabla \times \mathbf{F} = 0 \tag{1.82}$$

since the curl of any gradient is *always* zero. While (1.82) provides a test for the existence of a potential function, it does not help us obtain it. To obtain it, we must return to equation (1.80) and try to solve the three partial differential equations for V.

If a potential function does exist, then (1.77) and (1.78) together give

$$\frac{d}{dt}\left(\frac{1}{2}m\mathbf{v}\cdot\mathbf{v}+V\right)=0 \tag{1.83}$$

This immediately integrates to

$$\tfrac{1}{2}m\mathbf{v}\cdot\mathbf{v}+V=E \tag{1.84}$$

where E, the total energy, is, by construction, a constant. Equation (1.84) can be written as $E=T+V$, or the total energy is the sum of the kinetic and potential energies.

Without a doubt, the most important force in astronautics is gravity. Newton's law of gravity states that the force between two masses is proportional to the product of their masses, inversely proportional to the square of the distance between them, and acts along the line joining them. The force of gravity can then be written as

$$\mathbf{F}_g=-\frac{Gm_1m_2}{r^2}\frac{\mathbf{r}}{r} \tag{1.85}$$

The constant of proportionality is written G, the universal constant of gravitation, and the unit vector $-\mathbf{r}/r$ in (1.85) carries the direction information for the force as well as stating that it is a force of attraction. Now, if we form $\mathbf{F}_g\cdot\mathbf{v}$ and express the result in rectangular coordinates, we have

$$\mathbf{F}_g\cdot\mathbf{v}=-Gm_1m_2\frac{x\dot{x}+y\dot{y}+z\dot{z}}{(x^2+y^2+z^2)^{3/2}} \tag{1.86}$$

Some experimenting will show that this is the negative of the time derivative of

$$V=-Gm_1m_2(x^2+y^2+z^2)^{-1/2}=-\frac{Gm_1m_2}{r} \tag{1.87}$$

Notice that, like any potential-energy function, the newtonian gravity potential is only defined to within an arbitrary additive constant. Any other function $V'=V+C$, where C is constant, will also produce the same gravitational force (1.85) when the gradient of V' is calculated. This is also different from the familiar $V=mgh$, where h is the height above some reference level. In (1.87) write the radius $r=R_\oplus+h$, where $R_\oplus$ is the radius of the earth. Then expanding (1.87) using the binomial theorem *for small h*, we have

$$V=-\frac{Gm_1m_2}{R_\oplus+h}$$

$$\approx-\frac{Gm_1m_2}{R_\oplus}+\left[\frac{Gm_1}{R_\oplus^2}\right]m_2h+\cdots \tag{1.88}$$

where m_1 is the mass of the earth. If we drop the constant first term in (1.88), and realize that the bracketed quantity is g, the acceleration of a point mass near the surface of the earth, we recover the familiar $V = mgh$. However, this is only an approximation, valid if you stay very close to the surface of the earth. All fields of engineering do, *except* for astronautical engineering.

1.9   ANGULAR MOMENTUM

Another useful rearrangement of Newton's second law occurs if we take the cross product of $\mathbf{F} = m\mathbf{a}$ with an inertial radius vector $\mathbf{r}$. This yields

$$\mathbf{r} \times \mathbf{F} = \mathbf{r} \times m\ddot{\mathbf{r}}$$

$$= \frac{d}{dt}(m\mathbf{r} \times \dot{\mathbf{r}}) = \frac{d}{dt}\mathbf{H} \tag{1.89}$$

The first step above occurs since, using the chain rule for derivatives, $d/dt(\mathbf{r} \times \dot{\mathbf{r}}) = \mathbf{r} \times \ddot{\mathbf{r}} + \dot{\mathbf{r}} \times \dot{\mathbf{r}}$, and the last term is always zero. The *angular momentum* of the particle is then defined as

$$\mathbf{H} = m\mathbf{r} \times \dot{\mathbf{r}} \tag{1.90}$$

The other side of (1.89), $\mathbf{r} \times \mathbf{F}$, is termed the *torque,* or the *moment of the force,* about the origin. If the torque is written as $\mathbf{M} = \mathbf{r} \times \mathbf{F}$, then (1.89) becomes

$$\dot{\mathbf{H}} = \mathbf{M} \tag{1.91}$$

where, of course, the time derivative on the left must be taken with respect to an inertial frame.

We will find that (1.91) is the key to writing the equations of motion for rigid bodies. However, this concept is still quite useful in particle dynamics. In fact, an entire class of dynamics problems, *central-force problems,* are distinguished by the fact that the angular momentum $\mathbf{H}$ is constant. If the force on a particle always points toward or away from the origin of an inertial frame, then we may write $\mathbf{F} = F(x, y, z, t)\mathbf{r}/r$. Note that the magnitude of the force may depend on both position and time in a very general way, but the force always points along the radius vector. Then, the torque $\mathbf{M} = \mathbf{r} \times F\mathbf{r}/r = 0$, and (1.91) becomes

$$\frac{d}{dt}\mathbf{H} = 0 \tag{1.92}$$

This has as its immediate consequence the statement that $\mathbf{H}$ is constant throughout the motion of the particle.

Consider the plight of the space construction worker shown in Figure 1.12, who has lost contact with the space station. If her safety line is still attached to the station, she can influence her motion by putting a tension force $\mathbf{T}$ on the rope, or she can allow the tether to play out, in which case the tension is zero. In either case, since the tension force can be written as $\mathbf{T} = T\mathbf{r}/r$, her angular momentum will be constant. A first consequence of this is that her

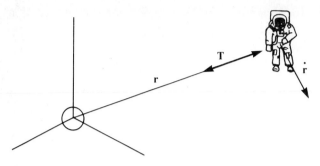

FIGURE 1.12
An astronaut tethered to a space station.

motion must always lie in the plane spanned by her initial position and velocity vectors at the moment she slipped. If she left the station with a substantial angular momentum, it might be quite dangerous for her to use the tether to pull herself in. If $\mathbf{H} = m\mathbf{r} \times \dot{\mathbf{r}}$ is constant, then as our astronaut decreases $\mathbf{r}$, her tangential velocity will increase. Both effects are general characteristics of central-force problems.

Another, less common use of the angular-momentum relationship is to eliminate a constraint force. Figure 1.13 shows the simple pendulum, with the forces acting on the pendulum bob being the tension T and the gravity mg. Since the constraint force always points back to the pivot point, this force will disappear when we calculate the torque

$$\mathbf{r} \times \mathbf{F} = \mathbf{r} \times (T\mathbf{r}/r + m\mathbf{g})$$
$$= -mgL \sin \theta \mathbf{s}_3 \tag{1.93}$$

Since the torque is not zero, the angular momentum

$$\mathbf{H} = m\mathbf{r} \times \dot{\mathbf{r}} = mL^2 \dot{\theta} \mathbf{s}_3 \tag{1.94}$$

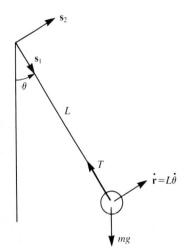

FIGURE 1.13
Angular-momentum calculation for the simple pendulum.

is not constant. However, if we apply (1.91), we obtain

$$\dot{\mathbf{H}} = mL^2\ddot{\theta}\mathbf{s}_3 = -mgL \sin\theta\mathbf{s}_3 \tag{1.95}$$

This is identical to the equation of motion of the pendulum, (1.49b), once the extra factor of L is removed from (1.95). In this case, since the motion of the system was purely rotational, and since the constraint force vanishes when the torque is calculated, angular momentum yields the complete equation of motion. While this is rarely the case, it can be a useful relationship, particularly in those cases when $\mathbf{H}$ is a constant of the motion.

1.10 SYSTEMS OF PARTICLES

Until now, we have concentrated on the motion of a single particle. When groups of particles interact, it is no longer possible to separately treat the motion of individual objects, and the particles are said to form a *system*. Obtaining the equations of motion for a system of particles is no harder than for a single particle, since each particle still obeys Newton's second law. However, we can now use Newton's third law for the first time to save half the labor of finding the forces between the masses in the system.

Consider the two-particle system shown in Figure 1.14, consisting of a shuttle orbiter and a satellite connected by a tether. The tether is to be considered a massless, inextensible string, although neither would be true in an actual tethered-satellite system. Each mass has two forces acting on it. The force of gravity acts between the center of the earth and the particle, with a magnitude proportional to the products of the masses of the particle and the earth, and points back along the direction to the center of the earth. Thus, the

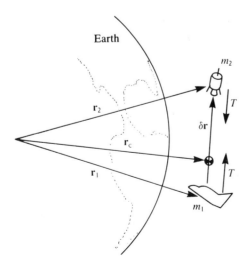

FIGURE 1.14
A satellite m_2 tethered to a shuttle orbiter m_1.

gravitational force on the orbiter m_1 is

$$\mathbf{F}_{g1} = -\frac{GM_\oplus m_1}{r_1^2}\frac{\mathbf{r}_1}{r_1} = -\frac{GM_\oplus m_1}{r_1^3}\mathbf{r}_1 \tag{1.96}$$

This is Newton's law of gravity in modern vector form. In addition, the tension in the tether will have magnitude T and will point along the direction between the orbiter and the satellite for m_1 and in the opposite direction for the satellite m_2. This is the promised application of Newton's third law. Notice that the tension T is not a constraint force, since it is not preventing anything from happening. In an actual tethered-satellite system, the tension would be the primary means of controlling the motion of the satellite.

If the orbiter and satellite are located by the position vectors $\mathbf{r}_1$ and $\mathbf{r}_2$, respectively, then their inertial accelerations are simply $\ddot{\mathbf{r}}_1$ and $\ddot{\mathbf{r}}_2$. The equations of motion can then be written as

$$m_1\ddot{\mathbf{r}}_1 = -\frac{GM_\oplus m_1}{r_1^3}\mathbf{r}_1 + T\frac{\mathbf{r}_2 - \mathbf{r}_1}{|\mathbf{r}_2 - \mathbf{r}_1|} \tag{1.97}$$

$$m_2\ddot{\mathbf{r}}_2 = -\frac{GM_\oplus m_2}{r_2^3}\mathbf{r}_2 - T\frac{\mathbf{r}_2 - \mathbf{r}_1}{|\mathbf{r}_2 - \mathbf{r}_1|} \tag{1.98}$$

where, of course, the quantity $(\mathbf{r}_2 - \mathbf{r}_1)/|\mathbf{r}_2 - \mathbf{r}_1|$ gives the direction of the tether force.

While the equations of motion (1.97) and (1.98) are correct, they may not be in the most advantageous form. In particular, the tether itself must be relatively short compared to the distances r_1 and r_2, or the physical strength of any actual material is quickly exceeded and the tether would break. So, we are actually only interested in equations of motion for the tether system for the case where $|\delta\mathbf{r}| \equiv |\mathbf{r}_2 - \mathbf{r}_1| \ll |\mathbf{r}_1|$ or $|\mathbf{r}_2|$.

Also, notice that if we add the two equations of motion, we obtain

$$m_1\ddot{\mathbf{r}}_1 + m_2\ddot{\mathbf{r}}_2 = -GM_\oplus\left(\frac{m_1\mathbf{r}_1}{r_1^3} + \frac{m_2\mathbf{r}_2}{r_2^3}\right) \tag{1.99}$$

and the tension force disappears completely. This is a general consequence of Newton's third law. If we take all the equations of motion for a system of particles and add them together, the *internal forces* cancel in pairs. If we define the position of the center of mass as the "weighted" average of the position vectors

$$\mathbf{r}_c = \frac{m_1\mathbf{r}_1 + m_2\mathbf{r}_2}{m_1 + m_2} \tag{1.100}$$

then (1.99) becomes

$$M\ddot{\mathbf{r}}_c = -GM_\oplus\left(\frac{m_1\mathbf{r}_1}{r_1^3} + \frac{m_2\mathbf{r}_2}{r_2^3}\right) \tag{1.101}$$

where $M = m_1 + m_2$ is the total mass of the system of particles. So, *the center of mass moves as if the total mass were concentrated at that point and all the* external *forces on the system acted at that point.* This is the reason that we may treat a space-shuttle orbiter, itself a system consisting of an enormous number of particles, as if it were a point mass. Another term used for the center of mass is the *center of gravity.* Near the earth's surface, where the gravitational force acts as if it were constant in magnitude and directed downward, the center of mass is also the effective point at which gravity acts. *That is* not *the case here.* When gravity changes its intensity with position, the center of mass and the center of gravity are no longer the same. In fact, the concept of a "center of gravity" becomes dangerous, and it is not often used in astronautics.

The disappearance of the tension force from (1.101) suggests that the position of the center of mass would itself be a much "nicer" coordinate for the system. We need another position vector to specify the position of both masses, and the relative separation vector $\delta\mathbf{r}$ suggests itself. In terms of $\mathbf{r}_c$ and $\delta\mathbf{r}$, the original position vectors are given by

$$\mathbf{r}_1 = \mathbf{r}_c - \frac{m_2}{M}\delta\mathbf{r} \qquad \mathbf{r}_2 = \mathbf{r}_c + \frac{m_1}{M}\delta\mathbf{r} \tag{1.102}$$

We already have an equation of motion for $\mathbf{r}_c$ and can obtain an equation of motion for $\delta\mathbf{r}$ by dividing (1.97) and (1.98) by m_1 and m_2, respectively, and subtracting. The result is

$$\delta\ddot{\mathbf{r}} = \ddot{\mathbf{r}}_2 - \ddot{\mathbf{r}}_1$$

$$= GM_\oplus\left(\frac{\mathbf{r}_2}{r_2^3} - \frac{\mathbf{r}_1}{r_1^3}\right) - T\frac{m_1 + m_2}{m_1 m_2}\frac{\delta\mathbf{r}}{|\delta\mathbf{r}|} \tag{1.103}$$

Equations (1.101) and (1.103) are an alternate set of equations of motion for the tethered-satellite problem. However, we still have not inserted the assumption that $\delta\mathbf{r}$ is small. That, however, will be left as a problem.

1.11 ENERGY FOR SYSTEMS OF PARTICLES

Begin with the equations of motion for a system of particles

$$\mathbf{F}_i = m_i\ddot{\mathbf{r}}_i \tag{1.104}$$

Then, as for a single particle, take the dot product of both sides of (1.104) with the inertial velocity $\dot{\mathbf{r}}_i$. The right side of the result will be

$$m_i\dot{\mathbf{r}}_i \cdot \ddot{\mathbf{r}}_i = \frac{d}{dt}\left(\frac{1}{2}m_i\dot{\mathbf{r}}_i \cdot \dot{\mathbf{r}}_i\right) \tag{1.105}$$

This is the derivative of the kinetic energy of particle i. But, if our system of particles really is a system, then we cannot expect to find an individual potential-energy function for each particle. The force on particle i, $\mathbf{F}_i$, will usually involve the coordinates of some or all of the other particles. It is this coupling that binds a group of particles together into an interacting system.

So, let us explore the possibility that the total energy of the *system* might be conserved. If we take each of equations (1.104), dot it with the appropriate inertial velocity $\dot{\mathbf{r}}_i$, and then add them together, we have

$$\sum_{i=1}^{N} \mathbf{F}_i \cdot \dot{\mathbf{r}}_i = \sum_{i=1}^{N} m_i \dot{\mathbf{r}}_i \cdot \ddot{\mathbf{r}}_i$$

$$= \frac{d}{dt} \sum_{i=1}^{N} \frac{1}{2} m_i \dot{\mathbf{r}}_i \cdot \dot{\mathbf{r}}_i = \frac{d}{dt} T \tag{1.106}$$

where T is the total kinetic energy of the system. The left side of (1.106) is again the rate at which the forces do work on the system. Even in this form, the work = energy equation (1.106) is a powerful result. For example, virtually all earthly dynamical systems include frictional forces among the $\mathbf{F}_i$, and these always act opposite to the direction of the velocity vectors $\dot{\mathbf{r}}_i$. This means that the dot products $\mathbf{F}_i \cdot \dot{\mathbf{r}}_i$ are negative for frictional forces, and this makes $\dot{T}$ negative if no other forces do positive work. Since the total kinetic energy T must be non negative, this says that all systems with frictional forces (and no other sources of energy input) must inevitably slow down and stop. Although in newtonian mechanics the natural state of a particle is motion (Newton's first law), it is easy to see how Aristotle could have believed that the natural state of a particle should be *rest*.

Now, let us explore the conditions under which the total energy of the system is constant. If the left side of (1.106) is also to be the perfect time derivative of the potential energy V:

$$\sum_{i=1}^{N} \mathbf{F}_i \cdot \dot{\mathbf{r}}_i = -\frac{d}{dt} V \tag{1.107}$$

then V, if it exists, must be a function *only of the position vectors* $\mathbf{r}_i$. If V involves the velocity components, then in expanding the right side of (1.107) with the chain rule we would obtain terms involving the accelerations $\ddot{\mathbf{r}}_i$, and these terms would have no counterparts on the right side of (1.107). Similarly, if the potential was a function of time, then the term $\partial V / \partial t$ on the right side of (1.107) would have no counterpart on the left side. So, the potential must be a function only of the position vectors.

This still does not mean that such a potential function V necessarily exists. Take (1.107) and expand the right side by the chain rule. Then, writing only the terms involving one particle,

$$\mathbf{F}_i \cdot \mathbf{v}_i = F_{xi}\dot{x}_i + F_{yi}\dot{y}_i + F_{zi}\dot{z}_i$$

$$= -\frac{\partial V}{\partial x_i}\dot{x}_i - \frac{\partial V}{\partial y_i}\dot{y}_i - \frac{\partial V}{\partial z_i}\dot{z}_i \tag{1.108}$$

If this is to be true for any velocity components $\dot{x}_i$, $\dot{y}_i$, $\dot{z}_i$, then the partial derivatives on the right must equal the force components on the left. This gives

$$F_{xi} = -\frac{\partial V}{\partial x_i} \qquad F_{yi} = -\frac{\partial V}{\partial y_i} \qquad F_{zi} = -\frac{\partial V}{\partial z_i} \qquad (1.109)$$

This can be put into the more modern form

$$\mathbf{F}_i = -\boldsymbol{\nabla}_i V \qquad (1.110)$$

Equation (1.110) is simply a shorthand for the three component equations (1.109), where $\boldsymbol{\nabla}_i$ is the vector gradient operator with respect to the coordinates of particle i:

$$\boldsymbol{\nabla}_i = \mathbf{i}\frac{\partial}{\partial x_i} + \mathbf{j}\frac{\partial}{\partial y_i} + \mathbf{k}\frac{\partial}{\partial z_i} \qquad (1.111)$$

in rectangular coordinates.

Notice that (1.110) states that we must be able to obtain *all* the forces in the system from *one* potential function. Forces for which this is true are called *conservative* forces. As we will see, gravity is a conservative force. Forces for which no potential function exists are called *nonconservative,* and friction is the most common example of this type of force. There is, however, an intermediate class of forces. Forces of *constraint* often do no work, since their function is to forbid motion in the direction of the constraint. Since this means that the velocity must be perpendicular to the constraint force, constraint forces often vanish when the dot product in (1.107) is calculated. But if the constraint surface moves with time, the constraint force does not vanish from (1.107). Consider the case of a ship at sea. If the ocean is calm, the buoyancy force **B** does no work, since it is perpendicular to the velocity. However, if the ship is caught by a tidal wave (the constraint surface moves with time), then the buoyancy force **B** can do quite considerable work on the ship, since the velocity of the ship will have a component along **B**. This is the distinction between working and nonworking constraint forces.

The determination of whether a force is conservative or nonconservative is made by attempting to solve the partial differential equations (1.109) for a potential function. If this is possible, then the force is, by definition, conservative. Energy is a useful quantity in its own right, but its true significance is not apparent here. In advanced mechanics, energy is the quantity used to formulate the equations of motion for the entire system. In the mechanics of Lagrange and Hamilton, it completely replaces the notion of force in obtaining the equations of motion.

1.12 ANGULAR MOMENTUM FOR SYSTEMS OF PARTICLES

Angular momentum is a useful concept for a single particle, especially in the case of central-force problems. It is also a valuable tool in the case of systems of particles. Begin with the equations of motion for a system of particles,

$$\mathbf{F}_i = m_i \mathbf{a}_i \qquad (1.112)$$

and, just as for a single particle, take the cross product of (1.112) with the position vector $\mathbf{r}_i$. However, we take the additional step of then adding the results together for all N particles. This results in the expression

$$\sum_{i=1}^{N} \mathbf{r}_i \times \mathbf{F}_i = \sum_{i=1}^{N} m_i \mathbf{r}_i \times \ddot{\mathbf{r}}_i$$

$$= \frac{d}{dt} \sum_{i=1}^{N} m_i \mathbf{r}_i \times \dot{\mathbf{r}}_i \qquad (1.113)$$

The last step above is possible since the extra term in the time derivative involves $\dot{\mathbf{r}}_i \times \dot{\mathbf{r}}_i$, which is always zero. We identify the sum in the last term as the *total angular momentum* $\mathbf{H}$ of the system, so

$$\mathbf{H} = \sum_{i=1}^{N} m_i \mathbf{r}_i \times \dot{\mathbf{r}}_i \qquad (1.114)$$

The other side of (1.113) is the total torque $\mathbf{M}$ on the system,

$$\mathbf{M} = \sum_{i=1}^{N} \mathbf{r}_i \times \mathbf{F}_i \qquad (1.115)$$

also referred to as the *moments of the forces* $\mathbf{F}_i$ *about the origin*. So, (1.113) is rewritten as

$$\dot{\mathbf{H}} = \mathbf{M} \qquad (1.116)$$

where the time derivative is, of course, taken with respect to an inertial frame.

The reason that angular momentum is a very useful concept for systems of particles is that the torque $\mathbf{M}$ never contains contributions from the *internal forces* within the system. Now, Newton's third law tells us that the force of particle i on particle j, $\mathbf{F}_{ij}$, must be equal and opposite to the force of particle j on particle i, $\mathbf{F}_{ji} = -\mathbf{F}_{ij}$. However, for almost all forces (only magnetism is an exception), internal forces in nature go one step further. The internal forces also act *along the line joining the two particles*. If an internal force $\mathbf{F}_{ij} = f_{ij}(\mathbf{r}_i - \mathbf{r}_j)$ lies along the vector joining the particles i and j, and $\mathbf{F}_{ji}$ is its negative, then these two internal forces have a contribution to the total torque given by

$$\mathbf{M}_{ij} = f_{ij}[\mathbf{r}_i \times (\mathbf{r}_i - \mathbf{r}_j) + \mathbf{r}_j \times (\mathbf{r}_j - \mathbf{r}_i)] = 0 \qquad (1.117)$$

Since internal torque contributions cancel in pairs, the total torque $\mathbf{M}$ can be calculated considering only the *external forces* on the system. If the system is isolated, and there are no external forces, then $\mathbf{M} = 0$ and the total angular momentum will be constant.

This rule is, in some sense, the companion to Newton's third law. That postulate basically states that there is nothing that an isolated system of particles can do which will cause its center of mass to accelerate. In colloquial terms, it says: "You can't lift yourself by your own bootstraps." The statement that the angular momentum of a system of particles is constant when the system is isolated is the rotational version of that same statement: "You can't spin yourself by your own bootstraps, either."

1.13   THE *N*-BODY PROBLEM

The fundamental problem in celestial mechanics is to calculate the orbits of a system of masses whose motion is influenced by their own gravitational fields. This is the famous *N-body problem.* For example, our own solar system consists of one star, nine planets, over fifty moons, tens of thousands of asteroids, and millions of comets. The description of the motion of this system is clearly important, but an exact solution to this problem has not been found in over three hundred years of study. However, it does possess all the usual conserved quantities and makes an excellent example system for that reason.

Figure 1.15 shows the N masses in a rectangular inertial reference frame. Their position vectors are $\mathbf{r}_i$, and their masses are m_i. Since we are using a simple rectangular inertial frame, the inertial accelerations are simply $m_i \ddot{\mathbf{r}}_i$. The force of gravity between mass m_i and m_j is proportional to the product of the masses and inversely proportional to the square of the distance between the two particles, $r_{ij} = |\mathbf{r}_i - \mathbf{r}_j|$. Gravity is an attractive force, so it acts along the unit vector $(\mathbf{r}_j - \mathbf{r}_i)/r_{ij}$ for particle i and along the negative of this direction for particle j.

So, writing Newton's second law for particle i, we have

$$m_i\ddot{\mathbf{r}}_i = \sum_{j\neq i}^{N} \frac{Gm_im_j}{r_{ij}^2}\frac{\mathbf{r}_j - \mathbf{r}_i}{r_{ij}} \tag{1.118}$$

where the sum extends from $j = 1$ to N, but not including particle i itself, since

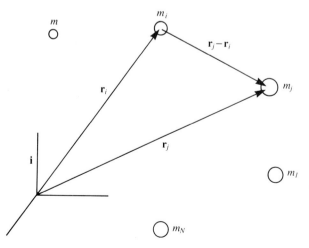

FIGURE 1.15
The *N*-body problem.

mass m_i cannot exert a force on itself. Since there is nothing special about particle i, (1.118) represents the equations of motion for the entire system when i is allowed to vary from 1 to N.

This is a dynamical system with $3N$ degrees of freedom, or a system of differential equations of order $6N$ when put into first-order form. The case of $N = 1$ is, of course, trivial, since Newton's first law covers it. The case of $N = 2$ is solvable and is the subject of the next chapter. When $N \geq 3$, no solution is known. However, the N-body problem possesses all the standard conserved quantities.

Begin by adding together all the equations of motion for this system:

$$\sum_{i=1}^{N} m_i \ddot{\mathbf{r}}_i = \sum_{i=1}^{N} \sum_{j \neq i}^{N} \frac{G m_i m_j}{r_{ij}^3} (\mathbf{r}_j - \mathbf{r}_i) \tag{1.119}$$

The imposing double sum on the right is zero, by Newton's third law. For every i and j, the force on particle i due to particle j is canceled by the equal and opposite force on particle j due to particle i. The right side of (1.119) is simply $M\ddot{\mathbf{r}}_c$, where $\mathbf{r}_c$ locates the center of mass of the N particles and M is the total mass. So (1.119) becomes

$$M\ddot{\mathbf{r}}_c = 0 \tag{1.120}$$

This integrates immediately to give

$$\mathbf{r}_c(t) = \mathbf{A}t + \mathbf{B} \tag{1.121}$$

where $\mathbf{A}$ is the constant velocity of the center of mass and $\mathbf{B}$ is its position at $t = 0$. These relations enable us to solve for the position and velocity of one mass if we know the positions and velocities of the others. Analytically, it is often convenient to place the origin of the coordinate frame at the center of mass, in which case $\mathbf{A} = \mathbf{B} = 0$. In our solar system, the center of mass is near the sun, but Jupiter and Saturn together can pull it outside the body of the sun. Observationally, the sun can be *seen,* while the center of mass cannot.

Next, let us try angular momentum. Take the cross product of each of equations (1.118) with the position vector $\mathbf{r}_i$ and sum to find

$$\sum_{i=1}^{N} m_i \mathbf{r}_i \times \ddot{\mathbf{r}}_i = \sum_{i=1}^{N} \sum_{j \neq i}^{N} \frac{G m_i m_j}{r_{ij}^3} \mathbf{r}_i \times (\mathbf{r}_j - \mathbf{r}_i) \tag{1.122}$$

Again, the expression on the right equals zero. The cross product of a vector with itself, $\mathbf{r}_i \times \mathbf{r}_i$, is always zero, while the term containing $\mathbf{r}_i \times \mathbf{r}_j$ is equal and opposite to the term containing the same vectors with their order exchanged. As the left side is $\dot{\mathbf{H}}$, the rate of change of the total angular momentum, (1.122) integrates to give the statement that

$$\mathbf{H} = \sum_{i=1}^{N} m_i \mathbf{r}_i \times \dot{\mathbf{r}}_i \tag{1.123}$$

is constant. This gives us three more constants of the motion, in addition to the six embodied in **A** and **B**.

The conservation of angular momentum depends on the fact that gravity *acts along the line joining the two objects*. This is also true for most other forces as well, including gas pressure and collisional forces. So, the solar system has conserved its angular momentum since the original cloud of gas and dust separated from a larger nebula and began to collapse. It is believed that as the cloud collapsed, the initial angular momentum would force it to assume the shape of a disk. This explains why the solar system is so very nearly flat, with the orbits of the major planets in nearly the same plane, with their spin axes nearly perpendicular to that plane, and with the orbits of their moon systems also nearly in that plane (Uranus is an exception to the last two statements).

Finally, let us check the N-body problem for energy conservation. Take each of the equations of motion (1.118) and sum their dot products with the velocity vectors $\dot{\mathbf{r}}_i$. The result is

$$\sum_{i=1}^{N} m_i \dot{\mathbf{r}}_i \cdot \ddot{\mathbf{r}}_i = \sum_{i=1}^{N} \sum_{j \neq i}^{N} \frac{Gm_i m_j}{r_{ij}^3} \dot{\mathbf{r}}_i \cdot (\mathbf{r}_j - \mathbf{r}_i) \tag{1.124}$$

The left side is always the derivative of the total kinetic energy T. To have conservation of the total energy, the right side must also be the derivative of one single function, the negative of the potential energy. Again examining the terms in the sum on the right in pairs, the vector parts are

$$\dot{\mathbf{r}}_i \cdot (\mathbf{r}_j - \mathbf{r}_i) + \dot{\mathbf{r}}_j \cdot (\mathbf{r}_i - \mathbf{r}_j) = -(\dot{\mathbf{r}}_j - \dot{\mathbf{r}}_i) \cdot (\mathbf{r}_j - \mathbf{r}_i) \tag{1.125}$$

and after some labor, the right side can be recognized as the time derivative of

$$-V = \sum_{i=1}^{N} \sum_{j>i}^{N} \frac{Gm_i m_j}{r_{ij}} \tag{1.126}$$

Equation (1.124) thus integrates to give the law of conservation of energy:

$$E = \sum_{i=1}^{N} \frac{1}{2} m_i \dot{\mathbf{r}}_i \cdot \dot{\mathbf{r}}_i - \sum_{i=1}^{N} \sum_{j>i}^{N} \frac{Gm_i m_j}{r_{ij}} \tag{1.127}$$

This completes the tally of known constants of the motion for the N-body problem. Much more could be desired. For example, while it is suspected that the solar system is stable, this has never been proven to be true. Nor has a complete solution ever been found. Approximate techniques were developed to handle the nearly circular orbits found throughout most of the solar system, and the digital computer has given us the ability to follow the evolution of a given system until time, patience, and computer budgets are exhausted. But none of these are a substitute for an actual solution. In this the N-body problem is a prototype for most of dynamics: The system equations of motion can be posed, but most physically interesting systems elude solution.

1.14 REFERENCES AND FURTHER READING

Unfortunately, most introductory dynamics texts simplify the material by restricting themselves to two dimensions. A notable exception to this trend is the book by Likins. (As a simple test, if you remember angular momentum as the vector which always comes out of the page, you probably had a two-dimensional dynamics course.) While this produces an easier text, too many of the problems we need to discuss in this book are three-dimensional.

Also, in this book we rely solely on the techniques of traditional newtonian mechanics. The advanced techniques of Lagrange and Hamilton are necessary to treat complex systems, such as coupled rigid bodies. The book by Meirovitch is an excellent introduction to these methods from the perspective of the astronautical engineer, while the book by Goldstein is the classic introduction to the advanced techniques from the point of view of the physicist.

Likins, P. W., *Elements of Engineering Mechanics*, McGraw-Hill, New York, 1973.
Meirovitch, L., *Methods of Analytical Dynamics*, McGraw-Hill, New York, 1970.
Goldstein, H., *Classical Mechanics*, Addison-Wesley, Reading, Mass., 1950.

1.15 PROBLEMS

1. Show that the two expressions (1.17) and (1.18) for the velocity of the astronaut in the space station are equivalent by constructing the rotation matrix R^{si} and transforming one expression into the other.

2. A reentry vehicle is flying through the atmosphere at velocity $\mathbf{V} = V\mathbf{b}_1$ at flight-path angle γ as shown in Figure 1.16. The vehicle keeps its velocity aligned with $\mathbf{b}_1$ as γ changes. Use the velocity rule to show that the inertial acceleration of the center of mass can be written as

$$\mathbf{A} = \dot{V}\mathbf{b}_1 + V\dot{\gamma}\mathbf{b}_2$$

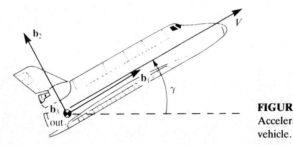

FIGURE 1.16
Acceleration calculation for a reentry vehicle.

3. A rotation matrix R^{si} can be thought of as a column matrix of the three unit vectors of the **s** frame written in their **i**-frame components:

$$R^{si} = [\mathbf{s}_1 \mathbin{\vdots} \mathbf{s}_2 \mathbin{\vdots} \mathbf{s}_3]$$

Use the velocity rule to show that a rotation matrix obeys the differential equation

$$\frac{{}^{\text{i}}d}{dt} R^{\text{si}} = \boldsymbol{\omega}^{\text{si}} \times R^{\text{si}}$$

where the cross product is interpreted as the cross product of $\boldsymbol{\omega}^{\text{si}}$ with each column of R^{si}.

4. The location of a vehicle is described by its longitude λ from Greenwich, its latitude δ, and its altitude H above the surface of the earth as shown in Figure 1.17. Of course, the earth itself rotates at angular velocity $\omega_{\oplus}$ about its polar axis. A coordinate frame $\mathbf{s}$ has its origin at the center of the earth, with the $\mathbf{s}_1$ vector in the plane of the equator, the $\mathbf{s}_2$ vector pointing at the vehicle, and $\mathbf{s}_3$ completing the right-handed set. Show that the angular velocity of the frame $\mathbf{s}$ is

$$\boldsymbol{\omega}^{\text{si}} = \dot{\delta}\mathbf{s}_1 + (\omega_{\oplus} + \dot{\lambda})\sin \delta\mathbf{s}_2 + (\omega_{\oplus} + \dot{\lambda}) \cos \delta\mathbf{s}_3$$

and show that the inertial velocity of the vehicle is

$$\mathbf{v} = -(R + H)(\omega_{\oplus} + \dot{\lambda}) \cos \delta\mathbf{s}_1 + \dot{H}\mathbf{s}_2 + (R + H)\dot{\delta}\mathbf{s}_3$$

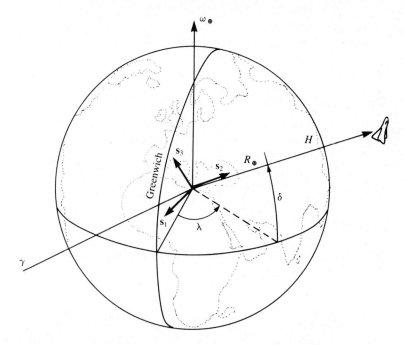

FIGURE 1.17
Flight vehicle located by altitude H, longitude λ, and latitude δ with respect to the rotating earth.

5. A future radio telescope might be constructed with a very large reflector and a separate spacecraft located at the focal point, a fixed distance R (perhaps several kilometers) from the main reflector. As the reflector changes its orientation, the two angles ϕ and θ shown in Figure 1.18 change with time. Write the angular

velocity of the **s** frame shown in the figure, and find an expression for the velocity the spacecraft needs to remain at the focal point.

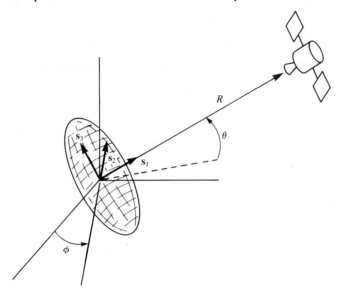

FIGURE 1.18
Receiver spacecraft stationkeeping at the focal point of a very large dish antenna.

6. An unfortunate astronaut, mass m, has lost his footing in the "low gravity" at the hub of the space station shown in Figure 1.19 and is about to "fall" down one of the spokes of the wheel. The station rotates at angular speed ω. Write equations of motion for the astronaut using coordinates x and y. On which side of the spoke should the ladder be located?

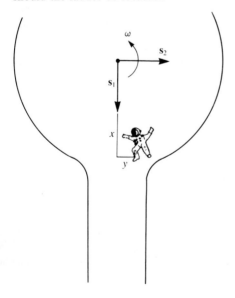

FIGURE 1.19
A "falling" astronaut in a rotating space station.

7. Rederive equations of motion for the baseball in the space colony considered in section 1.7. Use the cylindrical coordinates r, ϕ, z shown in Figure 1.20.

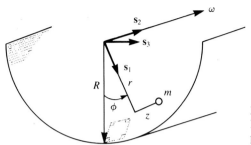

FIGURE 1.20
Polar coordinates for the baseball problem.

8. For the radio telescope considered in problem 5, find an expression for the acceleration of the receiver spacecraft in terms of the angles ϕ and θ and their first two derivatives. Express your answer in the unit vectors of the **s** frame.

9. For the vehicle considered in problem 4, show that the inertial acceleration of the spacecraft is

$$\mathbf{a} = \mathbf{s}_1\{(R + H)[-\ddot{\lambda}\cos\delta + 2(\omega_\oplus + \dot{\lambda})\dot{\delta}\sin\delta] - 2\dot{H}(\omega_\oplus + \dot{\lambda})\cos\delta\}$$
$$+ \mathbf{s}_2\{\ddot{H} - (R + H)[(\omega_\oplus + \dot{\lambda})^2\cos^2\delta + \dot{\delta}^2]\}$$
$$+ \mathbf{s}_3\{(R + H)[\ddot{\delta} + (\omega_\oplus + \dot{\lambda})^2\cos\delta\sin\delta] + 2\dot{H}\dot{\delta}\}$$

If the vehicle is subjected only to newtonian inverse-square-law gravity, write its equations of motion.

10. Show that the spherical pendulum of section 1.6 conserves energy, and write an expression for E. Also, show that this system does not conserve angular momentum but that the vertical component of **H** is constant.

11. Using the results of problems 4 and 9, show that the vehicle conserves total energy. Write an expression for E.

12. A system of particles is shown in Figure 1.21, along with an arbitrary point p. Define the angular momentum about point p as

$$\mathbf{H}^P = \sum_{i=1}^{N} m_i\mathbf{r}_i \times \dot{\mathbf{r}}_i$$

Calculate $\dot{\mathbf{H}}^P$, and substitute for $\mathbf{r}_i = \mathbf{R}_i - \mathbf{R}$, where $\mathbf{R}_i$ is the inertial position vector of particle i and **R** locates point p. Using Newton's second law, show that this becomes

$$\dot{\mathbf{H}}^P = \mathbf{M}^P + \ddot{\mathbf{R}} \times M\mathbf{r}_c$$

where the torque about point p is

$$\mathbf{M}^P = \sum_{i=1}^{N} \mathbf{r}_i \times \mathbf{F}_i$$

M is the total mass of particles, and $\mathbf{r}_c$ locates the center of mass of the system relative to point p. Argue that we can use the relationship

$$\mathbf{M}^P = \dot{\mathbf{H}}^P$$

in the cases where either (*a*) *p* is an inertial origin or (*b*) *p* is the system center of mass.

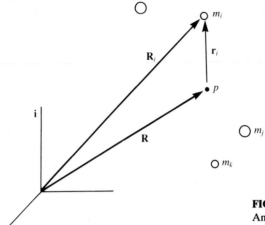

FIGURE 1.21
Angular momentum about an arbitrary point *p*.

13. For the tethered satellite considered in section 1.10, finish incorporating the assumption that the tether length δr is much less than the radius vectors $\mathbf{r}_1$ or $\mathbf{r}_2$. Use the binomial theorem to expand (1.101) to first-order in $\delta \mathbf{r}$, and show it becomes

$$M\ddot{\mathbf{r}}_c = -\frac{GM_\oplus M}{r_c^3}\mathbf{r}_c$$

Similarly, show that (1.103) for the relative position becomes

$$\delta\ddot{\mathbf{r}} = GM_\oplus\left(\frac{\delta\mathbf{r}}{\mathbf{r}_c^3} - \frac{3\mathbf{r}_c \cdot \delta\mathbf{r}}{r_c^5}\mathbf{r}_c\right) - T\frac{m_1 + m_2}{m_1 m_2}\frac{\delta\mathbf{r}}{|\delta\mathbf{r}|}$$

We will shortly see in Chapter 2 that the first equation states that the center of mass of the tether system follows an unperturbed two-body orbit. The bracketed term in the relative equation of motion above is called the *gravity-gradient term*.

CHAPTER

2

THE
TWO-BODY
PROBLEM

2.1 INTRODUCTION

The problem of correctly describing orbital motion has confronted the human race for at least 4 millennia. The planets visible to the naked eye excited wonder among early peoples (who seem to have looked at the night sky far more often than their modern descendants), and the association of these objects with gods in early religions increased the importance of the problem of predicting their motion. Claudius Ptolemy, in his second century A.D. book the *Almagest* (arabic for "the greatest"), cites observations made by Babylonian astronomers in the second millennium B.C. The system expounded by Ptolemy to explain these motions was centered on the supposedly immovable earth and was founded on the doctrine of the perfection of the circle. Under these assumptions, the basic system was capable of generating moderately accurate predictions (a few degrees) of planetary positions over intervals of several thousand years.

The ptolemaic scheme is shown in Figure 2.1. A large circle, the *deferent,* carries a smaller circle, the *epicycle,* with the planet placed on the epicycle. The earth is not located at the exact center of the deferent but is displaced slightly. Also, the angular rate of the center of the epicycle was not uniform as seen from the center of the circle. Rather, the speed was uniform as seen from the *equant,* a point displaced an equal amount from the earth but on the

41

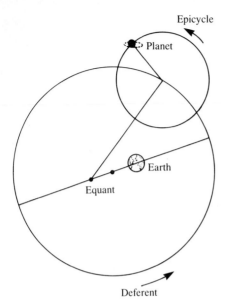

Epicycle

Planet

Equant

Earth

Deferent

FIGURE 2.1
The ptolemaic world scheme.

opposite side. These two features, a displaced earth with uniform motion seen from the equant, combine to produce a close approximation of the behavior of the actual elliptical orbit.

This system was considerably rearranged by Nicolaus Copernicus in 1543. Copernicus was led to place the sun at the center of his system by the realization that *all the epicycles in the ptolemaic system had a period of one year.* If the earth is nailed down, then our own orbital motion must be transferred to all other objects in the solar system to obtain a reasonable description of reality. This was not noticed earlier, since Ptolemy did not cite the *inertial* angular rate of the epicycles but rather their rate with respect to the line to the center of the deferent. Copernicus' contribution is thus basically a rearrangement of the ptolemaic scheme. He believed in the holiness of pure circular motion as devoutly as Ptolemy, and he even eliminated the ptolemaic equant in favor of a *small* epicycle to produce the same effect (Figure 2.2). Thus, Copernicus' system was the ptolemaic celestial machinery rearranged. It could definitely be argued that the rearrangement was less cumbersome, since the motion of each object was nearly described with a single circle. However, discarding the immovable earth seemed a great drawback of the theory to most of his contemporaries.

The contributions of Galileo and Kepler are the next major event in what has been called the *copernican revolution.* The Italian astronomer Galileo was the first to use the newly discovered telescope in astronomy. He discovered the craters of the moon (disproving the perfection of the heavenly spheres), the four large moons of Jupiter (disproving that "a center of motion cannot itself be in motion"), the phases of Venus and Mercury (showing that they circled

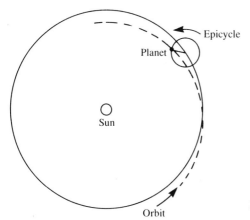

FIGURE 2.2
The copernican system.

the sun), and sunspots (proving that the sun should not be observed through a telescope without eye protection). All these phenomena can be accounted for in the ptolemaic system by appropriate rearrangements or modifications, but the psychological impact of these discoveries was immense. Galileo announced them as having *proved* the copernican system. Furthermore, he wrote in the language of his country, Italian, and not in the scholary language, Latin. This gave his discoveries a much wider audience at a time, during the High Renaissance, when the climate was ripe for the reception of new ideas.

Much has been made of Galileo's difficulties with the Catholic church. This event is often cited as a reaction of blind superstition to scientific truth. It is true that in 1632 Galileo was brought before the office of the Inquisition, forced to recant his belief in the copernican system, and sentenced to house arrest for the remaining 10 years of his life. However, it seems much more likely that his problems arose from the reaction of a powerful bureaucracy to Galileo's printed ridicule. The direct cause of contention, his *Dialogue on the Two Great World Systems,* was printed with the permission of the church and takes the form of a debate among three philosophers. These are Salvati, Galileo's mouthpiece; Sagredo, always quick to see the truth of Salvati's arguments; and Simplicito, an unreconstructed aristotelian. The disclaimers required by the censors were put into the mouth of Simplicito. This is the only argument Simplicito wins in the entire volume, and it did not pass unnoticed that while *simplicito* in Latin is the name of one of the most famous commentators on Aristotle, it means "idiot" in Italian. The book was written in Italian.

Johannes Kepler was a contemporary of Galileo. He lived a tempestuous life in Poland and Germany during the time of the religious wars between the Protestants and Catholics. Entering the service of Tycho Brahe, the greatest observational astronomer of the time, Kepler hoped for access to Tycho's data in order to work out the details of the copernican system, and also to prove certain ideas of his own. Kepler came into the possession of the observation

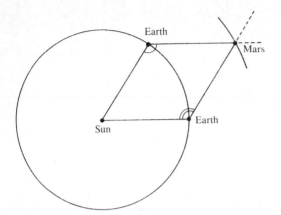

FIGURE 2.3
Kepler's method of triangulation.

books after Tycho's sudden death, and for the rest of his life fought a running battle in the courts with Tycho's heirs in order to keep them.

Using Tycho's data, Kepler set out to systematically map the solar system. Figure 2.3 shows his method of surveying the orbit of Mars. In the copernican system, Mars returns to the same point in its orbit every 687 days. Since this is not quite 2 years, the earth will not be in the same place in its orbit after one orbit of Mars. Taking the angle between the sun and Mars from Tycho's data, triangulation yields the absolute position of Mars. After many years of study and calculation, with many false starts and mistakes, Kepler was able to announce his three empirical laws of planetary motion:

1. The planets orbit the sun in elliptical orbits, with the sun at one focus of the ellipse.
2. The radius vector sweeps out equal areas in equal times.
3. The period of the orbit squared is proportional to the semimajor axis cubed.

These laws constitute a complete solution to the two-body problem of orbital motion. They enabled Kepler to publish predictions of planetary positions accurate to within 4 arcminutes, an improvement of over two orders of magnitude on the predictions of either the ptolemaic or copernican theories. Kepler finally made the break with the sacred perfect circle, but the real physical meaning of his three laws remained for another to discover.

Isaac Newton was a graduate student at Cambridge University when the plague struck London in 1665. Fleeing to his ancestral home of Woolsthorpe, Newton arguably spent the most productive 18 months in the history of humanity. During this time, he invented the differential and integral calculus and discovered the law of gravity and his three laws of motion. Using these, he was able to *derive* Kepler's three laws of planetary motion. Since the prevailing explanation of planetary motion was Descarte's vortex theory, Newton founded the discipline of fluid mechanics, showing that Descarte's explanation would not work. Turning his attention to light, Newton was able to formulate a particle theory which explained, for the first time, Snell's law of diffraction and the working of a prism. Realizing that a simple-lens telescope also dispersed light, producing colored halos around images, he invented and fabricated the first reflecting telescope.

Returning to his studies when the universities reopened, Newton received his doctorate in 1668. His doctoral dissertation *contained none of the discoveries cited above.* An attempt to publish his theory of color several years later received poor reviews, and for the next 19 years Newton sat on the greatest series of discoveries ever made by any single scientist. It remained for Edmund Halley, learning of Newton's discoveries during a visit, to squeeze the *Principia Mathematica Philosophia Naturalis* out of Newton, paying for its publication out of his own pocket.

Newton was elected president of the Royal Society, and was awarded with a knighthood, becoming Sir Isaac Newton. Along with the knighthood came an appointment as Master of the Royal Mint. This position was intended as a monetary reward and came complete with an underling to do all the work. Everyone was surprised, then, when Newton quit his job as a professor, moved into the Royal Mint, and took over direction of daily operation of that institution. (The ridges on the edge of coins were invented by Newton to stop people from cutting the coins down to obtain the precious metal.)

A whole generation of scientists and mathematicians grew up who built on Newton's discoveries but had never seen the great man at work. It apparently became enough of a joke that Leonard Euler sent Newton a problem that he (Euler) could not solve. According to the legend, Newton put it aside until his duties at the mint were finished, went home that night, and founded the calculus of variations. This field is the mathematical underpinnings of the modern discipline of optimal control. Euler was suitably impressed, and apparently no one asked Newton any more questions for the rest of his life.

In this chapter we will pose and solve the problem of two bodies in orbital mechanics. Orbital mechanics is itself a discipline with hundreds of years of history, and it is not possible in several chapters to include all the useful formulations and solved problems in this field. Rather, we will concentrate on the "classical" solution to the problem, and save applications in the field of astronautics for later chapters. The student wishing to delve further into orbital mechanics should consult the bibliography.

2.2 THE TWO-BODY PROBLEM

The simplest gravitational problem involves only two bodies: two point masses orbiting under their mutual gravitational attraction. Also, after 300 years, this is the only gravitational problem for which a closed-form solution has been found. We begin with the two masses and an inertial reference frame, shown in

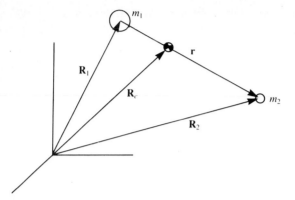

FIGURE 2.4
The two-body problem in an inertial frame.

Figure 2.4. The positions of the two masses are specified by position vectors $\mathbf{R}_1$ and $\mathbf{R}_2$, so the inertial accelerations are simply $\ddot{\mathbf{R}}_1$ and $\ddot{\mathbf{R}}_2$. Equating the gravitational force on each mass to $m_i\ddot{\mathbf{R}}_i$, we have the equations of motion for this system:

$$m_1\ddot{\mathbf{R}}_1 = -\frac{Gm_1m_2}{|\mathbf{R}_1 - \mathbf{R}_2|^3}(\mathbf{R}_1 - \mathbf{R}_2) \tag{2.1}$$

$$m_2\ddot{\mathbf{R}}_2 = -\frac{Gm_1m_2}{|\mathbf{R}_1 - \mathbf{R}_2|^3}(\mathbf{R}_2 - \mathbf{R}_1) \tag{2.2}$$

These equations, if put into component form, represent six second-order, nonlinear, coupled ordinary differential equations.

However, one-half of their solution is very easily found. If we add the two equations of motion, there results

$$m_1\ddot{\mathbf{R}}_1 + m_2\ddot{\mathbf{R}}_2 = 0 \tag{2.3}$$

Introduce the vector to the center of mass of the two objects, $\mathbf{R}_c$, and

$$\mathbf{R}_c = \frac{m_1\mathbf{R}_1 + m_2\mathbf{R}_2}{m_1 + m_2} \tag{2.4}$$

Then, (2.3) becomes

$$\ddot{\mathbf{R}}_c = 0 \tag{2.5}$$

This may be integrated twice to give

$$\mathbf{R}_c = \mathbf{R}_{c0} + \mathbf{V}_{c0}t \tag{2.6}$$

where the velocity vector of the center of mass, $\mathbf{V}_{c0}$, and its position at $t = 0$, $\mathbf{R}_{c0}$, are constant. This introduces six arbitrary constants into the solution, which is one-half of the twelve needed. So, at this point we are already in possession of half the solution to the equations of motion (2.1) and (2.2). The center of mass of the system moves in a straight line at constant velocity.

The remainder of the solution must concern the movement of mass 2 relative to mass 1. If we subtract (2.1) times m_2 from (2.2) times m_1 and introduce the position vector of mass 2 relative to mass 1, $\mathbf{r}$,

$$\mathbf{r} = \mathbf{R}_2 - \mathbf{R}_1 \qquad (2.7)$$

then we find

$$m_1 m_2 \ddot{\mathbf{r}} = -\frac{Gm_1 m_2(m_1 + m_2)}{r^3} \mathbf{r} \qquad (2.8)$$

Or, canceling the common factor $m_1 m_2$ and writing $\mu = G(m_1 + m_2)$, we have

$$\ddot{\mathbf{r}} = -\frac{\mu \mathbf{r}}{r^3} \qquad (2.9)$$

The gravitational parameter $\mu = G(m_1 + m_2)$ will, in the case of artificial satellites, be nearly equal to Gm_1. Even in the case of the solar system, it is not common to have two nearly equal masses in orbit about each other. In fact, the largest correction in the solar system is the case of earth's own moon. This parameter μ is used in place of the separate quantities G and m, since μ can be determined to far higher accuracy than either the gravitational constant G or the mass of the earth m. The problem lies in the fact that G can only be measured in exceedingly delicate laboratory experiments with known masses, while the product μ can be determined by accurate tracking of earth satellites.

Equation (2.9) specifies the motion of body 2 relative to body 1. It is a three-degree-of-freedom system, so it represents the other half of the two-body problem. Its solution will occupy us for the rest of this chapter.

2.3 ENERGY AND ANGULAR MOMENTUM

The relative equation of motion for the two-body problem,

$$\ddot{\mathbf{r}} = -\frac{\mu \mathbf{r}}{r^3} \qquad (2.10)$$

conserves both energy and angular momentum. To find the law of energy conservation, take the dot product of both sides of (2.10) with the relative velocity $\dot{\mathbf{r}}$:

$$\dot{\mathbf{r}} \cdot \ddot{\mathbf{r}} = -\frac{\mu}{r^3} \mathbf{r} \cdot \dot{\mathbf{r}} \qquad (2.11)$$

Now, the left side of this expression is the time derivative of $\frac{1}{2}\dot{\mathbf{r}} \cdot \dot{\mathbf{r}}$, the kinetic energy per unit mass. On the right side we can make further progress if we note that

$$\mathbf{r} \cdot \dot{\mathbf{r}} = r\dot{r} \qquad (2.12)$$

This bland statement conceals some sleight of hand. The dot product on the left is equal to the magnitude of $\mathbf{r}$ times the projection of the velocity vector $\dot{\mathbf{r}}$ along the position vector $\mathbf{r}$. Now, $\dot{\mathbf{r}}$ is the relative-velocity vector, and its projection on the position vector is the radial-velocity component. The radial velocity is, of course, the time derivative of the scalar radius vector r, so $\mathbf{r} \cdot \dot{\mathbf{r}} = r\dot{r}$.

With this substitution, (2.11) becomes

$$\frac{d}{dt}\frac{1}{2}v^2 = -\frac{\mu}{r^2}\dot{r}$$

$$= -\frac{d}{dt}\left(-\frac{\mu}{r}\right) \tag{2.13}$$

So, the right side of (2.11) is also the perfect time differential of something, the potential energy per unit mass:

$$V = -\frac{\mu}{r} \tag{2.14}$$

Integrating both sides of (2.13), we find the law of conservation of energy:

$$E = \tfrac{1}{2}v^2 - \frac{\mu}{r} \tag{2.15}$$

The total energy $E = T + V$ is constant, since it appears in (2.15) as the arbitrary constant when (2.13) is integrated. This is actually the total energy *per unit mass of satellite,* not the total mechanical energy of the two-body system. Working with the *specific* mechanical energy is actually an advantage, since the mass of an earth satellite may not be known (except that we are sure that it is much less than the mass of the earth). As a first application of the energy law, solve (2.15) for the speed v:

$$v = \left(2E + \frac{2\mu}{r}\right)^{1/2} \tag{2.16}$$

If the total energy E is negative, there is a maximum radius which will still permit a real-valued solution of (2.16) for the speed. If the satellite reaches this distance, it must stop and head back toward the primary body. On the other hand, if the total energy is positive, (2.16) permits real solutions for the speed at any radius r. The satellite is first able to get to infinity when the total energy is exactly zero. Then, (2.16) gives the *escape speed* as

$$v_{\text{esc}} = \sqrt{\frac{2\mu}{r}} \tag{2.17}$$

We have not yet shown that a satellite traveling at or above this speed must escape to infinity. This does turn out to be the case, however.

The equation of motion (2.10) also conserves angular momentum. Take its cross product with the radius vector **r** to find

$$\mathbf{r} \times \ddot{\mathbf{r}} = -\frac{\mu}{r^3} \mathbf{r} \times \mathbf{r} = 0 \tag{2.18}$$

since a vector crossed with itself is always zero. The two-body problem is an example of the class of dynamical systems called *central-force problems,* for which this will always happen. The left side of (2.18) is the time derivative of $\mathbf{r} \times \dot{\mathbf{r}}$, as application of the chain rule for derivatives will show. Then, (2.18) becomes the statement that $d/dt(\mathbf{r} \times \dot{\mathbf{r}}) = 0$, which integrates to become

$$\mathbf{r} \times \dot{\mathbf{r}} = \mathbf{H} \tag{2.19}$$

which again is a constant, since the vector **H** appears as an integration constant. As with the energy E, **H** is really the angular momentum per unit mass of satellite.

There are two immediate applications of (2.19). First, **H** is a vector which must be perpendicular to both the position vector **r** and the velocity vector $\dot{\mathbf{r}}$. Also, **H** is a constant vector. Thus, the *orbit lies within a plane,* the normal to which is the angular-momentum vector. This uses the direction information within the statement that **H** is constant. To interpret the information that the magnitude of **H** is constant, consider Figure 2.5. In a short interval of time dt the radius vector covers the area of the narrow triangle, which has area dA. Now, this triangle has base r and altitude $v_p \, dt$, where v_p is the velocity component perpendicular to **r**. So, the area of the triangle is

$$dA = \tfrac{1}{2} r v_p \, dt \tag{2.20}$$

Now, the magnitude of the angular-momentum vector is $|\mathbf{H}| = r v_p$. Comparing this to (2.20), we see that

$$H = 2\frac{dA}{dt} \tag{2.21}$$

Since H is constant, we have just proved Kepler's second law: The radius vector sweeps out equal areas in equal times.

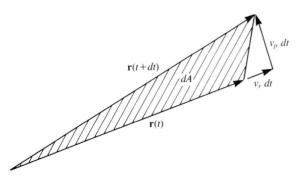

FIGURE 2.5
Kepler's law of areas.

Now, the energy law supplies us with an expression for the speed of the satellite as a function of radius. Also, we know that the orbit is planar, with the orbit plane fixed in inertial space. Finally, the law of areas tells us how fast the object moves along the orbit. However, we do not yet know the shape of the orbit. In fact, the shape of the orbit is the one remaining piece we need to complete the solution to the two-body problem.

2.4 THE ORBIT EQUATION

There are several ways to extract the shape of the orbit from the basic equations of motion of the two-body problem. All methods appear to be ad hoc, with no obvious direction apparent in the derivation. This is because there are no fixed procedures for solving a set of nonlinear differential equations. The sequence of operations which leads to a useful result is only discovered after trial and error. Indeed, no closed-form solution has ever been found in many important problems. So, with apologies for the nonintuitive nature of what follows, let us begin.

Start with the basic equation of motion (2.10), and take its post-cross product with the angular-momentum vector $\mathbf{H}$:

$$\ddot{\mathbf{r}} \times \mathbf{H} = -\frac{\mu}{r^3} \mathbf{r} \times \mathbf{H} \tag{2.22}$$

Since the angular-momentum vector is a constant, the left side of (2.22) is the time derivative of $\dot{\mathbf{r}} \times \mathbf{H}$. This would work, of course, if we crossed the equations of motion with *any* constant vector. The development to this point will only be useful if it leads to some form on the right-hand side which is also a perfect differential. The cross product on the right can be expanded as

$$\mathbf{r} \times \mathbf{H} = \mathbf{r} \times (\mathbf{r} \times \dot{\mathbf{r}})$$
$$= (\mathbf{r} \cdot \dot{\mathbf{r}})\mathbf{r} - (\mathbf{r} \cdot \mathbf{r})\dot{\mathbf{r}} \tag{2.23}$$

using the vector identity known as the *bac − cab* rule. Now, $\mathbf{r} \cdot \mathbf{r} = r^2$, and we remember that $\mathbf{r} \cdot \dot{\mathbf{r}} = r\dot{r}$, so (2.22) becomes

$$\frac{d}{dt}(\dot{\mathbf{r}} \times \mathbf{H}) = \frac{\mu}{r}\dot{\mathbf{r}} - \frac{\mu\dot{r}}{r^2}\mathbf{r} \tag{2.24}$$

The right side of the above is the time derivative of $\mu\mathbf{r}/r$. Equation (2.24) then can be written

$$\frac{d}{dt}\left(\dot{\mathbf{r}} \times \mathbf{H} - \mu\frac{\mathbf{r}}{r}\right) = 0 \tag{2.25}$$

So, crossing the equations of motion with $\mathbf{H}$ has led us to a form which can be integrated.

Integrating (2.25), we have

$$\dot{\mathbf{r}} \times \mathbf{H} - \mu \frac{\mathbf{r}}{r} = \mu \mathbf{e} \tag{2.26}$$

The vector **e** is, by construction, a constant vector. The factor of μ on the right is inserted to make the rest of the derivation somewhat simpler. The vector **e** must lie within the plane of the orbit, since the first term above is a vector perpendicular to **H**, and the second term is obviously within this plane. To learn the physical interpretation of **e**, however, it is necessary to finish the derivation of the orbit equation.

If we dot-multiply both sides of (2.26) with **r**, we obtain a scalar equation:

$$\mathbf{r} \cdot \dot{\mathbf{r}} \times \mathbf{H} = \mathbf{r} \cdot \mu \frac{r}{r} + \mu \mathbf{r} \cdot \mathbf{e} \tag{2.27}$$

Now, the dot and cross can be interchanged in the scalar triple product. This simplifies the first term to $\mathbf{r} \times \dot{\mathbf{r}} \cdot \mathbf{H} = H^2$. Introduce the angle v between the vectors **r** and **e** so that $\mathbf{r} \cdot \mathbf{e} = re \cos v$. The angle v is termed the *true anomaly*. Equation (2.27) then becomes

$$H^2 = \mu r + \mu r e \cos v \tag{2.28}$$

This can be solved for the radius-vector magnitude r as

$$r = \frac{H^2 / \mu}{1 + e \cos v} \tag{2.29}$$

This equation gives the behavior of r in terms of the angle v and constants of the orbit.

At this point we have reached the end of the derivation. It is usually an anticlimax, since (2.29) is not usually recognized for what it is: *the polar form of a conic section with the origin at one focus.* In other words, we have proved Kepler's first law: The orbit is a conic section with the primary object located at one focus. The vector **e** is called the *eccentricity vector,* and its scalar magnitude e is simply termed the *eccentricity* of the orbit. Notice in (2.29) that if $e < 1$, the radius r is bounded, while if $e \geq 1$, the denominator of the orbit equation can become zero, leading to infinite radius values. Also, (2.29) shows that r is smallest when $v = 0$, since the denominator is largest at this point. Since $v = 0$ when **r** and **e** are aligned, this means that the eccentricity vector points to the *perigee* of the orbit.

2.5 CONIC-SECTION GEOMETRY

Kepler, struggling with the unknown shape of Mars' orbit, lamented that if only the orbit were an ellipse, all the answers he needed could be found in the work of the Greek geometers. The orbit *is* a conic section, and the geometry

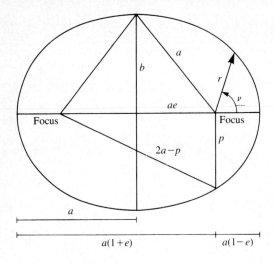

FIGURE 2.6
Ellipse geometry.

necessary to deal with it *was* developed by the ancient Greeks. Conic sections are so named because they can be generated by slicing a circular cone with a plane. There are five types of conic sections: the circle, ellipse, parabola, hyperbola, and straight line. All are possible orbits.

The geometry of the ellipse is shown in Figure 2.6. An ellipse is the locus of all points where the sum of the distances from two fixed points equals a constant. Conic sections have two foci (plural of focus[1]), the places where pins are stuck in the classic prescription for constructing an ellipse with two pins and a loop of string. From the pin-and-string method, it can be seen that the sum of the distances from each focus to any point on the periphery of the ellipse is constant. This constant sum is equal to the major diameter of the ellipse, as can be easily seen by imaging the string to lie along the major axis. It is more convenient to work with half this diameter, the *semimajor axis a*. The eccentricity *e*, already introduced, specifies the relative shape of the ellipse. If $e = 0$, the ellipse is a circle, a degenerate form of the ellipse with both foci at the center. If $e = 1$, the orbit becomes a parabola, an ellipse with one focus at the origin and the other at infinity. In the interval $0 \le e \le 1$ we have normal ellipses. If the eccentricity $e > 1$, the orbit is a hyperbola.

The distance from the center of the ellipse to either focus is ae. Drawing a triangle from each focus to the end of the minor axis and solving for the third side, the *semiminor axis b* is given by

$$b = \sqrt{a^2 - a^2 e^2} = a\sqrt{1 - e^2} \qquad (2.30)$$

---

[1] The focus gets its name from the fact that light or sound originating at one focus is reflected by an ellipse to the other focus. This property is used in some lasers to excite a laser rod with a flash tube, and is also occasionally used by playful architects to construct "whispering galleries."

since the hypotenuse has length a. The length of the chord from the focus vertically to the ellipse is of particular interest, since this is the radius H^2/μ given by (2.29) when v is 90°. This is the *parameter p*, whose classical name is the *semilatus rectum*. Drawing a triangle from each focus to the end of the latus rectum, the pythagorean theorem gives

$$(2ae)^2 + p^2 = (2a - p)^2 \tag{2.31}$$

This simplifies to

$$p = a(1 - e^2) \tag{2.32}$$

Evaluating (2.29) at $v = 90°$, we also have $p = H^2/\mu$, or using (2.32),

$$H = \sqrt{\mu p} = \sqrt{\mu a(1 - e^2)} \tag{2.33}$$

This gives us three different forms for the numerator of the conic-section expression (2.29).

The point furthest from the focus occupied by the earth is called the *apogee,* from the greek prefix *apo,* meeting "furthest," and the suffix *geos,* meaning "earth." Similarly, the closest point of the orbit to the earth is the *perigee.* Apogee occurs when the true anomaly $v = 180°$, while perigee occurs at $v = 0°$. The suffix on these terms is varied, within reason, depending on the name of the primary body. The closest and furthest points in an orbit about the sun are the *perihelion* and *aphelion,* respectively. The radius to perigee, r_p, and the radius to apogee, r_a, are found by using (2.32) in the equation of the ellipse and evaluating for the two special values of v. The results are

$$r_p = a(1 - e) \tag{2.34}$$

$$r_a = a(1 + e) \tag{2.35}$$

These are radii, not altitudes.

We now have two "languages" for describing the orbit: the geometric language of conic-section geometry and the dynamical language of position and velocity. We have already seen in (2.33) the geometric-language form of the angular momentum H, and a similar expression exists for the energy. Since energy is a constant, we can evaluate E at any convenient point on the orbit. At perigee, the radius is given by (2.34), while the speed at perigee is $v_p = H/r_p$, using the conservation of angular momentum. Then the energy is

$$E = \tfrac{1}{2}v_p^2 - \frac{\mu}{r_p}$$

$$= -\frac{\mu}{2a} \tag{2.36}$$

after substituting from (2.34) and (2.33) and simplifying the result. Notice that the total energy depends only on the semimajor axis of the orbit.

One final result is of extreme usefulness. We have already seen in (2.21) that the constant rate at which the radius vector sweeps out area is

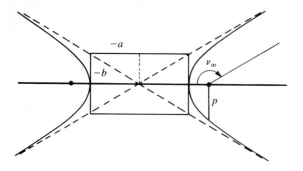

FIGURE 2.7
Geometry of the hyperbola.

$dA/dt = H/2$. The total area of an ellipse is $A = \pi ab$. So, the period of the orbit must be just $T = A/(dA/dt)$, and substituting from (2.30) and (2.33) we find

$$T = \frac{2\pi a^2 \sqrt{1 - e^2}}{\sqrt{\mu a}\sqrt{1 - e^2}} = \frac{2\pi}{\sqrt{\mu}} a^{3/2} \qquad (2.37)$$

This is Kepler's third law, as modified by Newton. The modification consists of the factor of μ in the denominator of (2.37). Notice that, like the energy, the period T of an elliptical orbit depends only on the semimajor axis a.

The same basic definitions work for the parabola and hyperbola as well, although there are some differences. In the parabola $a = \infty$ and $e = 1$, making (2.34) useless. Instead, the semilatus rectum p or perigee radius r_p is used to characterize the parabola. If a parabola is an ellipse whose apogee has been removed to infinity, then a hyperbola is an ellipse whose apogee has been carried through infinity and back around the other side (Figure 2.7). For gravity, only the branch of the hyperbola concave to the earth is physically possible. The radius reaches infinity when the denominator of the conic equation (2.29) goes to zero, giving the *asymptote* angle as

$$v_\infty = \cos^{-1} \frac{-1}{e} \qquad (2.38)$$

As $r \to \infty$, the satellite's speed approaches a constant. In the hyperbola, the semimajor axis a is negative, while the eccentricity e is greater than 1. Since by (2.36) positive-energy values occur for negative values of a, we have completed proving that a satellite with positive energy *must* escape.

2.6 KEPLER'S EQUATION

Kepler's second law is the key to calculating the position of a satellite in its orbit. Since the radius vector sweeps out equal areas in equal times, all we need is a convenient formula for the area of a pie-shaped wedge of an ellipse, with its apex at one focus. Kepler himself was the first to solve this problem, deriving the equation now bearing his name. Since it is difficult to operate with

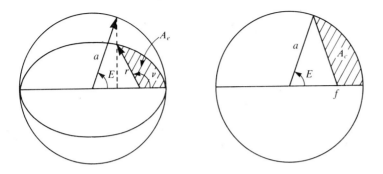

FIGURE 2.8
The eccentric anomaly E and Kepler's auxillary circle

an ellipse, Kepler introduced the auxiliary circle shown in Figure 2.8. An ellipse is, of course, simply a circle seen in projection. Any area in the ellipse is reduced by the ratio of the minor to major axes compared to the equivalent area in the auxiliary circle. This is obvious if we divide the area into narrow strips parallel to the minor axis. The equivalent area in the auxiliary circle has strips with the same base, but whose height has been increased by the ratio a/b.

Kepler also introduced the *eccentric anomaly E*, shown in Figure 2.8. Now, in the auxiliary circle, the area we wish to calculate is the singly shaded area in Figure 2.8. This area can be expressed as the difference between the sector and the triangle indicated in that figure. The area of the sector is simply $\frac{1}{2}a^2E$, where the eccentric anomaly is expressed in radians. The triangle has an altitude of $a \sin E$. The base of the triangle is ae, since the distance from the center to the focus is ae. Thus, the desired area A_c can be written

$$A_c = \tfrac{1}{2}a^2E - \tfrac{1}{2}a^2e \sin E$$
$$= \tfrac{1}{2}a^2(E - e \sin E) \tag{2.39}$$

This is, of course, the area calculated in the auxiliary circle, not the ellipse.

The corresponding area in the ellipse is then

$$A_e = \frac{b}{a}A_c = \frac{ab}{2}(E - e \sin E) \tag{2.40}$$

The total area of an ellipse is simply πab, and this is covered in one orbital period, given by Kepler's third law (2.37) as

$$T = 2\pi \sqrt{\frac{a^3}{\mu}} \tag{2.41}$$

The constant rate at which area is swept out is given by the total area divided by the period:

$$\frac{dA}{dt} = \frac{ab}{2}\sqrt{\frac{\mu}{a^3}} \tag{2.42}$$

Now, the area given in (2.40) should be the areal rate (2.42) times the time which has elapsed since perigee, $t - T_0$, where T_0 is the perigee time. Thus

$$M = E - e \sin E \tag{2.43}$$

where the *mean anomaly* is defined as

$$M = \sqrt{\frac{\mu}{a^3}}(t - T_0)$$

$$= n(t - T_0) \tag{2.44}$$

and the orbital *mean motion*, the angular frequency of the orbit, is given by

$$n = \sqrt{\frac{\mu}{a^3}} \tag{2.45}$$

Equation (2.43) is referred to as *Kepler's equation*. The use of the term "anomaly" for the angles M, E, and v is a holdover from the ptolemaic system, with its doctrine of the perfection of circular motion. To Ptolemy, any angle which did not increment uniformly with time was somehow wrong, or anomalous. However, the mean anomaly does increment linearly with time, so calling M an anomaly is itself anomalous.

However, to make (2.43) useful, we need to relate the eccentric anomaly E to the true anomaly v. Return to the triangles on the left of Figure 2.8. The distance between the center and the focus of the ellipse is ae, but this is also the sum of the bases of the two triangles. This gives

$$a \cos E - r \cos v = ae \tag{2.46}$$

and after substituting for r from the equation of a conic section (2.29), this becomes

$$\cos E = \frac{e + \cos v}{1 + e \cos v} \tag{2.47}$$

Similarly, the altitude of the triangle with vertex at the center of the orbit, $a \sin E$, must be the altitude of the triangle with vertex at the focus, $r \sin v$, when reduced by the factor b/a. This gives

$$\sin E = \frac{(1 - e^2)^{1/2} \sin v}{1 + e \cos v} \tag{2.48}$$

after simplification. Knowing both the sine and cosine of an angle allows either E or v to be determined without quadrant ambiguity. However, the most useful relation between these two angles is found by applying the half-angle identity for tangents from trigonometry. This leads to the result

$$\tan \tfrac{1}{2}v = \sqrt{\frac{1+e}{1-e}}\, \tan \tfrac{1}{2}E \tag{2.49}$$

which directly gives either E or v without quadrant confusion.

The form of Kepler's equation (2.43) is peculiar to circles and ellipses. In the case of parabolic and hyperbolic orbits, these results take slightly different forms. In the case of the parabola, Kepler's equation is

$$2\sqrt{\frac{\mu}{p^3}}\,(t - T_0) = \tan \tfrac{1}{2}v + \tfrac{1}{3}\tan^3 \tfrac{1}{2}v \tag{2.50}$$

which is a cubic equation for the quantity $\tan v/2$. For hyperbolas, the results analogous to (2.43) and (2.47) are

$$\sqrt{\frac{\mu}{(-a)^3}}\,(t - T_0) = e \sinh F - F \tag{2.51}$$

$$\cos v = \frac{e - \cosh F}{e \cosh F - 1} \tag{2.52}$$

Perhaps not surprisingly, these involve the *hyperbolic* trigonometric fuctions.

The fact that Kepler's equation takes different forms depending on the type of conic section has led several modern researchers to seek unified formulas for treating two-body motion. The *universal-variable* approach has been advocated by Battin, while Stiefel and Scheifele have discovered that the two-body problem can be transformed into a harmonic oscillator in a four-dimensional space. However, all such approaches still have buried within them a decision on the type of conic section. But it turns out that (2.43) for ellipses and (2.51) for hyperbolas are unified, since $E = iF$, where $i = \sqrt{-1}$.

2.7 THE CLASSICAL ORBITAL ELEMENTS

The two-body problem is a three-degree-of-freedom dynamical system, and a particular orbit is completely specified when the initial conditions $\mathbf{r}$ and $\mathbf{v}$ are specified at some initial time t_0. While this is true in a mathematical sense, it does not convey much information to mere mortals. Instead of citing the initial conditions, there are other quantities used to specify the orbit which enable the orbit to be directly visualized. The initial conditions $\mathbf{r}$ and $\mathbf{v}$ are six scalars, so any other system for specifying an orbit must also use six scalars. The most common such system is the six classical orbital elements, shown in Figure 2.9.

We have already encountered several of these quantities in previous sections. The semimajor axis a determines the size of the orbit, as well as specifying its period. The orbital eccentricity e determines the shape and type of conic section. The time of perigee passage T_0 gives a reference position of the satellite, fixing its position in the orbit at one instant of time. Together,

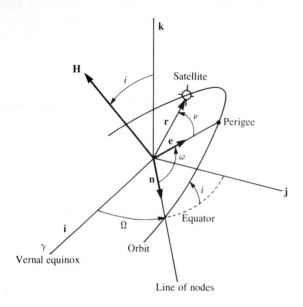

FIGURE 2.9
The classical orbital elements.

these three orbital elements form one group and completely determine the motion of the satellite within the orbital plane.

The second group of three orbital elements specifies the orientation of the orbit in space. As shown in Figure 2.9, these are the *right ascension of the ascending node Ω,* the *orbital inclination i,* and the *argument of the perigee ω.* The coordinate frame in the figure is also the standard frame for describing earth satellite orbits. The *xy* plane is the plane of the earth's equator, while the *z* axis is aligned with the earth's rotation axis. The *x* axis is tied to the *vernal equinox,* the point on the sky where the sun crosses the equator from south to north on the first day of spring. This point is traditionally denoted by the symbol γ. For orbits about the sun, the *xy* plane is usually the plane of the earth's orbit, the *ecliptic,* but the *x* axis still points to the vernal equinox. These reference frames differ by a rotation of about 23° about the *x* axis. This angle is the inclination of the earth's polar axis to its orbital plane, called the *obliquity of the ecliptic,* and is familiar from the usual mounting of earth globes. While the elements *a* and *e* refer to conic-section orbits, and therefore have been used in celestial mechanics only since the time of Kepler, many of these orientation angles are significantly older. For example, the vernal equinox γ is also termed the *first point of Aries,* the constellation of the Ram in the zodiac. Due to the precession of the earth's pole, it has been almost four thousand years since the vernal equinox was in this constellation, but the name has not changed in the interim. The Greek letter γ, used for this point on the sky, evolved from a Babylonian cuneiform symbol for the head of a ram.

(Due to precession of the earth's polar axis every 26,000 years, the frame in Figure 2.9 is actually in motion and therefore noninertial. Observing instruments refer their data to the instantaneous position of the equator and equinox, while theoretical work is done with respect to the equator and equniox of a standard epoch. Currently, the equator and equinox of the year 2000 are used. Small corrections are made to observational data to bring them into the inertial frame. To keep these corrections small, it is necessary to occasionally change the epoch of the standard reference frame. This will happen next in the year 2025, when the epoch will be changed to the year 2050.)

The right ascension of the ascending node (or *node* for short) is measured from the vernal equinox eastward along the equator to the point where the satellite's orbit crosses the equator from south to north. (Any angle measured in the plane of the equator from the vernal equinox is termed a *right ascension*. The angle similar to latitude is called the *declination*.) Notice that the line of nodes, the intersection of the equator and orbital planes, must be perpendicular to both the z axis and the angular-momentum vector **H**. The inclination of the orbit is measured from the equator plane to the orbit plane at the ascending node and lies in the range $0 \leq i \leq 180°$. Finally, the argument of perigee is measured from the line of nodes along the orbit (in the direction of motion) to the perigee point. This point is, of course, marked by the eccentricity vector **e**. For a low-altitude earth orbit, the inclination essentially sets the north and south limits of visibility of the earth's surface. If a satellite must see the entire surface of the earth, it must be placed in a *polar orbit,* or an orbit whose inclination is near 90°. Orbits with inclination less than 90° are termed *prograde*, since these satellites circle the earth in the same sense as the earth's rotation. Orbits with inclinations over 90° are termed *retrograde*. An orbit with $i = 0$ is termed an *equatorial* orbit. Notice also that a launch site at latitude δ cannot directly insert a satellite into an orbit with $i < \delta$.

Since both the initial conditions **r** and **v** at time t_0 and the six orbital elements a, e, T_0, Ω, i, and ω completely specify the orbit, it must be possible to transform one set of quantities to the other. In this section, consider the problem of calculating the elements given **r** and **v**. We can begin by calculating the orbital energy and angular momentum:

$$E = \tfrac{1}{2}v^2 - \frac{\mu}{r} \tag{2.53}$$

$$\mathbf{H} = \mathbf{r} \times \mathbf{v} \tag{2.54}$$

The energy immediately yields the semimajor axis

$$a = -\frac{\mu}{2E} \tag{2.55}$$

The eccentricity vector can also be calculated directly from the position and velocity vectors as

$$\mathbf{e} = \frac{1}{\mu}\left(\mathbf{v} \times \mathbf{H} - \frac{\mu\mathbf{r}}{r}\right) \tag{2.56}$$

and its magnitude furnishes the orbital eccentricity.

The line of nodes can be found by constructing a unit vector **n** along the direction of the ascending node. Since this line is perpendicular to both the unit vector **k** and the angular-momentum vector **H**, Figure 2.9 gives

$$\mathbf{n} = \frac{\mathbf{k} \times \mathbf{H}}{|\mathbf{H}|} \tag{2.57}$$

Since this vector lies in the equator plane, it must have the form

$$\mathbf{n} = \cos \Omega \mathbf{i} + \sin \Omega \mathbf{j} \tag{2.58}$$

With both $\sin \Omega$ and $\cos \Omega$ known, the angle Ω can be determined without quadrant ambiguity. The orbital inclination i is also simply obtained, since it is the angle between the angular momentum and z axis. This gives

$$\cos i = \frac{\mathbf{k} \cdot \mathbf{H}}{|\mathbf{H}|} \tag{2.59}$$

Since $0 \le i \le 180°$, the above equation determines i without ambiguity.

The argument of perigee is the angle between the node vector **n** and the eccentricity vector **e**. Its cosine can be found by calculating a dot product:

$$\cos \omega = \frac{\mathbf{n} \cdot \mathbf{e}}{|\mathbf{e}|} \tag{2.60}$$

However, this angle can lie in the range $0 \le \omega \le 360°$. Inspection of the figure shows that simply taking the inverse cosine of (2.60) is correct if the perigee lies above the equator plane. This is true if $\mathbf{e} \cdot \mathbf{k} > 0$. If **e** has a negative z component, the appropriate quadrant correction must be made.

The calculation of the true anomaly v (as a prelude to the calculation of the time of perigee passage) is quite similar. The true anomaly is the angle between the eccentricity vector and the radius vector. So

$$\cos v = \frac{\mathbf{e} \cdot \mathbf{r}}{|\mathbf{e}| \, |\mathbf{r}|} \tag{2.61}$$

Again, we are faced with possible quadrant problems. The quadrant obtained by simply taking the inverse cosine of (2.61) will be correct if the satellite is traveling from perigee toward apogee and incorrect if the satellite is moving from apogee toward perigee. By examining the sign of $\mathbf{r} \cdot \mathbf{v}$ this ambiguity can be resolved: If $\mathbf{r} \cdot \mathbf{v} > 0$, the satellite has not yet reached apogee, and $v < 180°$.

The true anomaly at time t_0 can now be traded for the eccentric anomaly according to

$$\tan \frac{E}{2} = \sqrt{\frac{1-e}{1+e}} \tan \frac{v}{2} \tag{2.62}$$

and then Kepler's equation

$$M = \sqrt{\frac{\mu}{a^3}} (t_0 - T_0) = E - e \sin E \tag{2.63}$$

will furnish the time of perigee passage directly.

There are several conditions under which calculation of the classical elements will fail. If the inclination of the orbit is 0 or 180°, then the node Ω is not defined, since the equator and orbital planes are then identical. This appears in (2.57) as a node vector of zero length. Also, if the orbit is perfectly circular, the eccentricity vector is of zero length, and it is impossible to determine *either* the argument of perigee or the true anomaly. Unfortunately, both nature and people persist in attempting to realize these conditions. Most orbits in the solar system are of small eccentricity and inclination, while a low-altitude earth satellite has very little choice but to be in a nearly circular orbit. The zero-inclination circular equatorial orbit is essential to the usefulness of geosynchronous communications satellites. Alternate forms of the classical orbital elements can be found which eliminate these difficulties.

Of course, if we can transform from a position and velocity vector $\mathbf{r}$ and $\mathbf{v}$ at a specified time t_0 to the set of orbital elements, then it is possible to perform the reverse transformation as well. If we change the time from t_0 to t before we perform the reverse calculation, then we will obtain the position and velocity at time t: $\mathbf{r}(t)$, $\mathbf{v}(t)$. The net effect of the two calculations is to obtain $\mathbf{r}(t)$ and $\mathbf{v}(t)$ given $\mathbf{r}(t_0)$ and $\mathbf{v}(t_0)$. But this is the complete solution, to which we will turn in the next section.

2.8 POSITION AND VELOCITY

Although we have calculated the classical elements from a position and velocity vector, this is only half the solution of the two-body problem. To obtain the complete solution, we must be able to calculate the position and velocity vectors at any other time. In a sense, this is the reverse of the process of calculating the elements from $\mathbf{r}$ and $\mathbf{v}$. However, it is common to use a slightly different formulation.

The calculation of position and velocity begins with the solution of Kepler's equation, (2.43),

$$M = \sqrt{\frac{\mu}{a^3}}\,(t - T_0)$$

$$= E - e \sin E \tag{2.64}$$

for the eccentric anomaly E. When we used this equation to calculate T_0 in section 2.7 there was no particular difficulty. Now, given the time t we can calculate the mean anomaly M in (2.64), but this equation is transcendental in the eccentric anomaly E: No closed-form solution for E is possible. Thus we must solve Kepler's equation by numerical methods.[1] Notice that if the eccentricity e is zero, the solution to Kepler's equation is trivial: $M = E$. Using

---

[1] There is an enormous literature on solving Kepler's equation, since it is one of the first transcendental equations in science which needed solution in wholesale quantities for practical purposes. Accurate ship navigation depended on accurate predictions of solar system bodies until only recently.

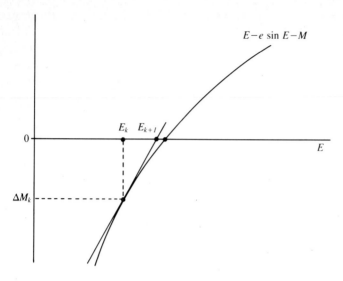

FIGURE 2.10
Iterative solution of Kepler's equation.

this as an approximation, rewrite (2.64) as

$$E = M + e \sin E$$
$$\approx M + e \sin M \tag{2.65}$$

which is a better approximation. This process can be continued to generate a series solution to Kepler's equation for small eccentricity.

For moderate eccentricities, the most common numerical technique is the *Newton-Raphson method*. Suppose we have an approximate value for E, say, E_k. When substituted into Kepler's equation, we will not obtain zero,

$$E_k - e \sin E_k - M = \Delta M_k \neq 0 \tag{2.66}$$

but will have some error ΔM_k. As shown in Figure 2.10, we can approximate the correction ΔE_k needed by assuming that the curve is locally a straight line, giving

$$\Delta E_k \approx -\frac{\Delta M_k}{dM/dE} \tag{2.67}$$

The necessary derivative is simply

$$\frac{dM}{dE} = 1 - e \cos E \tag{2.68}$$

Finally, the approximate correction obtained from (2.67) is added to obtain the next approximate value of the eccentric anomaly

$$E_{k+1} = E_k + \Delta E_k \tag{2.69}$$

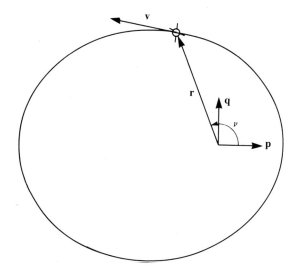

FIGURE 2.11
Orbital reference frame.

Since this is an iterative technique, we return to (2.66) and continue until the correction is negligible. If the initial guess is close enough, the Newton–Raphson technique converges very quickly to the actual answer.

Once the eccentric anomaly has been found, we can immediately obtain the true anomaly using (2.49),

$$\tan\frac{v}{2} = \sqrt{\frac{1+e}{1-e}}\,\tan\frac{E}{2} \tag{2.70}$$

and the scalar radius follows from the equation for a conic section:

$$r = \frac{a(1-e^2)}{1+e\cos v} \tag{2.71}$$

As shown in Figure 2.11, introduce a new coordinate frame tied to the orbit plane. The unit vector **p** points toward the perigee of the orbit, **q** is out the semilatus rectum, and **w** is along the orbit normal. In terms of these unit vectors the radius vector is simply

$$\mathbf{r} = r\cos v\mathbf{p} + r\sin v\mathbf{q} \tag{2.72}$$

To obtain the velocity vector, differentiate (2.72), realizing that the **p**, **q**, **w** frame is inertial:

$$\mathbf{v} = (\dot{r}\cos v - r\dot{v}\sin v)\mathbf{p} + (\dot{r}\sin v + r\dot{v}\cos v)\mathbf{q} \tag{2.73}$$

One new quantity, $r\dot{v}$, is the transverse-velocity component and can be found by remembering that the magnitude of the angular momentum is $H = r(r\dot{v}) = \sqrt{\mu p}$. This becomes

$$r\dot{v} = \sqrt{\frac{\mu}{p}}(1+e\cos v) \tag{2.74}$$

after substituting the conic-section expression. Also, $\dot{r}$ can be obtained by differentiating (2.71) to find

$$\dot{r} = \frac{pe \sin v \dot{v}}{(1 + e \cos v)^2} = r\dot{v}\frac{e \sin v}{1 + e \cos v}$$

$$= \sqrt{\frac{\mu}{p}}e \sin v \qquad (2.75)$$

after using (2.74). When these are inserted into (2.73), we have

$$\mathbf{v} = \sqrt{\frac{\mu}{p}}[-\sin v\mathbf{p} + (e + \cos v)\mathbf{q}] \qquad (2.76)$$

which gives the velocity vector in the $\mathbf{p}$, $\mathbf{q}$, $\mathbf{w}$ frame.

Although the orbital frame $\mathbf{p}$, $\mathbf{q}$, $\mathbf{w}$ is inertial, it is special to a particular orbit and is not the standard inertial frame. Returning to Figure 2.9, the position vector in standard inertial frame $\mathbf{r}_{ijk}$ is linked to the position vector $\mathbf{r}_{pqw}$ in the $\mathbf{p}$, $\mathbf{q}$, $\mathbf{w}$ frame (2.72) by a rotation matrix

$$\mathbf{r}_{ijk} = R\mathbf{r}_{pqw} \qquad (2.77)$$

The same rotation matrix will also transform the velocity vector (2.76) to the inertial frame. Using the elementary rotation matrices of section 1.4, this rotation matrix can be written as

$$R = R_3(-\Omega)R_1(-i)R_3(-\omega) \qquad (2.78)$$

We will actually cite this matrix in section 4.7, since the orbital elements Ω, i, and ω are just the Euler angles of rigid-body dynamics. However, remember from section 1.4 that a rotation matrix is just one set of unit vectors expressed in terms of their components in the other frame. The required rotation matrix is just

$$R = [\mathbf{p} \quad \vdots \quad \mathbf{q} \quad \vdots \quad \mathbf{w}] \qquad (2.79)$$

Since $\mathbf{p}$ points to perigee, it can be directly calculated by normalizing the eccentricity vector

$$\mathbf{p} = \frac{\mathbf{e}}{|\mathbf{e}|} \qquad (2.80)$$

while the orbit normal is the normalized angular momentum vector

$$\mathbf{w} = \frac{\mathbf{H}}{|\mathbf{H}|} \qquad (2.81)$$

The third unit vector is then $\mathbf{q} = \mathbf{w} \times \mathbf{p}$. Since $\mathbf{e}$ and $\mathbf{H}$ are originally calculated in the inertial-frame unit vectors, we obtain $\mathbf{p}$, $\mathbf{q}$, and $\mathbf{w}$ in the inertial frame, and (2.79) supplies the required rotation matrix. The rotation matrix can thus be easily calculated as a by-product of the original calculation of the orbital elements.

We can now take a set of initial conditions **r** and **v** at an initial time and find the orbital elements. Then, choosing a different time t, we can reverse the process and find the position and velocity at this new time. This is the complete solution to the problem of two gravitating bodies. The answer to any question we might ask about the two-body problem is implicit within it. However, there is still much more to the two-body problem, especially finding convenient answers to several frequently asked questions. We will consider some of these in later sections and chapters.

2.9 ORBIT DETERMINATION AND SATELLITE TRACKING

Although we now have a complete solution to the problem of two bodies, in actual practice there are additional complications. While working with the physical position and velocity vectors is possible during the mission analysis stage of a project, once the satellite is launched it becomes necessary to deal with the "real" world, and "real" data. Orbital mechanics is one of the most exact of sciences, but this exactness implies the need for similarly accurate data. And the data may not be directly related to the quantities we wish to know.

Perhaps the most benign case occurs when a rocket inserts a satellite into its initial orbit. A rocket carries an inertial-navigation system (see Chapter 6) which supplies the position and velocity vectors **r** and **v** in an inertial frame. Determining the orbit of the satellite is then a straightforward application of section 2.7, since the six components of position and velocity suffice to determine the six orbital elements. However, only rarely does the orbital analyst obtain data which are this directly related to the orbit.

For most of the history of orbital mechanics, objects were tracked by optical telescopes. A single observation consists of two angle measurements on the sky, and no information on the distance to the object is available. Since one observation consists of two scalar quantities (right ascension and declination), it takes a minimum of three such observations to determine an orbit. Even so, calculation of the orbit from three optical observations made from a moving observing site on the orbiting planet earth is not a simple task. Worse still, right ascension and declination are related back to the orbital elements by a complicated series of calculations. The problem had such importance for astronomy, however, that many solutions are available. One of the first is Laplace's method, while the first exact solution is due to Gauss.

With the advent of modern radar, the situation has changed somewhat. A radar typically measures two angles (usually azimuth and elevation), and by timing the return echo measures the range to the satellite as well. Knowing the position vector of the radar relative to the center of the earth, and the relative position vector from the radar to the satellite, some coordinate transformations and a vector addition will supply the position vector of the satellite. While a radar can also observe the range rate by measuring the Doppler shift of the returning echo, this does not supply complete information on the velocity vector. One often-used orbit-determinaton method, the Herrick–Gibbs method, takes several closely spaced position vectors from one radar site, runs a polynomial through these points, and differentiates the polynomial to obtain an approximate velocity vector. Many methods are available for determining an orbit from the two position vectors and the time interval which elapses between them. These methods are not only useful for processing radar data. Since many mission analysis problems are of the form "leave here now, arrive there then," they are also solvable by two position-vector–time-of-flight methods.

Radar has not eliminated the need for optical tracking. A radar pulse spreads out according to the inverse-square law for light until it hits the target, and then the return echo similarly spreads out on its return to the receiver. The intensity of the return echo is thus proportional to the inverse fourth power of the distance from the radar to the target. Radar tracking is thus limited to objects relatively close to the earth. Satellites in geosynchronous orbit are generally too far away to be tracked by radar, and their orbits must be determined from optical data.

The range of radar tracking can be extended by placing a transponder on the spacecraft. NASA deep-space craft are followed by range–range rate tracking, in which the range is observed by timing the (amplified) return echo, the range rate is obtained by the Doppler shift, and the angle data from the pointing of the antenna are (relatively) so inaccurate as to not be worth processing. In a sense, range–range rate data are the compliment of optical data: In the former method only the range information is processed; in the latter the range information is completely missing.

Orbits can be determined from Doppler-only data from telemetry receiving sites. In rendezvous, the orbit of the target may need to be determined from radar data taken from the active spacecraft, itself in orbit. Basically, there are at least as many orbit-determination methods as there are types of data from which an orbit may be determined. However, all such orbit-determination methods share one fact in common: Since the data are not perfectly accurate, the orbit calculated from the data is not perfectly accurate.

This fact has a profound influence on the tracking and navigation of spacecraft. One of the imperfectly determined orbital elements is the semimajor axis a. Since a determines the period through Kepler's third law, the period of the orbit is not perfectly known either. Taking differentials of Kepler's third law (2.37), we have the error in the period δT as

$$\delta T = \frac{3\pi}{\sqrt{\mu}} a^{1/2} \, \delta a \qquad (2.82)$$

in terms of the error δa in determining the true semimajor axis. Of course, we do not know the actual error δa, so we cannot fix this problem. Since our calculated period for the orbit is wrong, any predictions made from the orbital elements will depart further and further from reality as time increases. It is continually necessary to take new observations to control the growth of these errors.

Over time, a large body of observations is amassed on a particular orbiting object. Rather than continue to determine the orbit from the *minimum* number of observations, it seems reasonable that all the data, taken together, will produce a better orbit than any particular subset. The solution to this problem, also due to Gauss, is called the *method of least squares*. This technique finds an orbit passing *as close as possible to all the data.* In recent times, this work has been extended and applied to other fields as well, and the new discipline is termed *estimation theory*. The *Kalman filter* is a modern technique whose roots reach back to the work of Gauss.

However, the problem discussed above still remains. No data processing technique can produce perfection from data corrupted by unavoidable observational errors. Instead they produce an *estimate* of the true orbit: the "best" possible fit (in a statistical sense) to the true orbit. The error in the semimajor axis δa can be much smaller, but it will not be zero. A better estimate simply extends the time interval during which predictions can be trusted. Reality remains forever unobtainable.

2.10 REFERENCES AND FURTHER READING

Although the two-body problem is no more complex than other, equally important problems in astronautical engineering (e.g., the torque-free satellite), it is far older and has received far more attention. Generally, texts on orbital mechanics treat several different methods to achieve the same end, which can produce bewilderment on a first encounter. In this chapter we have studied the "classical" solution to the two-body problem and ignored the other methods.

Books on celestial mechanics also fall into two categories: the older "astronomy" books, which ignore maneuvers and concentrate on perturbation theory, and the newer "engineering" books, for which the reverse is usually true. An excellent introduction to orbital theory from the latter perspective is Bate, Mueller, and White, while the book by Danby is an excellent introduction from the astronomer's point of view. The newer universal-variable approach is covered in Battin's book, and a still more recent solution is found in the work of Stiefel and Scheifele. Battin's method is also covered in Bate, Mueller, and White.

The intricacies of perturbation theory are generally beyond the scope of this text. The classic introduction to this field from the astronomer's point of view is Brouwer and Clemence, while the newer work by Fitzpatrick concentrates on the application of these techniques to the problem of earth satellites.

Bate, R. R., D. D. Mueller, and J. E. White, *Fundamentals of Astrodynamics*, Dover, New York, 1971.

Danby, J. M. A., *Fundamentals of Celestial Mechanics*, Macmillan, New York, 1962.

Battin, R. H., *Astronautical Guidance*, McGraw-Hill, New York, 1964.

Stiefel, E. L., G. Scheifele, *Linear and Regular Celestial Mechanics*, Springer-Verlag, New York, 1971.

Brouwer, D., and G. M. Clemence, *Methods of Celestial Mechanics*, Academic, New York, 1961.

Fitzpatrick, P. M., *Principles of Celestial Mechanics*, Academic, New York, 1970.

2.11 PROBLEMS

1. Show that the speed of a satellite in a circular orbit is

$$v_c = \sqrt{\frac{\mu}{r}}$$

Compare this to escape velocity at the same radius. Calculate v_c and v_{esc} at $r = 6578$ km (200-km altitude) and at $r = 385,000$ km (the distance to the moon).

2. Use the energy relation to show that a satellite in a hyperbolic orbit arrives at $r = \infty$ with residual speed

$$v_\infty = \sqrt{\frac{-\mu}{a}} = \sqrt{2E}$$

3. Halley's comet last passed perihelion on Feb. 9, 1986. It has a semimajor axis $a = 17.9564$ AU and eccentricity $e = 0.967298$. [One astronomical unit (AU) is the distance between the earth and the sun.] Calculate the period of Halley's comet, and predict the date of the next return. Solve Kepler's equation, and calculate E, v, and the scalar radius vector r for your current date.

4. One extremely confusing thing about meeting orbital mechanics for the first time is that almost anything can be expressed in either the geometric language of the orbital elements or the dynamical language of energy, angular momentum, and so forth. Introduce the flight-path angle ϕ shown in Figure 2.12, the angle between the radius and velocity vectors. Express the tangential- and radial-velocity components in terms of the energy E, angular momentum H, the angle ϕ, and the radius r.

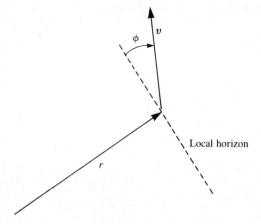

FIGURE 2.12
The flight-path angle ϕ.

5. It is also possible to express most formulas in terms of any of the anomalies. From Figure 2.8, show that the position vector is given by

$$\mathbf{r} = a(\cos E - e)\mathbf{p} + a\sqrt{1 - e^2} \sin E \mathbf{q}$$

in the unit vectors of the $\mathbf{p}$, $\mathbf{q}$, $\mathbf{w}$ frame. Also, show that the radius is

$$r = a(1 - e \cos E)$$

By differentiating the expression for $\mathbf{r}$, and substituting for $\dot{E}$ from the derivative of Kepler's equation (2.43), find an expression for the velocity vector $\mathbf{v}$ in terms of the eccentric anomaly E.

6. Continue the series solution to Kepler's equation begun in (2.65) to the second order in the eccentricity e. Expand any sines or cosines of small angles, and obtain a final answer involving only powers of e multiplying sines of multiples of the mean anomaly M.

7. Expand the solution to the two-body problem as a Taylor's series, obtaining

$$\mathbf{r}(t) = \mathbf{r}_0 + \mathbf{v}_0(t - t_0) - \frac{1}{2!}\frac{\mu}{r_0^3}\mathbf{r}_0(t - t_0)^2 + \cdots$$

If two position vectors $\mathbf{r}_0(t_0)$ and $\mathbf{r}_1(t_1)$ are available for an orbit, one of these position vectors must be traded for a velocity vector to determine the orbit. Solve for $\mathbf{v}_0$ above in terms of the two position vectors and the (assumed short) time interval $t_1 - t_0$. If the expansion is extended to the next order, is so simple a solution still possible?

8. A spaceship consisting of a payload m and a large solar sail, as shown in Figure 2.13, is in orbit about the sun. The sail is kept flat on to the sun and produces an acceleration from radiation pressure of

$$\mathbf{a}_1 = \frac{2SA}{cm}\frac{\mathbf{r}}{r^3}$$

where S is the solar constant, A is the sail area, and c is the speed of light. Show that the equation of motion of the spaceship is

$$\ddot{\mathbf{r}} = -\left(\mu - \frac{2SA}{cm}\right)\frac{\mathbf{r}}{r^3}$$

What types of conic sections are possible when μ is greater than $2SA/cm$? When they are equal? When μ is less than $2SA/cm$? In one of these cases the vehicle always escapes. Show that it will arrive at $r = \infty$ with residual speed

$$v_\infty = \left[v_0^2 - 2\left(\mu - \frac{2SA}{cm}\right)\frac{1}{r_0}\right]^{1/2}$$

where r_0 and v_0 are the initial distance and speed when the sail is unfurled.

9. Recall the equation of motion of the N-body problem (1.118), divided by m_i:

$$\ddot{\mathbf{r}}_i = -\sum_{j\neq i}^{N}\frac{Gm_j(\mathbf{r}_i - \mathbf{r}_j)}{|\mathbf{r}_j - \mathbf{r}_i|^3}$$

Assume that object $i = 1$ is the earth and body $i = 2$ is an earth satellite, while

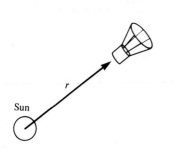

FIGURE 2.13
Solar-sail spacecraft.

$i = 3, 4, \ldots$ are the sun, moon, and so forth. Since the vector $\mathbf{r} = \mathbf{r}_2 - \mathbf{r}_1$ is the position vector of the satellite with respect to the earth, show that the equation of motion of the satellite is

$$\ddot{\mathbf{r}}_2 - \ddot{\mathbf{r}}_1 = \ddot{\mathbf{r}} = -\frac{\mu \mathbf{r}}{r^3} - \sum_{j=3}^{N} Gm_j \left(\frac{\mathbf{r}_2 - \mathbf{r}_j}{r_{2j}^3} - \frac{\mathbf{r}_1 - \mathbf{r}_j}{r_{1j}^3} \right)$$

The additional terms above are the perturbing acceleration of the rest of the solar system on the orbit of the earth satellite. Argue that the two-body problem is a good approximation when either (*a*) m_j is small, (*b*) $|\mathbf{r}_j|$ is very large, or (*c*) the satellite is "close" to the earth ($\mathbf{r}_{2j} \approx \mathbf{r}_{1j}$).

CHAPTER
3

EARTH
SATELLITE
OPERATIONS

3.1 INTRODUCTION

We have seen that the discipline of orbital dynamics was one of the first of the "exact" sciences. This field was developed for several hundred years by astronomers and applied to the motion of natural objects within the solar system and to multiple-star systems. However, much work remained to be done in this field when it became apparent that spaceflight was not just a dream but an actual possibility. For example, the concept of a *maneuver* is foreign to classical orbital dynamics.

In this chapter we will explore how orbital dynamics is applied to operation of earth satellites. After studying the two most common maneuver types, we will turn our attention to how launch and rendezvous are handled for orbiting spacecraft. An earth satellite does not perfectly follow the two-body dynamics of the last chapter, so we will look at the subject of orbital perturbations due to air drag and the oblate earth. Finally, in contrast to maneuver strategies with high-thrust chemical rockets, we will look at a simple spiral orbital transfer for a low-thrust vehicle.

3.2 THE HOHMANN TRANSFER

The problem of transfer between two coplanar circular orbits was first solved by Walter Hohmann in 1925. Although he performed his work before the advent of the discipline of optimal control, the Hohmann transfer is the optimal way to perform this maneuver, within wide limits. The major assumption in the Hohmann transfer is that the maneuvers are impulsive: They are assumed to occur instantaneously. This is usually a very good assumption, since most rockets are very high thrust devices, and burn times are short compared to the time the vehicle must coast.

Figure 3.1 shows two circular orbits with semimajor axes a_1 and a_2. Any orbit crossing both circular orbits could be used to transfer between them. An orbit which goes further from the earth than the outer orbit, however, would require extra energy to establish, since it would ascend further than necessary in the earth's gravity field. Also, the maneuver at the crossing point would require a considerable turning of the velocity vector. Conversely, an orbit which descends within the inner circular orbit would also be wasteful of energy, since it requires energy to slow down and fall inward toward the earth. The best transfer orbit, then, is an ellipse which just touches both circular orbits. This not only avoids the waste of energy to gain or lose "altitude" unnecessarily, but the velocity vectors of the circular orbit and the transfer orbit are aligned when they meet. This eliminates any need to burn fuel to simply turn the velocity vector.

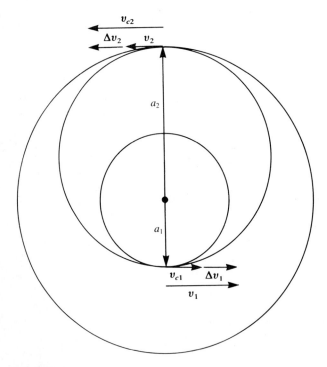

FIGURE 3.1
The Hohmann transfer.

The semimajor axis a of the transfer ellipse, is simply

$$a = \frac{a_1 + a_2}{2} \tag{3.1}$$

There will be two maneuvers required: one to leave the inner circular orbit and enter the transfer ellipse, the second to leave the transfer ellipse and enter the outer circular orbit. The magnitude of these maneuvers can be found by simply subtracting the speeds of the two orbits at their point of tangency. The speed in a circular orbit is $v_c = \sqrt{\mu/r}$, while the speed in the elliptical orbit can be found from the energy law, (2.15) and (2.36),

$$v^2 = \mu \left(\frac{2}{r} - \frac{1}{a} \right) \tag{3.2}$$

using the appropriate value of the radius vector. The first-maneuver velocity change is given by

$$\Delta v_1 = \sqrt{\frac{2\mu}{a_1} - \frac{2\mu}{a_1 + a_2}} - \sqrt{\frac{\mu}{a_1}} \tag{3.3}$$

since the first maneuver occurs at $r = a_1$, and inserting (3.1) for the semimajor axis of the elliptical orbit. Similarly, the second maneuver is given by

$$\Delta v_2 = \sqrt{\frac{\mu}{a_2}} - \sqrt{\frac{2\mu}{a_2} - \frac{2\mu}{a_1 + a_2}} \tag{3.4}$$

since the second maneuver occurs at $r = a_2$.

The order of the two subtractions in (3.3) and (3.4) has been chosen to give positive velocity changes in an outward transfer, since both maneuvers are made to increase the speed of the spacecraft. At the first maneuver the speed of the vehicle must be increased to enable it to "gain altitude" in the gravitational field of the primary body. As the spacecraft approaches the second-maneuver point, it will be moving too slow to remain at $r = a_2$. If its velocity is not increased, it will fall back toward the central object. So, in an outward transfer, both maneuvers are speed increases. Conversely, in an inward transfer both maneuvers are made to slow the vehicle down.

The time required to perform the Hohmann transfer is simply one-half the period of the transfer ellipse. That is, the time between maneuvers will be given by

$$\Delta t = \pi \sqrt{\frac{a^3}{\mu}} \tag{3.5}$$

For a Hohmann transfer from a low earth parking orbit to a geosynchronous orbit (an orbit with a period of 1 day), the time interval is about 5.3 h. A Hohmann transfer from the orbit of the earth to that of Mars, on the other hand, will require over 258 days. Although the Hohmann transfer minimizes the total velocity requirement for the trip (under most conditions), it also maximizes the time required for the transfer. This is not a great difficulty in the transfer to geosynchronous orbit. In fact, the second maneuver is often delayed by an integral number of transfer orbit periods to allow the satellite to drift into a position near its planned location above the earth's surface. This is out of the question for a Mars expedition, of course.

3.3 INCLINATION-CHANGE MANEUVERS

A change in the orbit plane is one of the most expensive of all possible maneuvers. The reason for this can be seen from Figure 3.2, where a change from one circular orbit to another with a different orbit plane, but with the same period, is made. The maneuver, of course, must occur where the two orbits intersect at the relative line of nodes. The velocity-change requirement is found by subtracting the velocity vectors of the two orbits. Since both orbits have the same speed v, the required velocity change is given by

$$\Delta v = 2v \sin \frac{\Delta i}{2} \tag{3.6}$$

This is easily shown by slicing the isosceles velocity triangle into two right triangles.

The velocity change given by (3.6) can be very large. An inclination change of 60° requires a $\Delta v = v$, or the full orbital velocity itself. This is obviously not acceptable since a 60° plane change for a low earth-orbiting satellite requires a rocket large enough to have launched the payload in the first place. Of course, the booster required to launch *that* payload would be huge.

But there is a cheaper way to obtain large inclination changes. We can escape from the circular orbit to infinity for a maneuver cost of $\Delta v_\infty =$

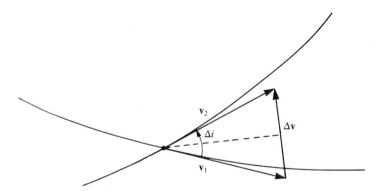

FIGURE 3.2
Direct inclination-change maneuver.

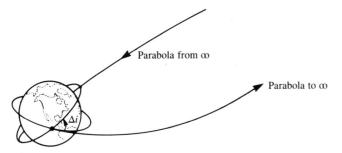

FIGURE 3.3
Inclination change by escape and return.

$(\sqrt{2}-1)v$, which is just the difference between escape speed and circular orbital speed. Since the outgoing parabolic orbit will arrive at infinity with zero velocity, we can change our orbit plane at will for zero cost. At the same time we can head back toward the earth on another parabolic orbit, tangent to the second circular orbit in Figure 3.3. A second maneuver of the same size as the first will place us into the new circular orbit. The total cost of these two maneuvers is

$$\Delta v_{tot} = 2(\sqrt{2}-1)v \qquad (3.7)$$

This is independent of the magnitude of the inclination change. Equating (3.6) and (3.7), the escape-to-infinity option is less expensive when $\Delta i > 48.9°$. The drawback of this method, of course, is that it takes an infinite time to traverse the two parabolic orbits.

Thus, at 60° of plane change, it would require a full satellite booster to perform the orbit-plane change. At about 50° of plane change, the satellite is capable of being an interplanetary spaceship. Of course, it is vastly preferable to obtain the correct inclination during the original launch. By varying the launch azimuth, satellite orbits with inclinations larger than the latitude of the launch site can be obtained by direct-ascent trajectories, and no inclination change is necessary. Very small inclination changes may be required to precisely obtain a certain specified inclination for an orbit, but these are trim maneuvers to remove errors introduced by the booster guidance system or to eliminate natural orbit perturbations. So, with one exception, there is no need to change the orbital plane of a satellite once it is in orbit.

The one exception occurs when an orbital inclination *less* than the latitude of the launch site is desired. Launch sites are placed as close to the equator as national pride and international politics will allow [the European Space Agency (ESA) launch site is *on* the equator], but it still may be necessary to launch into orbits with lower inclinations. The major example of this is a launch into a geosynchronous, equatorial orbit. A satellite orbiting with the same angular rate as the earth's rotation will appear to hang stationary against the sky *only if* the orbital inclination is zero. In this case, a

Hohmann transfer must also be performed, since a direct ascent to an orbit at 6.6 earth radii is very inefficient. The inclination-change maneuver could be performed at the first Hohmann maneuver, when leaving the parking orbit, or it could be performed at the second Hohmann maneuver at high altitude. Since the orbital velocity in the parking orbit will be about 7 km/s, while approaching the geosynchronous orbit the required velocity is less than 2 km/s, it is far cheaper to postpone the inclination change to the time of the second Hohmann burn. By adding vectorially the two required velocity changes, and performing the resultant Δv as one burn, considerable additional savings can be realized.

3.4 LAUNCH TO RENDEZVOUS

The launch of a spacecraft to rendezvous with another object already in orbit (e.g., a space station) demands the most critical launch timing of any mission. Orbit-plane changes are very expensive, and performing them after achieving orbit would greatly reduce the useful payload. Alternately, flying the launch vehicle some distance before turning into the required orbital plane (a *dogleg maneuver*) is also very expensive, since the velocity achieved by the booster before the turn is largely wasted. The only efficient way to rendezvous with another orbiting spacecraft is to launch when the space-station *orbit* passes directly overhead as seen from the launch site. The station itself does not need to pass over the launch site at the moment of launch, since in-orbit differences can be eliminated with phasing maneuvers.

Figure 3.4 shows the orbit plane projected onto the earth's surface. Assume that the orbit has inclination i and right ascension of the node Ω. The launch site has latitude l. From the figure it can be seen that there will be two launch opportunities per day as the launch site rotates with the earth under the

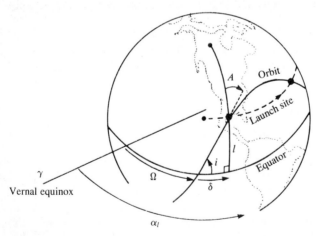

FIGURE 3.4
Coplanar launch to rendezvous.

orbital plane. The first chance comes when the right ascension of the launch site is $\alpha_1 = \Omega + \delta$, while the second will occur when $\alpha_2 = \Omega + 180° - \delta$. To find the angle δ, it is convenient to use spherical trigonometry. In this discipline, a triangle consisting of arcs of great circles on a sphere can be solved in the same manner as plane trigonometry. The only major difference is that both the sides as well as the angles of the triangle are measured as angles.

The spherical triangle shown in Figure 3.4 is a right triangle, since the vertical side, the latitude of the launch site, is perpendicular to the equator. Applying one of Napier's relations for a right spherical triangle, we have

$$\sin \delta = \tan l \cot i \tag{3.8}$$

This relation furnishes δ without sign ambiguity, since $-90° \leq \delta \leq 90°$.

Then, the right ascension of the launch site is given by

$$\alpha = \alpha_g + \lambda_E + \omega_\oplus(t - t_0) \tag{3.9}$$

where α_g is the right ascension of Greenwich at time t_0, λ_E is the east longitude of the launch site, and $\omega_\oplus$ is the inertial rotation rate of the earth. Contrary to popular opinion, the earth only rotates once every 24 h with respect to the *sun*. Since the earth orbits the sun once per year, it travels almost 1°/day along its orbit. However, this means that the earth completes one rotation with respect to the stars in slightly less than one solar day. Precisely, the inertial rotation period of the earth is $23^h56^m4.0905^s$, which is also termed *one sidereal day*. The angle α is termed the *sidereal time* at the launch site by astronomers, who insist upon measuring right ascensions in terms of hours, minutes, and seconds. It is not a time, however, but an angle. The right ascension of Greenwich, α_g, can be found from the *American Ephemeris and Nautical Almanac*.

Adding the angles along the equator in Figure 3.4, the two launch times are given by

$$t_1 = t_0 + \frac{\Omega + \delta - \alpha_0 - \lambda_E}{\omega_\oplus} \tag{3.10}$$

$$t_2 = t_0 + \frac{\Omega + 180° - \delta - \alpha_0 - \lambda_E}{\omega_\oplus} \tag{3.11}$$

This gives the local time of launch as a function of the target-orbit elements and the two constants α_0 and t_0 specifying the earth's inertial orientation.

The launch azimuth A is also of interest. Using another of Napier's rules,

$$\cos i = \cos l \sin A \tag{3.12}$$

or, solving for the azimuth,

$$\sin A = \frac{\cos i}{\cos l} \tag{3.13}$$

This is the inertial azimuth of the launch. Since the launch vehicle already possesses eastward velocity due to the earth's rotation, the apparent azimuth with respect to the earth's surface will differ somewhat from this.

Coplanar launch to rendezvous is perhaps the most restrictive of the criteria which lead to "launch windows." Generally, a rendezvous mission must be launched within at most a few seconds of the precise time. The requirement is so strict that the precession of the target-orbit plane, which causes the node Ω to change by several degrees per day, must be included in the accurate launch-time calculation (see section 3.7).

Now, since the launch must occur when the station's *orbit* passes over the launch site, we have little control over where the station will be *within* the orbit. It is not necessary, however, to launch directly into the same orbit as the space station. If the orbiter is launched into an orbit with a smaller semimajor axis, then it will have a shorter period and will catch up with the higher space station. This is termed a *chase orbit*. Once the proper angular relationship is obtained, the orbiter can execute a small Hohmann transfer to enter the station orbit at the position of the station itself. Many other options are possible, including elliptical chase orbits and using several different orbits in the process of arriving at the station. Rendezvous usually takes 1 to 2 days. This is ordinarily not a problem, especially since the crew will have already put in a full day by the time they reach orbit.

3.5 RELATIVE MOTION AND RENDEZVOUS

The best executed series of maneuvers will not exactly achieve the desired final orbit. Even if it were possible to execute the maneuvers exactly, it would not be desirable to execute the final maneuver of a Hohmann transfer only a few meters from the station. To avoid the possibility of a collision, some standoff distance is necessary. So, an approaching shuttle will find itself in close proximity to the space station, but not exactly at the station. The shuttle is then faced with the problem of closing the intervening distance to rendezvous and doing this in such a way as to comply with operational and safety constraints in the vicinity of the station. A similar problem arises for operations in the vicinity of the station itself, where nearby free-flying platforms must be visited and their relative motion controlled to prevent the possibility of collision. There is a simple solution for the relative motion of two satellites when one is in a circular orbit. Known in astronautics as the *Clohessy-Wiltshire equations*, they were first derived by the celestial mechanician Hill in 1878 for a slightly different purpose.

Consider the two satellites shown in Figure 3.5. The first, which we will call the station, follows a circular orbit at radius r_0 and mean motion $n = (\mu/r_0^3)^{1/2}$. Take the plane of the station's orbit as the reference plane, and describe the position of the second satellite, the orbiter, by the cylindrical coordinates δr, $\delta \theta$, and δz. With respect to the coordinate axes shown in the figure, the position, velocity, and acceleration vectors of the orbiter are

$$\mathbf{r} = (r_0 + \delta r)\mathbf{e}_r + r_0\,\delta\theta\,\mathbf{e}_\theta + \delta z\,\mathbf{e}_z \tag{3.14}$$

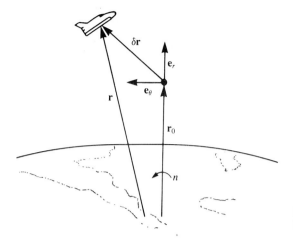

FIGURE 3.5
Dynamics of two close satellites.

$$\frac{{}^e d}{dt}\mathbf{r} = \delta\dot{r}\,\mathbf{e}_r + r_0\,\delta\dot{\theta}\,\mathbf{e}_\theta + \delta\dot{z}\,\mathbf{e}_z \tag{3.15}$$

$$\frac{{}^e d^2}{dt^2}\mathbf{r} = \delta\ddot{r}\,\mathbf{e}_r + r_0\,\delta\ddot{\theta}\,\mathbf{e}_\theta + \delta\ddot{z}\,\mathbf{e}_z \tag{3.16}$$

We have assumed that the relative coordinates are small. But the coordinate axes themselves rotate with the angular velocity of the orbit, $\boldsymbol{\omega} = n\mathbf{k}$, so the inertial acceleration of the shuttle is

$$\frac{{}^i d^2}{dt^2}\mathbf{r} = \frac{{}^e d^2}{dt^2}\mathbf{r} + 2\boldsymbol{\omega}\times\frac{{}^e d}{dt}\mathbf{r} + \boldsymbol{\omega}\times(\boldsymbol{\omega}\times\mathbf{r}) \tag{3.17}$$

Also, the orbiter is itself following a two-body orbit, so its gravitational acceleration is given by

$$\begin{aligned}
\mathbf{a}_g &= -\frac{\mu\mathbf{r}}{r^3} \\
&= -\frac{\mu[(r_0 + \delta r)\mathbf{e}_r + r_0\,\delta\theta\,\mathbf{e}_\theta + \delta z\,\mathbf{e}_z]}{(r_0^2 + 2r_0\,\delta r + \delta r^2 + r_0^2\,\delta\theta^2 + \delta z^2)^{3/2}}
\end{aligned} \tag{3.18}$$

Expanding the denominator by the binomial theorem and assuming small relative coordinates, this becomes

$$\mathbf{a}_g \approx -\frac{\mu r_0 \mathbf{e}_r}{r_0^3} - \frac{\mu}{r_0^3}(-2\delta r\,\mathbf{e}_r + r_0\,\delta\theta\,\mathbf{e}_\theta + \delta z\,\mathbf{e}_z) \tag{3.19}$$

Finally, equating the inertial acceleration of the orbiter to the gravitational acceleration (Newton's second law) and remembering the definition of the orbital mean motion n, we have the relative equations of motion

$$\delta\ddot{r} - 2nr_0\,\delta\dot{\theta} - 3n^2\,\delta r = 0 \tag{3.20}$$

$$r_0\,\delta\ddot{\theta} + 2n\,\delta\dot{r} = 0 \tag{3.21}$$

$$\delta\ddot{z} + n^2\,\delta z = 0 \tag{3.22}$$

These are valid for small displacements in the radial and out-of-plane directions but remain correct for any magnitude change in the in-track coordinate $\delta\theta$.

These are a set of linear, constant-coefficient differential equations and can be solved by the eigenvalue-eigenvector approach. However, in this case that would be unnecessarily cumbersome. The δz equation (3.22) is a simple harmonic oscillator and is decoupled from the in-plane motion of the orbiter. Its solution is given by

$$\delta z(t) = \delta z_0 \cos nt + \frac{\delta\dot{z}_0}{n}\sin nt \tag{3.23}$$

in terms of the initial conditions δz_0, $\delta\dot{z}_0$ at $t = 0$. The $\delta\theta$ equation (3.21) can be rewritten as

$$\frac{d}{dt}(r_0\,\delta\dot{\theta} + 2n\,\delta r) = 0 \tag{3.24}$$

which immediately integrates to give

$$\delta\dot{\theta} = \delta\dot{\theta}_0 + \frac{2n}{r_0}(\delta r_0 - \delta r) \tag{3.25}$$

where the constant of integration has again been evaluated in terms of the initial values δr_0, $\delta\dot{\theta}_0$ at $t = 0$. When this is substituted into the radial equation (3.20), we have

$$\delta\ddot{r} + n^2\,\delta r = 4n^2\,\delta r_0 + 2nr_0\,\delta\dot{\theta}_0 \tag{3.26}$$

This again is a simple harmonic oscillator, only this time with a forcing term. Its complete solution consists of the homogeneous solution

$$\delta r_h = A\cos nt + B\sin nt \tag{3.27}$$

and a constant particular solution

$$\delta r_p = 4\delta r_0 + \frac{2}{n}r_0\delta\dot{\theta}_0 \tag{3.28}$$

Finally, evaluating the constants A and B in terms of the initial conditions, the complete radial solution $\delta r_h + \delta r_p$ becomes

$$\delta r(t) = -\left(\frac{2}{n}r_0\,\delta\dot{\theta}_0 + 3\delta r_0\right)\cos nt + \frac{\delta\dot{r}_0}{n}\sin nt + 4\delta r_0 + \frac{2}{n}r_0\,\delta\dot{\theta}_0 \tag{3.29}$$

To obtain the last piece of the solution, return to (3.25) and substitute the radial solution to find

$$\delta\dot\theta = \left(-3\delta\dot\theta_0 - \frac{6n\,\delta r_0}{r_0}\right) + \left(\frac{6n\,\delta r_0}{r_0} + 4\delta\dot\theta_0\right)\cos nt - \frac{2\delta\dot r_0}{r_0}\sin nt \quad (3.30)$$

This integrates to give the final part of the solution

$$\delta\theta(t) = \delta\theta_0 - \left(3\delta\dot\theta_0 + \frac{6n\,\delta r_0}{r_0}\right)t$$

$$+ \left(\frac{4\delta\dot\theta_0}{n} + \frac{6\delta r_0}{r_0}\right)\sin nt + \frac{2\delta\dot r_0}{nr_0}\cos nt - \frac{2\delta\dot r_0}{nr_0} \quad (3.31)$$

So, the position solution is given by (3.23), (3.29), and (3.31). Taking derivatives of these equations will give the behavior of the relative-velocity components

$$\delta\dot r(t) = (2r_0\,\delta\dot\theta_0 + 3n\,\delta r_0)\sin nt + \delta\dot r_0\cos nt \quad (3.32)$$

$$\delta\dot z(t) = -z_0 n\sin nt + \dot z_0\cos nt \quad (3.33)$$

where $\delta\dot\theta$ is given by (3.30).

Define the relative position vector $\delta\mathbf{r}^T = \{\delta r,\ r_0\,\delta\theta,\ \delta z\}$, and the relative-velocity vector $\delta\mathbf{v}^T = \{\delta\dot r,\ r_0\,\delta\dot\theta,\ \delta\dot z\}$. Then, the solution can be put into the matrix form

$$\delta\mathbf{r}(t) = \Phi_{rr}\,\delta\mathbf{r}(t=0) + \Phi_{rv}\,\delta\mathbf{v}(t=0) \quad (3.34)$$

$$\delta\mathbf{v}(t) = \Phi_{vr}\,\delta\mathbf{r}(t=0) + \Phi_{vv}\,\delta\mathbf{v}(t=0) \quad (3.35)$$

where the four matrices Φ are given by

$$\Phi_{rr} = \begin{bmatrix} 4-3\cos\psi & 0 & 0 \\ 6(\sin\psi - \psi) & 1 & 0 \\ 0 & 0 & \cos\psi \end{bmatrix} \quad (3.36)$$

$$\Phi_{rv} = \begin{bmatrix} \dfrac{1}{n}\sin\psi & \dfrac{2}{n}(1-\cos\psi) & 0 \\ \dfrac{2}{n}(\cos\psi - 1) & \dfrac{4}{n}\sin\psi - \dfrac{3}{n}\psi & 0 \\ 0 & 0 & \dfrac{1}{n}\sin\psi \end{bmatrix} \quad (3.37)$$

$$\Phi_{vr} = \begin{bmatrix} 3n\sin\psi & 0 & 0 \\ 6n(\cos\psi - 1) & 0 & 0 \\ 0 & 0 & -n\sin\psi \end{bmatrix} \quad (3.38)$$

$$\Phi_{vv} = \begin{bmatrix} \cos\psi & 2\sin\psi & 0 \\ -2\sin\psi & -3+4\cos\psi & 0 \\ 0 & 0 & \cos\psi \end{bmatrix} \quad (3.39)$$

where $\psi = nt$. This is the complete Clohessy–Wiltshire solution in two matrix-vector equations.

The solution can be interpreted in terms of the usual two-body problem. The out-of-plane motion δz reflects a slightly different orbital plane and is oscillatory with the period of the original orbit. The radial motion also oscillates with the period of the orbit, and this motion is a slight eccentricity of the orbiter relative to the circular station orbit. Finally, the angular solution for $\delta\theta$ also contains oscillations if an eccentricity is present, but in addition contains a term proportional to the time t. This term produces an in-track drift when the orbiter and the station have slightly different orbital periods. Figure 3.6 shows some typical relative trajectories over several orbital periods.

This solution can be used for many purposes involving proximity operations of two spacecraft. For example, the space station will have several unmanned platforms flying in formation with the main station. The immediate vicinity of a manned spacecraft is a relatively dirty environment, contaminated by leakage and rocket exhausts, and is not a good place to put delicate optics. A manned station cannot provide very fine pointing for telescopes and other instruments, since movement of the crew will disturb the orientation of the station. For the same reason, a manned station cannot actually provide pure "0g," both because of crew movement and because the large size of the station will induce slight gravitational accelerations at points on the station far from the center of mass.

Unmanned platforms eliminate these difficulties, but they must be kept close to the main station if they are to be conveniently visited. To keep the platform in the vicinity of the station, the term proportional to the time in

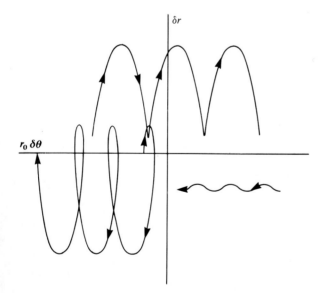

FIGURE 3.6
Spacecraft relative orbits.

(3.31) must be suppressed. This means that

$$\delta\dot{\theta}_0 + \frac{2n}{r_0}\,\delta r_0 = 0 \tag{3.40}$$

to prevent long-term drift. Various possible relative trajectories are possible, some keeping the same distance between the platform and station and some allowing this distance to vary.

The solution (3.34) can also be used to calculate rendezvous trajectories. Pick the intended maneuver time as $t = 0$, and chose the arrival time T. If the desired position components are known at T, then (3.34) becomes

$$\delta\mathbf{v}(t = 0) = (\Phi_{rv})^{-1}[\delta\mathbf{r}(T) - \Phi_{rr}\,\delta\mathbf{r}(t = 0)] \tag{3.41}$$

where the starting position $\delta\mathbf{r}(t = 0) = \{\delta r,\, r_0\,\delta\theta,\, \delta z\}^T$ is known at $t = 0$, but the initial relative-velocity vector $\delta\mathbf{v}(t = 0) = \{\delta\dot{r},\, r_0\delta\dot{\theta},\, \delta\dot{z}\}^T$ at $t = 0$ is not known. (The orbiter's velocity vector at $t = 0$ is indeed known, but it is not the correct one, or we would not be performing a maneuver.) If the orbiter is trying to rendezvous with the station itself, then the desired final position components will be zero (or nearly zero). If a visit is planned to a coorbiting platform, the final position coordinates will be the position of the platform at the intercept time T.

Equation (3.41) gives the relative velocity required to arrive at the desired final position at time T. Subtracting this velocity from the orbiter's velocity before the maneuver will produce the required maneuver (magnitude and direction) at $t = 0$. A similar subtraction at the final time will produce the second required maneuver. However, the final time T is arbitrary, and we can vary T over a reasonable range to search for the minimum total-velocity-change solution. So, repeating the solution (3.41) for many different times T and calculating the total Δv for each case should reveal the minimum energy-rendezvous solution.

3.6 DECAY LIFETIME

Satellites in low earth orbits have finite lifetimes due to the effect of atmospheric drag. Although the atmosphere at several hundred kilometers of altitude is an almost perfect vacuum by earthly standards, the satellite must travel at very high speed through this low-density medium for years. The effects of drag are cumulative and eventually become significant. A satellite in a low orbit suffers an acceleration due to drag given by

$$a_d = \frac{1}{2}\frac{C_d A}{m}\,\rho v^2 \tag{3.42}$$

where C_d is the ballistic coefficient, A is the presented area of the satellite, and m is its mass. These three quantities are not normally determinable separately, so they are grouped into the single quantity

$$B^* = \frac{C_d A}{m}$$

(3.43)

The other quantities in (3.42) are ρ, the atmospheric density, and v, the satellite velocity.

The first effect of air drag on a satellite orbit is to circularize the orbit. This occurs since the effect of drag is much more pronounced near perigee, when the satellite is deepest in the atmosphere. The satellite is slowed at this point, lowering the apogee height until it nearly equals the perigee height. We will assume that the orbit has circularized and study the behavior of the semimajor axis. With a circular orbit, all the air drag force is in the tangential direction. For simplicity, we will assume that the atmosphere is exponential (see section 9.2), with air density given by

$$\rho = \rho_0 e^{-(r-R_e)/h}$$

(3.44)

where ρ_0 is a fictitious base density of the atmosphere, R_e is the radius of the earth, and h is called the *atmospheric scale height*. At the top of the atmosphere, h is approximately 6 to 8 km.

So, assume a circular orbit, where $a = r$ and the satellite velocity is given by

$$v = \sqrt{\frac{\mu}{r}}$$

(3.45)

Air drag will change the specific energy of the orbit, since the drag acceleration a_d does work on the satellite at a rate of $\mathbf{a}_d \cdot \mathbf{v} = -v a_d$. The negative sign is necessary because drag opposes the velocity. Remember that the energy and semimajor axis a are related by $E = -\mu/2a$, so by calculating the rate of change of the energy and equating this to the work rate due to air drag, we obtain

$$\frac{dE}{dt} = \frac{\mu}{2a^2}\frac{da}{dt} = -v a_d$$

(3.46)

Substituting in the above from (3.42) to (3.45), the equation of motion for the semimajor axis becomes

$$\frac{da}{dt} = -\sqrt{\mu a}\, B^* \rho_0 e^{-(a-R_e)/h}$$

(3.47)

The variables in this equation can be separated to yield the two definite integrals

$$\int_{a_0}^{a} \frac{e^{(a-R_e)/h}}{\sqrt{a}}\, da = -\sqrt{\mu}\, B^* \rho_0 \int_{t_0}^{t} dt$$

(3.48)

We are interested in finding an approximate solution to (3.48), without, of course, totally destroying the physical significance of the problem by making excessively brutal simplifications. Let us begin by changing variables on the left side of (3.48) to altitude above the surface, $H = a - R_e$, to obtain

$$\int_{H_0}^{H} \frac{e^{H/h}}{\sqrt{R_e + H}} \, dH = -\sqrt{\mu} \, B^* \rho_0 (t - t_0) \qquad (3.49)$$

The radical in the denominator is proportional to the satellite velocity, which does not greatly increase during the last few hundred kilometers of decay. We can thus approximate this quantity by the root of R_e alone. This avoids expanding the exponential function, which is not desirable since the air density *does* vary enormously during a decay trajectory. With this simplification, the integral above can be performed by simple processes to yield

$$\frac{h}{\sqrt{R_e}} (e^{H/h} - e^{H_0/h}) = -\sqrt{\mu} \, B^* \rho_0 (t - t_0) \qquad (3.50)$$

This may be solved to give the behavior of altitude with time as

$$H(t) = h \ln \left[e^{H_0/h} - \frac{\sqrt{\mu R_e} \, B^* \rho_0}{h} (t - t_0) \right] \qquad (3.51)$$

A typical satellite in a low earth orbit might have an initial altitude of from 20 to 40 scale heights. In this case, the first term in the logarithm is huge and dominates for a very long time. Thus (3.51) reduces to the statement $H = H_0$. However, eventually the linear term will become comparable to the first term, and then the altitude decreases very quickly indeed. An estimate of the time to elapse to decay can be obtained by setting the altitude in (3.51) to zero. Since the terminal descent is very rapid, this introduces little error. The satellite reaches zero altitude when the argument of the logarithm function equals unity. Interpreting the time difference $t - t_0$ as the time to elapse to decay t_d, we easily find

$$t_d = \frac{h}{\sqrt{\mu R_e} \, B^* \rho_0} (e^{H_0/h} - 1) \qquad (3.52)$$

As expected, the satellite lifetime is inversely proportional to both B^* and ρ_0, only one of which is under the control of the engineer. However, the satellite lifetime is an exponentially growing function of initial orbital altitude. So, rather than try to design streamlined satellites, the easiest way to extend the lifetime of a low earth satellite is to place it in a slightly higher orbit. An increase of H_0 by one scale height (6 to 8 km) will extend the satellite lifetime by almost a factor of 3.

In the analysis above, the most questionable assumption was the model for the air density. At any given moment, the air density is approximately exponential, but the base density of the atmosphere ρ_0 and the scale height h may change drastically in a short period of time. This can occur during a solar

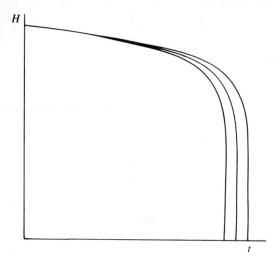

H

t

FIGURE 3.7
Altitude versus time during final decay.

storm. Also, they change with the phase of the 11-year sunspot cycle. This makes long-term decay predictions very inaccurate.

Figure 3.7 shows three decay trajectories as a function of time. They differ only in the scale height *h* for the atmosphere, and these differed by only 5%. At the beginning, it is *extremely* difficult to tell these trajectories apart. So, even if *h* were a dependable constant of the atmosphere, it would still be very difficult to predict the final plunge into the atmosphere. This fact renders even an accurate orbital fit in the predecay environment of very doubtful value in accurately predicting the final decay time. In practice, it is seldom possible to predict the actual decay date until about a week before decay, and the final decay orbit cannot be predicted until a few hours in advance.

3.7 EARTH OBLATENESS EFFECTS

Besides air drag, the most pronounced departure from two-body motion for a low earth satellite is caused by the fact that the earth is not perfectly spherical. Because the earth rotates, it bulges at the equator by almost 20 km. The extra mass distributed around the earth's equator leads to two major effects in the orbit of a low earth satellite.

The extra mass about the equator leads to a torque on the orbit, as shown in Figure 3.8. When the satellite is at the points in its orbit furthest from the equator, the extra force due to the bulge causes a net torque on the orbit. When the satellite is crossing the equator, the extra mass simply increases the inward radial acceleration. Averaged over the entire orbit, the effects of this torque produce gyroscopic effects in the motion of the orbit. The bulge causes the orbit plane to precess at the rate

$$\dot{\Omega} = -\frac{3nJ_2R_\oplus^2}{2a^2(1-e^2)^2}\cos i \qquad (3.53)$$

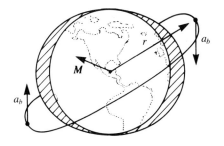

FIGURE 3.8
The earth's equatorial bulge.

where a, e, and i are the usual orbital elements, n is the mean motion, $R_\oplus$ is the radius of the earth, and J_2 is a dimensionless number that characterizes the departure of the earth from a sphere. It has the value $J_2 = 0.001082$ for the earth. The *regression of the nodes*, as this effect is called, causes the orbit plane to precess exactly like the motion of a top.

The second major effect is less obvious from the top analogy, but it occurs in that system as well. The elliptic orbit also rotates in its own plane at a rate

$$\dot{\omega} = -\frac{3nJ_2R_\oplus^2}{2a^2(1-e^2)^2}\left(\tfrac{5}{2}\sin^2 i - 2\right) \tag{3.54}$$

This effect is termed the *advance of the perigee*. However, the sign of $\dot{\omega}$ can be either positive or negative depending on the inclination.

These effects cause the two-body elements to increase or decrease linearly with time:

$$\omega = \omega_0 + \dot{\omega}t \tag{3.55}$$

$$\Omega = \Omega_0 + \dot{\Omega}t \tag{3.56}$$

In contrast, the semimajor axis a, the eccentricity e, and the inclination i only undergo small periodic changes as the satellite orbits the earth.

There are two very important applications of these oblateness effects. Low-orbit weather satellites (the Nimbus series) and earth-resources satellites (the Landsats) are launched into orbits where the node regresses at precisely 360°/year, or slightly less than 1°/day. As shown in Figure 3.9, this locks the satellite's orbital plane into a constant relationship to the sun, and such satellites are called *sun-synchronous*. Since such satellites are usually launched around midday local time, this ensures that forever afterward half the satellite's orbit will be in sunlight and that the satellite will not end up orbiting over the earth's "twilight zone." It also has the effect of ensuring that all photographs taken by the satellite are taken at the same local time at the area being photographed. This is a great aid in photointerpretation, since the user does not have to deal with the effects of different solar illumination. In fact, the different satellites of the Landsat series are known by their time of day: the 10 A.M. satellite, the 2 P.M. satellite, and so forth. Equating the nodal secular rate to the desired value supplies one relation between a, e, and i. If we wish

88 SPACEFLIGHT DYNAMICS

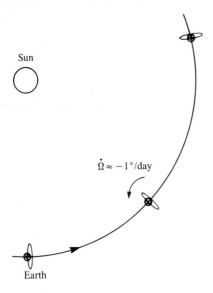

Sun

$\dot{\Omega} \approx -1°/day$

Earth

FIGURE 3.9
Sun-synchronous orbits.

our sun-synchronous satellite orbit to be circular, then we find a relation between a and i. It is extremely convenient that for the earth, at least, sun-synchronous orbits are nearly polar. This makes possible an orbit combining global coverage and constant illumination angle in one satellite.

The second application concerns the rate of advance or regression of the line of apsides, the line joining perigee and apogee. Setting $d\omega/dt$ to zero, we find that the orbit does not rotate when

$$\tfrac{5}{2}\sin^2 i - 2 = 0 \tag{3.57}$$

This yields the solution

$$i_c = \sin^{-1}\frac{2}{\sqrt{5}} \tag{3.58}$$

The two angles $i_c = 63.4349°$ or $i_c = 116.5650°$ are called the *critical inclinations*. They do not depend on the value of J_2 for the planet being orbited. An orbit at these inclinations will not rotate in its own plane. This type of orbit is used extensively by the Soviet Union for its high-eccentricity Molyniya communications satellites. Substantial portions of the Soviet Union are so far north that satellites in geosynchronous orbit are not visible. Molyniya satellites are placed in highly eccentric orbits with a period of about 12 h. Since the apogee must be in the northern hemisphere to be useful for high-latitude communications, the Russians need $\omega = 270°$. Furthermore, if ω changes with time, apogee would migrate into the southern hemisphere, and the satellite would be useless for an extended period of time. By launching these satellites at the precise value of the critical inclination, this problem can be avoided.

3.8 LOW-THRUST ORBIT TRANSFER

At the opposite end of the spectrum from impulsive maneuvers are the trajectories which must be used with very low thrust vehicles. Virtually all propulsion systems with very high exhaust velocities have very low absolute thrusts. Such vehicles will be capable of boosting very large payloads into high-velocity trajectories, with far lower fuel requirements than conventional chemical rockets. However, they may take a considerable time to achieve the required high velocity.

Consider the problem of transfer from a low circular orbit to a higher circular orbit with a very low thrust vehicle. The fact that we wish to increase the semimajor axis of the orbit a means that the total two-body energy $E = -\mu/2a$ must also change. Taking the derivative of this relation with respect to time, we find

$$\frac{dE}{dt} = \frac{\mu}{2} a^{-2} \frac{da}{dt} \tag{3.59}$$

Now, the specific energy E can only change if the low-thrust propulsion system does work on the vehicle. The rate at which the propulsion system performs work on the spacecraft is given by the usual relation $dE/dt = \mathbf{A} \cdot \mathbf{v}$. Here, instead of the absolute force $\mathbf{F}$, we must use the specific force $\mathbf{F}/m = \mathbf{A}$, the acceleration of the vehicle.

To make the most efficient use of the low-thrust propulsion system, we want to maximize the rate at which the energy of the orbit increases. This means that the dot product $\mathbf{A} \cdot \mathbf{v}$ must be maximized. Since we will assume that the vehicle acceleration is already as high as possible, and since the velocity of the spacecraft is beyond our immediate control, the dot product is maximized when the vehicle acceleration and velocity vectors are aligned. This means that the spacecraft must accelerate along the forward direction in its circular orbit.

If the vehicle has a very low acceleration, then the orbit is likely to remain nearly circular. The resulting trajectory will then be a slow outward spiral. In this case, we may substitute for the vehicle velocity the velocity in the instantaneous circular orbit $v = \sqrt{\mu/a}$ to find

$$\frac{dE}{dt} = A \left(\frac{\mu}{a} \right)^{1/2} = \frac{\mu}{2} a^{-2} \frac{da}{dt} \tag{3.60}$$

This yields an equation of motion for the semimajor axis of the orbit as

$$\frac{da}{dt} = \frac{2}{\sqrt{\mu}} a^{3/2} A \tag{3.61}$$

The variables a and time t are easily separated, yielding the two definite integrals

$$\int_{a_0}^{a} \frac{da}{a^{3/2}} = \frac{2}{\sqrt{\mu}} A \int_{t_0}^{t} dt \tag{3.62}$$

The integrations are simple, and give one form of the solution as

$$a_0^{-1/2} - a^{-1/2} = \frac{A}{\sqrt{\mu}}(t - t_0) \tag{3.63}$$

In this form it is easy to calculate the time required to escape from the earth. The orbit becomes parabolic when $a \to \infty$. This occurs in the finite time

$$t_{esc} = t - t_0 = \frac{1}{A}\sqrt{\frac{\mu}{a_0}} \tag{3.64}$$

Actually, the assumption that the orbit remains circular will break down as the spacecraft approaches escape speed. However, (3.64) does provide a useful approximation for the spiral escape time.

To complete the solution, (3.63) could be solved for the explicit dependence of the semimajor axis as a function of time. It is more useful, however, to solve for the time required to complete a given transfer:

$$t - t_0 = \frac{\sqrt{\mu}}{A}(a_0^{-1/2} - a^{-1/2}) \tag{3.65}$$

This gives the time the spacecraft must accelerate as a function of the radii of the starting and ending circular orbits. The total velocity change expected of the spacecraft will be the time the engines must run multiplied by the acceleration of the spacecraft, or

$$\Delta v_{tot} = A(t - t_0) = \sqrt{\frac{\mu}{a_0}} - \sqrt{\frac{\mu}{a}} \tag{3.66}$$

This is equal to the difference in the circular orbital speeds of the initial and final orbits and is independent of the acceleration of the spacecraft.

As an example, consider a low-thrust transfer from a circular parking orbit at 200 km altitude ($a_0 = 6578$ km) to geosynchronous orbit at $a = 42,164$ km. If the spacecraft has an acceleration of 0.01 m/s^2 (about one "milligee," or a thousandth of the acceleration gravity), the transfer can be completed in about 5.45 days. The total velocity-change requirement is given by (3.66) as 4.71 km/s. This is somewhat higher than the 3.9 km/s required to perform a Hohmann transfer to the same final orbit. The difference between the spiral Δv and the Hohmann value is termed the *kinematic inefficiency* of low-thrust orbit transfer. However, this extra Δv is bought at a great decrease in total fuel required for the low-thrust vehicle.

Of course, if the spacecraft's acceleration were only 0.001 m/s^2, the transfer would require 54 days, while the total Δv would remain unchanged. This forces the vehicle to spend a very long time within the Van Allen belts (see Chapter 9), with the attendant radiation damage to the vehicle and payload electronics and solar cells. Possibly long transit times across the Van Allen belts are one drawback of very low thrust orbit transfer vehicles. However, the advantages they offer in terms of delivering much larger payloads to high orbit makes them attractive, if the trip times are reasonable.

3.9 REFERENCES AND FURTHER READINGS

In this chapter we have seen examples of both maneuver strategies and perturbation theory. Maneuvers are discussed in most astronautically oriented texts on orbital theory. For example, Bate, Mueller, and White and Thomson both discuss the Hohmann transfer and other impulsive maneuvers. The emphasis in determining a sequence of maneuvers is to minimize the total Δv requirement. To prove that the Hohmann transfer is optimal would take us into the discipline of optimal control. Since maneuvering fuel must first be brought up from the earth's surface, it is quite "expensive," and minimizing the total Δv not only saves launch costs but also maximizes the payload delivered to the final orbit. Further examples of optimal maneuvers are found in the problems in this chapter and in Chapter 11.

Perturbation theory can be applied both to natural forces (e.g., air drag and the earth's oblateness) and to low-thrust orbit transfer. Both Danby and Fitzpatrick discuss oblateness and air-drag perturbations on a satellite orbit. Low-thrust orbit transfer is again usually studied within the discipline of optimal control, not to minimize fuel usage but to minimize total flight time of these vehicles.

Bate, R. R., D. D. Mueller, and J. E. White, *Fundamentals of Astrodynamics,* Dover, New York, 1971.
Thomson, W. T., *Introduction to Space Dynamics,* Wiley, New York, 1961.
Danby, J. M. A., *Fundamentals of Celestial Mechanics,* Macmillan, New York, 1962.
Fitzpatrick, P. M., *Principles of Celestial Mechanics,* Academic, New York, 1970.

3.10 PROBLEMS

1. A geosynchronous orbit has a period of $23^h56^m4.09^s$, or one sidereal day. Calculate the radius of a geosynchronous orbit. Starting from a parking orbit at $r_1 = 6578$ km and an inclination of 28°, calculate the two maneuvers required to reach geosynchronous, equatorial orbit. Do the required 28° inclination change at apogee as one burn by combining it with the second Hohmann transfer maneuver.

2. A lunar module (LM) lifts off from the lunar surface and flies a powered trajectory to its burnout point at 30 km altitude, as shown in Figure 3.10. The velocity vector of the LM is parallel to the lunar surface at burnout. It then coasts halfway around the moon, where it must rendezvous with the Apollo command module (CM) in a 250-km circular orbit. The mass of the moon is 0.0123 times the mass of the earth, and the radius of the moon is 1740 km.

   (*a*) What is the burnout speed of the LM in kilometers per second?
   (*b*) What is the magnitude of the maneuver required to rendezvous with the command module in kilometers per second?
   (*c*) What is the coast time for the LM? To assure a quick rendezvous, it is desirable that the LM and CM arrive at the rendezvous point *together* (or at least close). Where must the CM be in relation to the LM at burnout? Cite your answer as either a time differential or an angle differential of the CM ahead or behind of the LM burnout point. (This sets the "launch window" for the LM takeoff.)

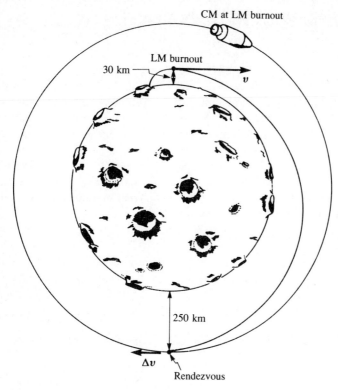

CM at LM burnout

LM burnout

30 km

v

250 km

Δv

Rendezvous

FIGURE 3.10
Lunar Module ascent trajectory.

3. A low-thrust orbit transfer vehicle is in a circular orbit and is attempting to change its orbital inclination i. The radius of the orbit is a, and the vehicle has a constant (low!) acceleration $\mathbf{A}$. As shown in Figure 3.11, consider the thrust profile where

$$\mathbf{A} = \begin{cases} -A\mathbf{w} & -90° < v < 90° \\ A\mathbf{w} & 90° < v < 270° \end{cases}$$

where $\mathbf{w}$ is the orbit normal vector.

(a) Write $\mathbf{r}$ in terms of the in-plane unit vectors $\mathbf{p}$ and $\mathbf{q}$, and evaluate the specific torque on the orbit, $\mathbf{M} = \mathbf{r} \times \mathbf{A}$. As the thrust is low, integrate this over one orbit and show that the change in the angular-momentum vector is

$$\Delta \mathbf{H} = \int_0^T \mathbf{M}\, dt = 4Aa \sqrt{\frac{a^3}{\mu}}\, \mathbf{q}$$

[*Hint*: Show $dv/dt = (\mu/a^3)^{1/2}$.]

(b) Argue that, since the change in inclination per orbit is small, the change in the inclination is

$$\Delta i \approx \frac{|\Delta \mathbf{H}|}{H} = \frac{4a^2 A}{\mu}$$

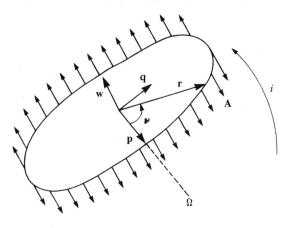

FIGURE 3.11
Low-thrust inclination change.

4. A satellite leaves a parking orbit at inclination i and executes a Hohmann transfer to geosynchronous equatorial orbit, as shown in Figure 3.12. Part of the required inclination change Δi_1 is performed during the first maneuver, and the remainder $\Delta i_2 = i - \Delta i_1$ is done during the second maneuver. If the speeds in the circular orbits are v_{c1} and v_{c2}, respectively, and the perigee and apogee speeds in the Hohmann transfer ellipse are v_p and v_a, show that the total Δv for both maneuvers is

$$\Delta v = (v_{c1}^2 + v_p^2 - 2v_{c1}v_p \cos \Delta i_1)^{1/2} + [v_{c2}^2 + v_a^2 - 2v_{c2}v_a \cos (i - \Delta i_1)]^{1/2}$$

Write the condition that minimizes the total required Δv as a function of Δi_1, the amount of inclination change performed during the first maneuver. Argue that $\Delta i_1 = 0$ (as in problem 1) is *not* optimal.

5. When the orbit of a satellite is projected back onto the rotating earth, the resulting curve is called a *ground trace*. As shown in Figure 3.13, this often looks nearly sinusoidal for a low-altitude circular orbit. Show that tracks for sequential orbits are separated by a longitude difference of

$$\Delta \lambda = \frac{2\pi \omega_\oplus a^{3/2}}{\sqrt{\mu}}$$

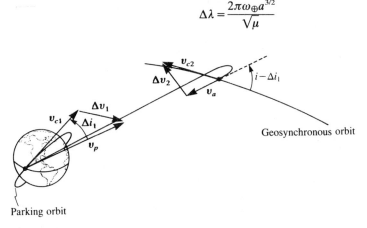

Geosynchronous orbit

Parking orbit

FIGURE 3.12
Hohmann transfer with two inclination changes.

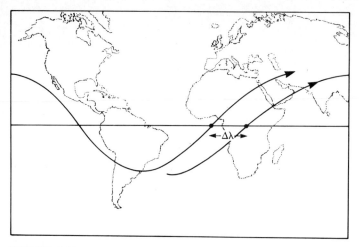

FIGURE 3.13
Ground-trace separation at the equator.

at the equator, where $\omega_\oplus$ is the angular rotation rate of the earth. Also, show that the satellite can see a swath of angular width

$$\theta = 2 \cos^{-1} \frac{R_\oplus}{a}$$

measured from the center of the earth. Compare θ and $\Delta\lambda$ numerically from $a = 1R_\oplus$ to $a = 3R_\oplus$, where $R_\oplus$ is the radius of the earth. For overlapping coverage, we wish that $\theta > \Delta\lambda$. Are there any altitude bands in this range of a for which this is not possible?

6. Geosynchronous satellites can be restationed using very small maneuvers. The satellite is initially in a circular orbit at semimajor axis a_0, and a small in-track maneuver changes its speed from v_0 to $v_0 + dv$, and the semimajor axis changes to $a_0 + da$. Taking differentials of the energy relation for two-body orbits, (2.15) and (2.36), show that

$$dv \approx \frac{\mu}{2v_0 a_0^2} da$$

Using the mean-motion definition (2.45), argue that the satellite now drifts in longitude at a rate

$$\dot{\lambda} \approx -\tfrac{3}{2}\sqrt{\mu}\, a_0^{-5/2}\, da$$

for small da and dv. If the satellite is to be moved *eastward* along the equator, what is the direction of the initial maneuver? Explain, in words, *why* this should be so.

7. The Hohmann transfer is not always optimal in terms of total Δv. Calculate the total velocity change from a low earth orbit (say, 1.03 earth radii) to the orbit of the moon at 60 earth radii. Compare this to the three-maneuver strategy of (a) escaping to infinity on a parabolic orbit, (b) a zero Δv maneuver to approach the moon's orbit on another tangent parabola, and (c) circularizing in the moon's orbit. Use at least five significant figures. Can you locate the crossover point (the ratio a_2/a_1) where the escape option becomes cheaper?

CHAPTER
4

RIGID-BODY DYNAMICS

4.1 INTRODUCTION

Until now we have treated the dynamics of particles: masses treated as if they are dimensionless points. However, any actual object has physical size and can thus rotate as well as translate. If the individual atoms in an object are fixed in relation to each other, the object is termed a *rigid body*. As you will see by the naming conventions throughout this chapter, much of rigid-body dynamics is due to the great mathematician Leonard Euler. Of course, our purpose in laying the foundations of this field is to apply rigid-body dynamics to the rotational motion of satellites and other space vehicles.

A rigid body can be thought of as a collection of many individual particles of mass, with very many constraints. In particular, none of the distances between individual particles can change as the body moves and rotates. The first question to be resolved is how many degrees of freedom a rigid body possesses. Consider building a rigid body particle by particle. The first mass particle can move freely, without limitations. Thus, the first particle possesses three degrees of freedom associated with the translational motion of the rigid body as a whole. The placement of the second mass particle is constrained by the fixed distance between it and the first particle. The second particle must thus lie somewhere on the surface of a sphere centered on the first, and two more degrees of freedom are obtained. The third particle must lie at given distances from both the first and second particles, and in general its motion is constrained to lie on a circle whose axis is the line joining the first and second particles. The third particle thus adds one additional degree of freedom. The placement of the fourth particle is subject to three constraints, since its distance from each of the first three particles is known. Its position is thus completely specified, and no additional degrees of freedom are gained with the fourth or subsequent mass particles.

A rigid body is thus a dynamical system with, in general, six degrees of freedom. Three degrees of freedom are associated with the translational motion of some given point in the body, usually the center of mass, while three degrees of freedom describe the rotational motion of the system. The equations of motion for the translational degrees of freedom are easily found. Write Newton's law of motion for each of the particles in the system

$$\mathbf{F}_i = \mathbf{f}_{ie} + \sum_{j \neq i}^{N} \mathbf{f}_{ij} = m_i \mathbf{a}_i \tag{4.1}$$

where the force $\mathbf{F}_i$ on particle i has been broken into the *external* force on this particle, $\mathbf{f}_{ie}$, and the *internal* forces, where the force on particle i due to particle j is $\mathbf{f}_{ij}$. If we sum all these equations of motion we have

$$\sum_{i=1}^{N} \mathbf{f}_{ie} + \sum_{i=1}^{N} \sum_{j \neq i}^{N} \mathbf{f}_{ij} = \sum_{i=1}^{N} m_i \mathbf{a}_i \tag{4.2}$$

where the sums extend over the rigid body. Now, because of Newton's third law, for every force $\mathbf{f}_{ij}$ there is an equal but opposite force $\mathbf{f}_{ji}$. Thus, the internal forces sum exactly to zero on the left side of (4.2).

Since a rigid body contains very many particles (e.g., atoms), we commit very little error in replacing the summations with integrals. Define the total external force on the rigid body as

$$\mathbf{F}_e = \int d\mathbf{f}_{ie} = \sum_{i=1}^{N} \mathbf{f}_{ie} \tag{4.3}$$

where $d\mathbf{f}_{ie}$ is the differential force on a small mass element and the integral extends over the rigid body. Also, define the position of the center of mass as

$$\mathbf{r}_c = \frac{1}{M} \int \mathbf{r}_i \, dm = \frac{1}{M} \sum_{i=1}^{N} m_i \mathbf{r}_i \tag{4.4}$$

where M is the total mass of the body and dm is the mass of a small volume element of the body. Then, (4.2) becomes

$$\mathbf{F}_e = M \frac{d^2}{dt^2} \mathbf{r}_c \tag{4.5}$$

That is, the center of mass behaves as if all the mass of the rigid body were concentrated at that point, and the total external force acted there. Thus, three of the six component equations of motion for a rigid body are simply the familiar equations of motion for a mass particle. In fact, this is the major reason we can treat extended objects as if they were particles when considering the dynamics of their center of mass. The other three equations of motion for the rotational degrees of freedom will be the major topic for this chapter.

4.2 THE CHOICE OF ORIGIN

In the first section we saw the special status of the center of mass of the rigid body for formulating the translational equations of motion. It is natural to inquire if this point has similar advantages for the formulation of the rotational equations of motion. Now, the rotational motion of the body is in large part described by

$$\mathbf{M} = \dot{\mathbf{H}} \tag{4.6}$$

where $\mathbf{M}$ is the applied torque and $\mathbf{H}$ is the total angular momentum. The time derivative on the right of (4.6) must be taken with respect to inertial space. However, in the calculation of $\mathbf{H}$ we have a choice of the origin about which to calculate the moments. As shown in Figure 4.1, it is possible to use a point O in the calculation of $\mathbf{H}$ which may not be an inertial origin. In this section we will ask if there are any points other than inertial origins having advantages for the calculation of $\mathbf{H}$.

Let us begin with the statement of Newton's second law, $\mathbf{dF} = dm\,\ddot{\mathbf{r}}$ for a small mass element dm, and take moments about point O in Figure 4.1. Then, the total moment (torque) about point O is given by

$$\mathbf{M}^o = \int (\mathbf{d} + \boldsymbol{\rho}) \times \mathbf{dF} = \int (\mathbf{d} + \boldsymbol{\rho}) \times \ddot{\mathbf{r}}\, dm \tag{4.7}$$

where the integrals extend over the entire rigid body. Now, the inertial acceleration $\ddot{\mathbf{r}}$ can be written as

$$\ddot{\mathbf{r}} = \ddot{\mathbf{R}} + \ddot{\mathbf{d}} + \ddot{\boldsymbol{\rho}} \tag{4.8}$$

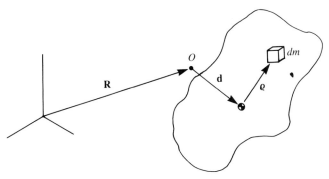

FIGURE 4.1
Calculation of the angular momentum.

where $\mathbf{R}$ locates point O, $\mathbf{d}$ gives the offset of point O from the center of mass, and $\boldsymbol{\rho}$ locates the mass element dm with respect to the center of mass. When this is inserted in the right side of (4.7) and expanded, we have

$$\mathbf{M}^o = \int (\mathbf{d} + \boldsymbol{\rho}) \times (\ddot{\mathbf{d}} + \ddot{\boldsymbol{\rho}}) \, dm + \int \mathbf{d} \times \ddot{\mathbf{R}} \, dm + \int \boldsymbol{\rho} \, dm \times \ddot{\mathbf{R}} \qquad (4.9)$$

The first term on the right is the rate of change of the angular momentum about the point O. That is, it is the time derivative of

$$\mathbf{H}^o = \int (\mathbf{d} + \boldsymbol{\rho}) \times (\dot{\mathbf{d}} + \dot{\boldsymbol{\rho}}) \, dm \qquad (4.10)$$

The last term in (4.9) is zero, since this integral effectively computes the position of the center of mass seen from the center of mass:

$$\int \boldsymbol{\rho} \, dm \equiv 0 \qquad (4.11)$$

In the second integral on the right side of (4.9) it is possible to remove the quantity $\mathbf{d} \times \ddot{\mathbf{R}}$, since this quantity does not depend on the variables of integration. The remaining integral $\int dm = \mathrm{M}$, the total mass of the body. So, (4.9) simplifies to the statement that

$$\mathbf{M}^o = \dot{\mathbf{H}}^o + M\mathbf{d} \times \ddot{\mathbf{R}} \qquad (4.12)$$

The last term above accounts for the completely arbitrary choice of the point O. We have not specified anything about the motion of O, either with respect to the inertial frame or with respect to the body. However, this extra term vanishes for two important special cases. First, if point O is itself an inertial origin, then $\ddot{\mathbf{R}} \equiv 0$. This case would arise, for example, if the rigid body were pivoted about point O. The second special case occurs when $\mathbf{d} = 0$. The point O now coincides with the center of mass of the rigid body, and the extra term in (4.12) again vanishes. So, if point O is *either* an inertial origin *or* the center of mass,

$$\mathbf{M}^o = \dot{\mathbf{H}}^o \qquad (4.13)$$

It is always possible to use one of these two special cases to formulate the equations of motion of a rigid body. In particular, in satellite dynamics it is usually necessary to use the center of mass, since the nearest inertial origin may be far away and inconveniently located. The time derivative in (4.13) must still be calculated with respect to inertial space.

Now, return to the calculation of the torque about point O and assume O is the center of mass. Separate the force $\mathbf{dF}$ on the mass element dm into its external component $\mathbf{df}_{ex}$ and those internal forces impressed by other parts of the rigid body, $\mathbf{df}_{in}$. Then, the torque becomes

$$\mathbf{M}^o = \int \boldsymbol{\rho} \times \mathbf{df}_{ex} + \int \boldsymbol{\rho} \times \mathbf{df}_{in} \qquad (4.14)$$

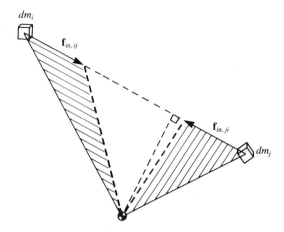

FIGURE 4.2
Cancellation of internal torques.

Now, Newton's third law still holds for the forces between any two mass elements. These internal forces must be equal and opposite. However, almost all physical forces go one step further, and the force between any two particles will act along the line joining the particles. The only exception to this is the magnetic force. However, this is not the force responsible for keeping rigid bodies rigid. From Figure 4.2, whenever the internal forces act along the line joining the two particles, the internal torques cancel in pairs. The magnitude of the component torques are twice the areas of the two shaded triangles. The triangles have equal bases (the two forces) and the same altitude, so their areas are equal. The cross products in (4.14) have a magnitude equal to the length of the perpendicular lever arm (the altitude of the triangles) times the magnitude of the force (the bases of the triangles). But this is simply twice the area of the shaded triangles, so the cross products are equal in magnitude. The directions are opposite, since the two forces have opposite directions.

The conclusion is that internal forces are not capable of exerting a torque on a rigid body. Just as you cannot lift yourself by your bootstraps, you cannot spin yourself by your bootstraps, either. The torque on the body then becomes

$$\mathbf{M}^o = \int \boldsymbol{\rho} \times d\mathbf{f}_{ex} \tag{4.15}$$

involving only the *external* forces on the body.

As an example, consider the top shown in Figure 4.3. The external forces acting on the top are gravity and the reaction force of the tabletop. We are interested in the motion of the top assuming that the pivot point is fixed, so this will be our choice for the origin O. Only gravity will then contribute to the total torque on the top. Since the force on the element of mass dm is $d\mathbf{f}_{ex} = dm\,\mathbf{g}$, equation (4.15) becomes

$$\mathbf{M}^o = \int \boldsymbol{\rho} \times g \, dm = \int \boldsymbol{\rho} \, dm \times \mathbf{g} \tag{4.16}$$

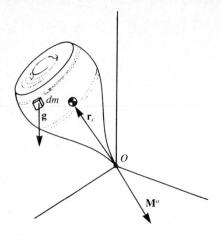

FIGURE 4.3
Gravitational torque on a top.

since **g** is a constant vector. Then, since the origin O is not the center of mass, $\int \boldsymbol{\rho} \, dm = M\mathbf{r}_c$, where $\mathbf{r}_c$ locates the center of mass. We have

$$\mathbf{M}^o = \mathbf{r}_c \times M\mathbf{g} \tag{4.17}$$

This states that gravity acts on the top as if all the mass were concentrated at the center of mass. In fact, the center of mass is sometimes referred to by the confusing term "center of gravity." The two points are the same *only* when gravity can be treated as a constant force across the rigid body. We will see several examples where this is not the case.

4.3 ANGULAR MOMENTUM AND ENERGY

A rigid body consists of a large number of individual particles. The total angular momentum will be the sum of the angular momentum for each mass element, so the inertial angular momentum of a rigid body can be written as

$$\mathbf{H} = \int \mathbf{r} \times \mathbf{v} \, dm \tag{4.18}$$

where the integration extends over the entire object and **v** is the inertial velocity of the mass element dm about the center of mass. While (4.18) is correct, it is extremely inconvenient to use, since the position vectors **r** of the mass elements change with time in the inertial frame. It is greatly to our advantage to do these integrals *in a frame of reference tied to the body,* since the mass distribution is static in such a frame. This is shown in Figure 4.4. Now, the velocity of the mass element about the center of mass is due solely to the rotation of the body,

$$\mathbf{v} = \boldsymbol{\omega} \times \mathbf{r} \tag{4.19}$$

where **ω** is the inertial angular velocity of the rigid body. Inserting (4.19) into

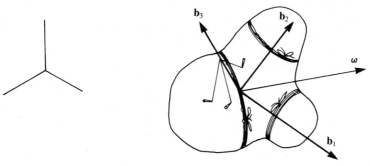

FIGURE 4.4
Frame of referenceing tied to a rigid body.

(4.18), we find

$$\mathbf{H} = \int \mathbf{r} \times (\boldsymbol{\omega} \times \mathbf{r})\, dm \qquad (4.20)$$

We have now implicitly committed ourselves to evaluating the integrals (4.20) in a *body frame of reference*, since the limits on the integral are simple only in a frame where the mass distribution does not change with time. However, the expression (4.20) still yields the inertial angular momentum, although it will be expressed in the unit vectors of a noninertial frame.

It is further desirable to somehow remove the angular-velocity vector from within the integrals (4.20). This would separate (4.20) into a portion depending on the mass distribution within the body and a portion which gives its rotational velocity. The mass integrals could then be performed once and for all time. Write the vectors $\mathbf{r}$ and $\boldsymbol{\omega}$ in their body-frame components:

$$\mathbf{r} = x\mathbf{b}_1 + y\mathbf{b}_2 + z\mathbf{b}_3 \qquad (4.21)$$

$$\boldsymbol{\omega} = \omega_1\mathbf{b}_1 + \omega_2\mathbf{b}_2 + \omega_3\mathbf{b}_3 \qquad (4.22)$$

Then, either working out the double cross product directly or remembering the *bac − cab* vector identity

$$\mathbf{A} \times (\mathbf{B} \times \mathbf{C}) = \mathbf{B}(\mathbf{A} \cdot \mathbf{C}) - \mathbf{C}(\mathbf{A} \cdot \mathbf{B})$$

the expression for the angular momentum becomes

$$\mathbf{H} = \int \left\{ \begin{matrix} \mathbf{b}_1[\omega_1(y^2 + z^2) - \omega_2 xy - \omega_3 xz] \\ \mathbf{b}_2[-\omega_1 yx + \omega_2(x^2 + z^2) - \omega_3 yz] \\ \mathbf{b}_3[-\omega_1 zx - \omega_2 zy + \omega_3(x^2 + y^2)] \end{matrix} \right\} dm \qquad (4.23)$$

We have not yet factored the $\boldsymbol{\omega}$ vector from this expression, but it should be removable, since it is not a function of the variables of integration x, y, and z or the limits of the integration. The major question is what type of object would be left behind when $\boldsymbol{\omega}$ is factored from the integrals. It cannot be another vector, since a dot product will yield a scalar as a result (which **H** is not), while a cross product would not have the ω_1 component present in the $\mathbf{b}_1$ line of (4.23). After some experimenting, (4.23) can be recognized as the product of a *matrix* and a column vector

$$\mathbf{H} = \int \begin{bmatrix} y^2 + z^2 & -yx & -zx \\ -xy & x^2 + z^2 & -zy \\ -xz & -yz & x^2 + y^2 \end{bmatrix} dm \begin{bmatrix} \omega_1 \\ \omega_2 \\ \omega_3 \end{bmatrix} \qquad (4.24)$$

The integral of a matrix is the matrix of integrals of individual elements. Thus, all information on the mass distribution within the rigid body is contained within the *moment of inertia matrix*

$$I = \left\{ \begin{array}{ccc} \int (y^2 + z^2)\, dm & -\int yx\, dm & -\int zx\, dm \\ -\int xy\, dm & \int (x^2 + z^2)\, dm & -\int zy\, dm \\ -\int xz\, dm & -\int yz\, dm & \int (x^2 + y^2)\, dm \end{array} \right\} \qquad (4.25)$$

This matrix is symmetric and need only be obtained once for a given rigid body.

The angular momentum can then be written as the matrix-vector product

$$\mathbf{H} = I\boldsymbol{\omega} \qquad (4.26)$$

which must, we recall, be evaluated in the unit vectors of the body frame. The first thing to be noticed about (4.26) is that *the angular momentum* **H** *and the angular velocity* **ω** *are not generally aligned*. Consider the wheel and axle shown in Figure 4.5. The $\mathbf{b}_3$ axis is aligned with the wheel axis, and the angular-velocity vector **ω** must also lie in the $\mathbf{b}_3$ direction. However, unless

$$\int zx\, dm = \int zy\, dm = 0 \qquad (4.27)$$

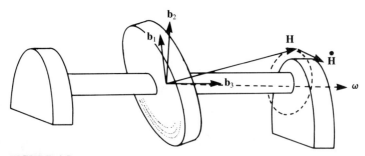

FIGURE 4.5
Unbalanced wheel and axle.

the angular momentum **H** will not lie completely along the $\mathbf{b}_3$ direction. Its components in the body frame will be given by (4.26), which will yield a constant expression for **H** if **ω** is a constant vector.

However, as shown in Figure 4.5, the vector **H** will only be constant seen from the body frame. Seen from the inertial frame, $\dot{\mathbf{H}} \neq 0$, and this latter statement implies that there must be a torque **M** on the system. Such a torque is undesirable in the wheel of a vehicle, since it leads to excessive wear in the wheel bearing and tire surface. When you pay an auto mechanic to "dynamically balance" your new set of tires, you are paying for equation (4.27) to be satisfied.

The kinetic energy of a rigid body can also be calculated as if it were a system of particles. Thus,

$$T = \int \tfrac{1}{2}\mathbf{v} \cdot \mathbf{v} \, dm$$

$$= \int \tfrac{1}{2}(\mathbf{\omega} \times \mathbf{r}) \cdot (\mathbf{\omega} \times \mathbf{r}) \, dm \qquad (4.28)$$

after (4.19) is used. Expanding (4.28) into its components, we have

$$T = \tfrac{1}{2} \int \left(\omega_2^2 z^2 - 2\omega_2\omega_3 yz + \omega_3^2 y^2 \right.$$

$$\left. + \omega_3^2 x^2 - 2\omega_1\omega_3 xz + \omega_1^2 z^2 + \omega_1^2 y^2 - 2\omega_1\omega_2 xy + \omega_2^2 x^2 \right) dm \quad (4.29)$$

When this expression is split into its nine separate terms, the ω_i components can be brought outside the integrations, and the integrals then are recognizable as elements of the moment of inertia matrix (4.25). The expression for T can then be written

$$T = \tfrac{1}{2}(\omega_1^2 I_{11} + \omega_2^2 I_{22} + \omega_3^2 I_{33} + 2\omega_1\omega_2 I_{12} + 2\omega_1\omega_3 I_{13} + 2\omega_2\omega_3 I_{23}) \quad (4.30)$$

This can be put into the form

$$T = \tfrac{1}{2}\mathbf{\omega} \cdot \mathbf{H} = \tfrac{1}{2}\mathbf{\omega} \cdot I\mathbf{\omega} \qquad (4.31)$$

In this last form the expression for the kinetic energy is reminiscent of the familiar $T = \tfrac{1}{2}mv^2$ of particle dynamics, with the angular velocity playing the role of the linear velocity and the moment of inertia matrix being the analogue of the mass for rotational systems.

There are three major systems of notation for writing the expressions for the angular momentum and kinetic energy. This occurs because there are three names for the construct in (4.25). Until now we have called this rectangular array of numbers a matrix. It is also termed a *second-rank tensor,* and alternately a *dyadic.* These latter two terms are based on several additional properties of the moment of inertia matrix, principally the way it transforms from one coordinate frame to another. You should not be intimidated by the terms tensor or dyadic, since one deals with the object in (4.25) according to the rules of matrix multiplication. Note also, for your comfort, that a scalar is termed a *zero-rank tensor,* while a vector is a *first-rank tensor.*

However, the notations do differ depending on which system a particular author advocates. The dyadic equivalents of (4.26) and (4.31) are

$$\mathbf{H} = I \cdot \boldsymbol{\omega} \tag{4.32}$$

$$T = \tfrac{1}{2}\boldsymbol{\omega} \cdot I \cdot \boldsymbol{\omega} \tag{4.33}$$

The extra "dot" in (4.33) arises from the way dyads are written in component form. To keep track of where a particular element of I belongs, dyadic notation appends *two* unit vectors to each element. The tensor scheme uses subscripted indices for the same purpose, while in the matrix method the position of an element can be seen at a glance. In the matrix notation scheme, the vector $\boldsymbol{\omega}$ is written as a column matrix. Then, (4.26) and (4.31) are written as matrix equations

$$[\mathbf{H}] = [I][\boldsymbol{\omega}] \tag{4.34}$$

$$T = \tfrac{1}{2}[\boldsymbol{\omega}]^T[I][\boldsymbol{\omega}] \tag{4.35}$$

since a dot product is produced by transposing the first vector into a row matrix. The notation we have used is, of course, the tensor scheme. We will follow this throughout the book.

4.4 THE PRINCIPAL-BODY-AXIS FRAME

We have already seen that in three-dimensional dynamics the angular velocity $\boldsymbol{\omega}$ and the angular momentum $\mathbf{H}$ will not necessarily be aligned. However, under what conditions are they aligned? If $\mathbf{H}$ and $\boldsymbol{\omega}$ are parallel, then one is a scalar multiple of the other,

$$\mathbf{H} = I\boldsymbol{\omega} = \lambda\boldsymbol{\omega} \tag{4.36}$$

where λ is a scalar. We still do not know if (4.36) admits of any solutions at all. Write the vector $\lambda\boldsymbol{\omega}$ as $\lambda[\mathbf{1}]\boldsymbol{\omega}$, where $[\mathbf{1}]$ is the identity matrix. Then, (4.36) can be written

$$\{I - \lambda[\mathbf{1}]\}\boldsymbol{\omega} = 0 \tag{4.37}$$

Equation (4.37) obviously admits the *trivial* solution of $\boldsymbol{\omega} = 0$, but this is not of interest.

Equation (4.37) is an example of an *eigenvalue problem* in mathematics. We are interested in nontrivial vectors $\boldsymbol{\omega}$ satisfying (4.37). Now, if $\boldsymbol{\omega}_1$ is such a solution, then any scalar multiple of this will also be a solution. Hence, we are really only interested in the *direction* of the vectors $\boldsymbol{\omega}$ satisfying (4.37). Now, from linear algebra, it is well known that (4.37) will only have nontrivial solutions when the determinant of the coefficient matrix is zero itself. That is, we must determine λ from

$$|I - \lambda[\mathbf{1}]| = 0 \tag{4.38}$$

This is termed the *characteristic equation*. Since the moment of inertia matrix is of dimension 3, equation (4.38) will yield a cubic equation for λ when the determinant is expanded.

The characteristic equation will always yield three positive values for the scalar λ for any physically reasonable moment of inertia matrix. This is necessary since a negative λ value implies, via (4.36), that **H** and **ω** are in opposite directions. Such a body would spin ever faster when braked, making an infinite source of energy available. A matrix with all three eigenvalues λ_i positive and real is termed a *positive definite matrix*. However, the moment of inertia matrix has more structure than this. Since the I matrix is also symmetric, linear algebra tells us that the vectors ω_i corresponding to the three λ_i are *perpendicular*. So, there are only three possible axes in the rigid body about which **H** and **ω** are parallel. In all other orientations, the angular velocity and angular momentum do not line up. The three special λ_i are termed the *principal moments of inertia,* while the unit vectors ω_i are called the *principal axes of inertia.*

Once the three λ_i values are found from the cubic equation (4.38), the principal-axis vectors can be easily found. If we insert one λ_i into (4.37), we will have a set of three equations in three unknowns, the three components of ω_i. However, these three equations will be singular, since through (4.38) we explicitly arranged for this to happen. If we arbitrarily choose one component of the **ω** vector to be, say, 1, then any two of these three equations can be used to solve for the other two components. The resulting vector may then be normalized, as only the *direction* of the principal axis is of interest.

Since these three unit vectors are mutually perpendicular, they can be used as the basis vectors for a body frame of reference. This special body frame is called the *principal-axis frame.* In this frame, the moment of inertia tensor takes a particularly simple form. Since (4.36) must be true when each of the unit vectors of this special frame are inserted, the moment of inertia tensor becomes

$$I = \begin{pmatrix} \lambda_1 & 0 & 0 \\ 0 & \lambda_2 & 0 \\ 0 & 0 & \lambda_3 \end{pmatrix} \qquad (4.39)$$

in the principal-axis frame. Note that although (4.39) is diagonal, **H** and **ω** will only be aligned when the body rotates about one of the three principal axes.

The transformation from the principal-axis frame to the original body frame is, of course, performed with a rotation matrix. To see how the moment of inertia tensor transforms under a rotation, write the angular-momentum relation with respect to two body frames **a** and **b**:

$$\mathbf{H}^a = I^a \boldsymbol{\omega}^a \qquad \mathbf{H}^b = I^b \boldsymbol{\omega}^b \qquad (4.40)$$

Superscripts refer to the frame in which the components are expressed; the two equations refer to the same rigid body in the same rotation state. Now, since simple vectors transform with a rotation matrix (see section 1.4), we can write

$$\mathbf{H}^{b} = R^{ab}\mathbf{H}^{a} \qquad \boldsymbol{\omega}^{b} = R^{ab}\boldsymbol{\omega}^{a} \tag{4.41}$$

where R^{ab} is the rotation matrix which transforms from the **a** frame to the **b** frame. Inserting (4.41) into (4.40), one obtains

$$\mathbf{H}^{b} = R^{ab}\mathbf{H}^{a} = I^{b}R^{ab}\boldsymbol{\omega}^{a} \tag{4.42}$$

Or, rearranging,

$$\mathbf{H}^{a} = (R^{ab})^{-1}I^{b}R^{ab}\boldsymbol{\omega}^{a} = I^{a}\boldsymbol{\omega}^{a} \tag{4.43}$$

By comparing the two forms above, we find that the moment of inertia tensor transforms according to

$$I^{a} = (R^{ab})^{-1}I^{b}R^{ab} \tag{4.44}$$

In fact, this is the relation proving that the moment of inertia matrix *is* a tensor or a dyadic. It takes two rotation matrices to transform a second-rank tensor under a rotation. (There is a simple progression here. Zero-rank tensors are scalars and require no rotation matrices to transform them, while first-rank tensors are vectors and require one rotation matrix to transform their coordinates.)

The rotation matrix R^{ab} transforming from the original body frame to the principal-axis body frame is already available. One definition of a rotation matrix is that it is the array of the three unit vectors of one reference frame *expressed in components of the other frame.* The three unit vectors $\boldsymbol{\omega}_i$ are the three axis vectors of the principal-axis frame, and the eigenvector calculation (4.37) will produce these vectors in terms of their components in the original body frame. (Their components in the principal-axis frame itself, are, of course, trivially written down.) All we need do to find the rotation matrix transforming from the original body frame to the principal-axis frame is to assemble the three $\boldsymbol{\omega}_i$ vectors by columns into the matrix R^{ab}. The principal-axis frame has significant advantages for doing dynamics of rigid bodies. Since this frame can always be found, we will usually work in this frame from now on.

4.5 THE PARALLEL-AXIS THEOREM

It is also necessary to learn how the moment of inertia matrix changes when we change the origin of the body frame. Figure 4.6 shows a rigid body with the original set of axes centered on the center of mass. A second set of axes $\mathbf{b}_1'$, $\mathbf{b}_2'$, and $\mathbf{b}_3'$ are parallel to the first, but the origin has been displaced by an amount $\Delta\mathbf{r}$ from the center of mass. We wish to calculate the moment of inertia matrix about this new origin.

The old body-frame coordinates x, y, and z are related to the new coordinates x', y', and z' by

$$x = x' + \Delta x \qquad y = y' + \Delta y \qquad z = z' + \Delta z \tag{4.45}$$

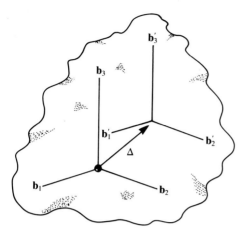

FIGURE 4.6
Body frames for moment of inertia calculation.

Consider the calculation of a diagonal element of the moment of inertia tensor. The first diagonal element with respect to the *new* origin is given by

$$I'_{xx} = \int (y'^2 + z'^2)\, dm$$

$$= \int (y^2 + z^2)\, dm + (\Delta y^2 + \Delta z^2) \int dm - 2\Delta y \int y\, dm - 2\Delta z \int z\, dm \qquad (4.46)$$

The first integral is the first diagonal element of the moment of inertia matrix about the old origin. The second integral yields M, the total mass of the body. Finally, since we have put the origin of the unprimed frame at the body center of mass, the last two integrals are zero. Thus

$$I'_{xx} = I_{xx} + M(\Delta y^2 + \Delta z^2) \qquad (4.47)$$

The expressions for the other two diagonal elements are similar.

An off-diagonal moment of inertia element becomes

$$I'_{xy} = -\int x'y'\, dm$$

$$= -\int xy\, dm - \Delta x\, \Delta y \int dm + \Delta x \int y\, dm + \Delta y \int x\, dm \qquad (4.48)$$

Again, the last two integrals are zero, while the second yields the total mass of the object. Hence,

$$I'_{xy} = I_{xy} - M\Delta x\, \Delta y \qquad (4.49)$$

with similar expressions for the other off-diagonal elements. Combining both types of elements, the old and new moment of inertia matrices are related by

$$I' = I + M \begin{bmatrix} \Delta y^2 + \Delta z^2 & -\Delta y\, \Delta x & -\Delta z\, \Delta x \\ -\Delta x\, \Delta y & \Delta x^2 + \Delta z^2 & -\Delta z\, \Delta y \\ -\Delta x\, \Delta z & -\Delta y\, \Delta z & \Delta x^2 + \Delta y^2 \end{bmatrix} \qquad (4.50)$$

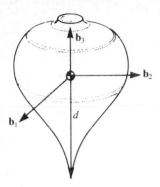

FIGURE 4.7
Pivot moment of inertia calculation for a top.

The diagonal elements always increase in this transformation, while the off-diagonal elements may either increase or decrease.

Consider the top shown in Figure 4.7. Assume the moment of inertia matrix about the center of mass is given by

$$I = \begin{bmatrix} A & 0 & 0 \\ 0 & A & 0 \\ 0 & 0 & C \end{bmatrix} \qquad (4.51)$$

Two moments of inertia are equal, since the top is symmetric. To study motion of the top in its usual use, we will need the moment of inertia matrix evaluated about the pivot point. The displacements from the center of mass to the pivot are $\Delta x = \Delta y = 0$ and $\Delta z = -d$. Straightforward calculation then gives

$$I' = \begin{bmatrix} A + Md^2 & 0 & 0 \\ 0 & A + Md^2 & 0 \\ 0 & 0 & C \end{bmatrix} \qquad (4.52)$$

Although we have assumed that one body frame has its origin at the center of mass, it is not difficult to construct a transformation between any two body frames, neither of which is a center of mass frame. One need only transform from the first frame to the center of mass frame and from there to the desired second frame. Two transformations similar to (4.50) are necessary. The major use of the parallel-axis theorem is in the calculation of the moment of inertia matrix of a large vehicle from information on its component parts. The moment of inertia matrices of shells, plates, solid shapes, and so forth, may be found tabulated in most introductory dynamics texts. Of course, the moment of inertia matrix of the entire vehicle is the sum of the moment of inertia matrices of the component parts of the spacecraft, *once these have been transformed to the center of mass of the spacecraft* as the new origin and *rotated into a common reference frame*. The only alternate to constructing the moment of inertia tensor by the parallel-axis theorem, rotations, and addition is to build the vehicle and measure the moment of inertia matrix by spin testing. This usually occurs when it is too late to make major modifications in the configuration, and is very difficult for physically large systems.

4.6 EULER'S EQUATIONS

The equations of motion for a rigid body can be found by remembering that the applied torque must equal the rate of change of the angular momentum:

$$\mathbf{M} = \dot{\mathbf{H}} \tag{4.53}$$

However, the expression we have for the angular momentum of a rigid body,

$$\mathbf{H} = I\boldsymbol{\omega} \tag{4.54}$$

is usable only in the components of a body frame, where I is constant. We must remember that (4.54) is expressed in body-frame components, while we need to perform an inertial-frame derivative for (4.53). The body frame has great advantages, not the least of which is that the moment of inertia matrix is constant in this frame. So, it would seem that (4.53) should be expressed in the unit vectors of the body frame.

Remember that the time derivative of a vector with respect to the inertial frame is related to the time derivative with respect to the body frame by

$$\frac{^id}{dt}\mathbf{A} = \frac{^bd}{dt}\mathbf{A} + \boldsymbol{\omega} \times \mathbf{A} \tag{4.55}$$

Here, $\boldsymbol{\omega}$ is the angular velocity of the body frame, which is just the $\boldsymbol{\omega}$ used in the calculation of the angular velocity (4.54). When we apply (4.55) to the angular momentum, (4.53) becomes

$$\mathbf{M} = \frac{^bd}{dt}I\boldsymbol{\omega} + \boldsymbol{\omega} \times I\boldsymbol{\omega} \tag{4.56}$$

In the rotating frame, the moment of inertia matrix is a constant, so

$$\mathbf{M} = I\dot{\boldsymbol{\omega}} + \boldsymbol{\omega} \times I\boldsymbol{\omega} \tag{4.57}$$

where the dot in the first term on the right of (4.57) is the body-frame time derivative of $\boldsymbol{\omega}$.

Now, the moment of inertia matrix is simplest when expressed in the principal-axis frame. Assume the moment of inertia matrix is

$$I = \begin{bmatrix} A & 0 & 0 \\ 0 & B & 0 \\ 0 & 0 & C \end{bmatrix} \tag{4.58}$$

in the principal-axis frame. Also, let the $\boldsymbol{\omega}$ vector have components $\boldsymbol{\omega} = \omega_1\mathbf{b}_1 + \omega_2\mathbf{b}_2 + \omega_3\mathbf{b}_3$ in the principal-axis body frame. Then, (4.57) can be expressed in the principal-axis body frame components as

$$M_1 = A\dot{\omega}_1 + (C - B)\omega_2\omega_3$$
$$M_2 = B\dot{\omega}_2 + (A - C)\omega_1\omega_3 \tag{4.59}$$
$$M_3 = C\dot{\omega}_3 + (B - A)\omega_1\omega_2$$

These are *Euler's equations*. They are three coupled, nonlinear, first-order differential equations, and they constitute one-half of the rotational equations of motion for a rigid body.

4.7 ORIENTATION ANGLES

Euler's equations are three differential equations for the components of the angular-velocity vector as seen from the body frame. If these equations are solved, we still do not have a solution for the orientation of the body. To obtain the second half of the dynamics, then, it will be necessary to introduce three angles specifying the orientation of the body and to obtain three more differential equations for these angles in terms of $\boldsymbol{\omega}$. Although it takes three angles to specify a general rotation of a rigid body, it is an unfortunate fact that any possible system of three angles becomes singular in some particular orientation. There are two systems of orientation angles commonly used: the classical Euler angles and the pitch, roll, and yaw angles. These two systems are singular in different situations, and difficulties can often be avoided by choosing one or the other at the outset.

The classical Euler angles ϕ, θ, and ψ are shown in Figure 4.8. These are also the right ascension of the ascending node, the inclination, and the argument of perigee used to specify the orbit in celestial mechanics. One transforms from the inertial **i** frame to the body frame **b** by rotating about the **k** axis by ϕ, then rotating about the node vector **n** by θ, and finally rotating about the $\mathbf{b}_3$ vector by ψ. The total rotation matrix can be written as

$$R^{\mathbf{ib}} = R_3(\psi)R_1(\theta)R_3(\phi) \tag{4.60}$$

where R_1 and R_3 are the elementary rotation matrices about the x and z axes, respectively. This matrix is simply the matrix used to rotate from the inertial

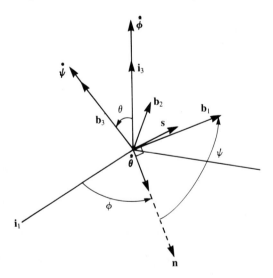

FIGURE 4.8
The classical Euler angles.

frame to the orbit frame in the two-body problem. Completing the multiplications in (4.60), we have the rotation matrix as

$$R^{ib} = \begin{bmatrix} c\phi c\psi - s\phi c\theta s\psi & s\phi c\psi + c\phi c\theta s\psi & s\theta s\psi \\ -c\phi s\psi - s\phi c\theta c\psi & -s\phi s\psi + c\phi c\theta c\psi & s\theta c\psi \\ s\theta s\phi & -c\phi s\theta & c\theta \end{bmatrix} \quad (4.61)$$

where c indicates cosine and s indicates sine. Since this is a rotation matrix, its inverse is its transpose. Since the Euler angles ϕ, θ, and ψ are also the two-body orbital elements Ω, i, and ω, the rotation matrix needed in section 2.8 is just the inverse of (4.61).

However, at the moment, we need three more expressions to complete the equations of motion for a rigid body. Since Euler's equations will supply us with the three components of the angular-velocity vector ω in the body frame, we can find the other three equations by expressing the angular velocity in terms of the Euler angles and writing this in the body-frame unit vectors. Now, angular velocities add, so the angular velocity of the body frame with respect to the inertial frame is given by

$$\omega = \dot{\phi}\mathbf{k} + \dot{\theta}\mathbf{n} + \dot{\psi}\mathbf{b}_3 \quad (4.62)$$

This result is, however, in a mixed set of basis vectors. By inspection of Figure 4.8,

$$\mathbf{n} = \cos\psi\mathbf{b}_1 - \sin\psi\mathbf{b}_2 \quad (4.63)$$

since the node vector always lies in the $\mathbf{b}_1\mathbf{b}_2$ plane. The $d\phi/dt$ component can be resolved into a $\mathbf{b}_3$ component and a component in the $\mathbf{b}_1\mathbf{b}_2$ plane as

$$\dot{\phi}\mathbf{k} = \dot{\phi}\cos\theta\mathbf{b}_3 + \dot{\phi}\sin\theta\mathbf{s} \quad (4.64)$$

Now, the $\mathbf{s}$ vector lies at $90°$ to the nodal vector $\mathbf{n}$, so

$$\mathbf{s} = \sin\psi\mathbf{b}_1 + \cos\psi\mathbf{b}_2 \quad (4.65)$$

Combining (4.62) to (4.65), we can write the ω vector in its body-frame components as

$$\omega_1 = \dot{\phi}\sin\theta\sin\psi + \dot{\theta}\cos\psi \quad (4.66)$$

$$\omega_2 = \dot{\phi}\sin\theta\cos\psi - \dot{\theta}\sin\psi \quad (4.67)$$

$$\omega_3 = \dot{\psi} + \dot{\phi}\cos\theta \quad (4.68)$$

Equations (4.66) to (4.68) are the second half of the equations of motion of a rigid body. Their left sides are obtained from the solution to Euler's equations. Then, they become three coupled differential equations for the Euler angles. Their solution will furnish the orientation of the body as a function of time. Of course, there are cases when the three Euler equations and the three Euler angle rate equations must be solved together. If the torque components M_i appearing in Euler's equations are functions of the inertial orientation of the body, then it is not possible to separate the solution of these six first-order differential equations. They would have to be solved together.

The classical Euler angles are singular when $\theta = 0$. In this event, it is not possible to tell the difference between the angles ϕ and ψ, since the line separating them, the line of nodes, is not defined. Unfortunately, the dynamics tends to produce reasonable values of the sum $d\phi/dt + d\psi/dt$ by moving the undefined node at high rate and adding two values of $d\phi/dt$ and $d\psi/dt$ which are very large, nearly equal, and of opposite sign. This can cause severe numerical difficulties at best and can produce singularities in the equations of motion if θ goes exactly to zero. One solution to this difficulty is to switch to another set of orientation angles when this occurs. The most commonly used alternate set are the orientation angles used in aircraft dynamics: roll, pitch, and yaw.

The aircraft orientation angles are shown in Figure 4.9. An airplane is free to change its orientation as it desires, but the mathematical transformation from the inertial frame to the body frame requires some convention on the order of the three rotations. Starting from the inertial frame, the standard order is first yaw about the vertical axis ψ_1, then pitch about the new wing axis ψ_2, and finally roll about the new longitudinal axis ψ_3. The complete rotation matrix from the inertial to the body frame is then

$$R^{ib} = R_1(\psi_3)R_2(\psi_2)R_3(\psi_1) \tag{4.69}$$

After tedious calculation, the rotation matrix itself is

$$R^{ib} = \begin{bmatrix} c\psi_2 c\psi_3 & c\psi_1 s\psi_3 + s\psi_1 s\psi_2 c\psi_3 & s\psi_1 s\psi_3 - c\psi_1 s\psi_2 c\psi_3 \\ -c\psi_2 s\psi_3 & c\psi_1 c\psi_3 - s\psi_1 s\psi_2 s\psi_3 & s\psi_1 c\psi_3 + c\psi_1 s\psi_2 s\psi_3 \\ s\psi_2 & -s\psi_1 c\psi_2 & c\psi_1 c\psi_2 \end{bmatrix} \tag{4.70}$$

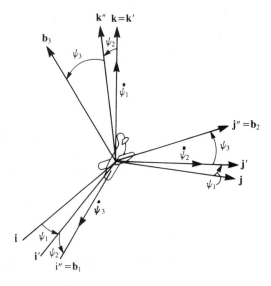

FIGURE 4.9
The yaw, pitch, and roll orientation angles.

where c indicates cosine and s indicates sine. Again, however, to supplement Euler's equations we need the expression for the angular-velocity vector written in the body unit vectors.

The total angular velocity is just the sum of the three component angular velocities:

$$\boldsymbol{\omega} = \dot{\psi}_1 \mathbf{k} + \dot{\psi}_2 \mathbf{j}' + \dot{\psi}_3 \mathbf{b}_1 \tag{4.71}$$

in a mixed set of unit vectors. Now, by inspection of Figure 4.9,

$$\mathbf{j}' = \cos \psi_3 \mathbf{b}_2 - \sin \psi_3 \mathbf{b}_3 \tag{4.72}$$

For the $\mathbf{k}$ vector, some care is necessary

$$
\begin{aligned}
\mathbf{k} &= -\sin \psi_2 \mathbf{i}'' + \cos \psi_2 \mathbf{k}'' \\
&= -\sin \psi_2 \mathbf{b}_1 + \cos \psi_2 \sin \psi_3 \mathbf{b}_2 + \cos \psi_2 \cos \psi_3 \mathbf{b}_3
\end{aligned} \tag{4.73}
$$

Then, inserting (4.72) and (4.73) into (4.71), we have the three components of the $\boldsymbol{\omega}$ vector in the body frame:

$$\omega_1 = \dot{\psi}_3 - \dot{\psi}_1 \sin \psi_2 \tag{4.74}$$

$$\omega_2 = \dot{\psi}_2 \cos \psi_3 + \dot{\psi}_1 \cos \psi_2 \sin \psi_3 \tag{4.75}$$

$$\omega_3 = -\dot{\psi}_2 \sin \psi_3 + \dot{\psi}_1 \cos \psi_2 \cos \psi_3 \tag{4.76}$$

The roll, pitch, and yaw system becomes singular for a pitch angle of $\psi_2 = 90°$. This corresponds to a vertical flight orientation for an aircraft, an uncommon orientation which cannot be maintained for a long period of time. However, these angles are not singular near a condition of level flight, while the classical Euler angles are singular in this orientation. The choice of an appropriate system can thus depend on the particular orientations of interest.

However, there are situations where a vehicle can assume any attitude at all, and either set of orientation angles could experience difficulties. In this case, other methods are available to specify the orientation of the body frame. Perhaps the easiest is to remember that a rotation matrix is simply the matrix of unit vectors of one reference frame *expressed in the components of the other frame*. That is,

$$R^{\mathrm{ib}} = [\mathbf{b}_1 \ \vdots \ \mathbf{b}_2 \ \vdots \ \mathbf{b}_3] \tag{4.77}$$

where the $\mathbf{b}_i$ are written in their inertial-frame components. Then, applying the time-derivative rule,

$$\frac{^{\mathrm{i}}d}{dt}\mathbf{b}_i = \frac{^{\mathrm{b}}d}{dt}\mathbf{b}_i + \boldsymbol{\omega} \times \mathbf{b}_i \tag{4.78}$$

But, the $\mathbf{b}_i$ unit vectors are constants in the $\mathbf{b}$ frame, so (4.78) simplifies to

$$\dot{\mathbf{b}}_i = \boldsymbol{\omega} \times \mathbf{b}_i \tag{4.79}$$

This yields nine first-order equations of motion for the components of the body-frame rotation matrix. This method is often used in a vehicle computer, as in a strapdown navigation unit. All nine equations are, of course, not needed. For example, $\mathbf{b}_3 = \mathbf{b}_1 \times \mathbf{b}_2$ can be used to eliminate three of them.

4.8 THE SIMPLE TOP

As an example of obtaining the equations of motion for a rigid body, consider the simple top shown in Figure 4.10. To apply Euler's equations, we need to write the torque on the top in its body-frame components. Assume that the pivot point of the top is fixed. Then this point is a fixed point in an inertial frame, and we can use it as the origin. Also, the constraint forces from the tabletop have no torques about the pivot point, and only the gravity torque needs to be included. We have already calculated this torque in an earlier section. Resolving it into its body-frame components, we find

$$\mathbf{M} = mgd \sin \theta \mathbf{n}$$
$$= \mathbf{b}_1(mgd \sin \theta \cos \psi) - \mathbf{b}_2(mgd \sin \theta \sin \psi) \qquad (4.80)$$

For a symmetric top, the moments of inertia are A, A, and C. Then, Euler's equations can be written

$$M_1 = mgd \sin \theta \cos \psi = A\dot{\omega}_1 + (C - A)\omega_2\omega_3$$
$$M_2 = -mgd \sin \theta \sin \psi = A\dot{\omega}_2 + (A - C)\omega_1\omega_3 \qquad (4.81)$$
$$M_3 = 0 = C\dot{\omega}_3 + (A - A)\omega_1\omega_2$$

Since the top is symmetric, the last of these immediately reduces to the statement that ω_3 is a constant.

These are only half the equations of motion, however. We normally expect one second-order differential equation (or two first-order differential equations) for each degree of freedom in the system. The top has three

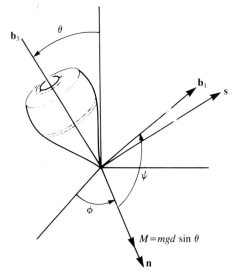

$$M = mgd \sin \theta$$

FIGURE 4.10
The simple top.

degrees of freedom, so equations (4.81) are not complete. The other half of the equations of motion can be taken to be the Euler-angle rate equations

$$\omega_1 = \dot{\phi} \sin \theta \sin \psi + \dot{\theta} \cos \psi$$

$$\omega_2 = \dot{\phi} \sin \theta \cos \psi - \dot{\theta} \sin \psi \qquad (4.82)$$

$$\omega_3 = \dot{\psi} + \dot{\phi} \cos \theta$$

From the third equation above, we learn that the conserved quantity ω_3 is not the "spin" of the top $\dot{\psi}$ but also includes a contribution due to the top's precessional motion. Equations (4.81) and (4.82) constitute six first-order differential equations for the motion of the top. Since the torque depends on two of the Euler angles, they must be solved together.

While there is nothing wrong with the above derivation, an alternate procedure is possible in the case of a symmetric body which often leads to simpler equations of motion. We were originally driven to Euler's equations by the necessity of using an inertial derivative in the statement that $\mathbf{M} = d\mathbf{H}/dt$, and by the fact that the moment of inertia matrix is (usually) only constant in a body frame. However, the nodal frame $\mathbf{c}$, with unit vectors $\mathbf{n}$, $\mathbf{s}$, and $\mathbf{b}_3$, also has the property that the moment of inertia is constant. The $\mathbf{b}$ and $\mathbf{c}$ frames differ only by the rotation of the top about $\mathbf{b}_3$, and the top is symmetric about this axis.

Equation (4.80) shows that the torque is considerably simpler in the nodal frame. Write the angular-velocity vector of the $\mathbf{c}$ frame as

$$\boldsymbol{\omega}^{ic} = \dot{\theta}\mathbf{n} + \dot{\phi} \sin \theta \mathbf{s} + \dot{\phi} \cos \theta \mathbf{b}_3 \qquad (4.83)$$

while the inertial angular velocity of the body frame is

$$\boldsymbol{\omega}^{ib} = \boldsymbol{\omega}^{ic} + \dot{\psi}\mathbf{b}_3$$

$$= \dot{\theta}\mathbf{n} + \dot{\phi} \sin \theta \mathbf{s} + (\dot{\psi} + \dot{\phi} \cos \theta)\mathbf{b}_3 \qquad (4.84)$$

Then, an alternate expression for $\mathbf{M} = d\mathbf{H}/dt$ is obtained if we calculate the inertial derivative in the $\mathbf{c}$ frame:

$$\mathbf{M} = \frac{{}^c d}{dt} I\boldsymbol{\omega}^{ib} + \boldsymbol{\omega}^{ic} \times I\boldsymbol{\omega}^{ib} \qquad (4.85)$$

This has the same advantages as Euler's equations, since the $\mathbf{c}$-frame derivative of the moment of inertia matrix I is still zero.

Now, performing the indicated calculations in (4.85), we have

$$M_{\mathbf{n}} = mgd \sin \theta$$

$$= A\ddot{\theta} + (C - A)\dot{\phi}^2 \sin \theta \cos \theta + C\dot{\psi}\dot{\phi} \sin \theta$$

$$M_{\mathbf{s}} = 0$$

$$= A\ddot{\phi} \sin \theta + 2A\dot{\theta}\dot{\phi} \cos \theta + C\dot{\theta}(\dot{\psi} + \dot{\phi} \cos \theta) \qquad (4.86)$$

$$M_{\mathbf{b}_3} = 0 = C\frac{d}{dt}(\dot{\psi} + \dot{\phi} \cos \theta)$$

Equations (4.86) are three second-order differential equations for the motion of the top. The third equation repeats the conservation law we found earlier. However, these equations are somewhat simpler in form, since the torque components on the left side are less complicated.

As an application of these equations, consider the familiar (but special) case of θ constant. In this case, we have $\dot{\theta} = d^2\theta/dt^2 = 0$. Now, this is an *assumption* about the motion, and it may not represent a possible solution to (4.86); and even if possible, it may not be a stable motion. To see if it is permitted, substitute the assumption that θ is constant into the equations of motion. The second of equations (4.86) immediately reduces to the statement that $d^2\phi/dt^2 = 0$. Thus, if θ is constant, the precession rate $\dot{\phi}$ must be constant. The first of equations (4.86) then can be solved for the precession rate

$$\dot{\phi} = -\frac{C\dot{\psi}}{(C-A)\cos\theta} \qquad (4.87)$$

This gives the precession rate necessary to maintain constant θ with a given spin rate. Notice that in this case *neither* the top's angular velocity *nor* its angular momentum are lined up with the symmetry axis $\mathbf{b}_3$.

4.9 REFERENCES AND FURTHER READING

Although rigid-body dynamics is a standard part of any dynamics course, the best coverage of this field is to be found in textbooks on spacecraft rotational motion. Kane, Likins, and Levinson is especially noteworthy for its innovative and exhaustive treatment of rotational kinematics, while Hughes' work takes a more conventional approach. Meirovitch has a comprehensive treatment of rotational dynamics using both the classical and advanced techniques of lagrangian dynamics.

Kane, Thomas R., Peter W. Likins, and D. A. Levinson, *Spacecraft Dynamics*, McGraw-Hill, New York, 1983.
Hughes, Peter C., *Spacecraft Attitude Dynamics*, Wiley, New York, 1986.
Meirovitch, L., *Methods of Analytical Dynamics*, McGraw-Hill, New York, 1970.

4.10 PROBLEMS

1. The angular-velocity component expressions for the Euler-angle rates $\dot{\phi}$, $\dot{\theta}$, and $\dot{\psi}$,

$$\omega_1 = \dot{\phi}\sin\theta\sin\psi + \dot{\theta}\cos\psi$$

$$\omega_2 = \dot{\phi}\sin\theta\cos\psi - \dot{\theta}\sin\psi$$

$$\omega_3 = \dot{\psi} + \dot{\phi}\cos\theta$$

are three linear equations in the three angular rates. Solve for these angular rates, and show that difficulties occur for $\theta \to 0$.

2. The expression $\mathbf{H} = I\boldsymbol{\omega}$ is still correct when expressed in the unit vectors of an inertial frame, although the moment of inertia tensor is no longer constant. Show that the moment of inertia matrix obeys

$$\frac{{}^i d}{dt} I = \boldsymbol{\omega} \times I = \begin{bmatrix} 0 & -\omega_3 & \omega_2 \\ \omega_3 & 0 & -\omega_1 \\ -\omega_2 & \omega_1 & 0 \end{bmatrix} [I]$$

and that the statement $\mathbf{M} = \dot{\mathbf{H}}$ becomes

$$\mathbf{M} = (\boldsymbol{\omega} \times I)\boldsymbol{\omega} + I\dot{\boldsymbol{\omega}}$$

in the inertial frame.

3. In a Frisbee satellite deployment from the shuttle payload bay, shown in Figure 4.11, a spring pushes the satellite from below while it rotates about a socket joint attached to one wall. The socket is only partial, and the satellite floats free before it moves very far. The shuttle's mass is large enough that it may be treated as inertial compared to the satellite. Show, while the spring is pushing the satellite, that the velocity and acceleration of the center of mass are

$$\mathbf{v} = \omega r \mathbf{b}_1 \qquad \mathbf{a} = \dot{\omega} r \mathbf{b}_2$$

where ω is the angular velocity of the satellite and r is its radius. Then, show that the equation of motion of the center of mass is

$$m\mathbf{a} = \mathbf{F}_s - \mathbf{F}_p$$

where $\mathbf{F}_s$ is the spring force and $\mathbf{F}_p$ is the pivot reaction force. Also, show that the rotational equation of motion is

$$I\dot{\omega} = rF_p$$

where I is the $\mathbf{b}_3$ moment of inertia of the satellite. Combine these two results to eliminate the socket force to find

$$\left(m + \frac{I}{r^2}\right)\mathbf{a} = \mathbf{F}_s$$

or the equivalent rotational form

$$(I + mr^2)\dot{\omega} = rF_s$$

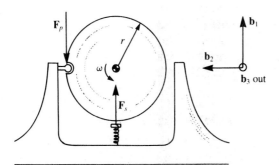

FIGURE 4.11
Frisbee satellite deployment.

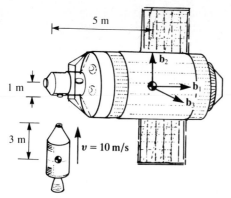

FIGURE 4.12
Nonaxial docking.

One difficulty with the Frisbee launch technique is that the separation speed and rotational speed are linked by $v = \omega r$. Low separation speeds (desirable) thus imply low rotational speeds (not so desirable). Usually the satellite must spin itself up further after separation.

4. After a Frisbee launch, a satellite must be spun up from, 0.05 rad/s to 1 rad/s. The moment of inertia of the satellite is $3400 \text{ kg} \cdot \text{m}^2$ about the spin axis, and paired thrusters are located at 2 m from the spin axis. They must be pulsed and give an impulse of $20 \text{ N} \cdot \text{s}$ per pulse per thruster. How many pulses must be commanded to achieve the desired final spin rate?

5. After having demolished the radial docking port on his first try, Mario Jones, famed racecar driver turned astronaut, is approaching the Skylab space station's bottom port for his second docking attempt at $\mathbf{v} = 10 \text{ m/s } \mathbf{b}_2$. As shown in Figure 4.12, calculate the following:

(a) The velocity of the center of mass.
(b) The angular momentum about the center of mass. Docking is successful (i.e., the wreckage stays together). At the moment of contact, two forces arise within the system which the astronauts find hard to ignore but which you can ignore after you explain *why* they can be ignored. All spacecraft attitude control devices fail at docking. Calculate:
(c) The moment of inertia tensor of the docked configuration.
(d) The angular velocity of the docked configuration. The spacecraft masses are $m_a = 15,000 \text{ kg}$ and $m_{sl} = 50,000 \text{ kg}$ for the Apollo and Skylab, and their moment of inertia tensors are

$$I_a = \begin{bmatrix} 20 & 0 & 0 \\ 0 & 12 & 0 \\ 0 & 0 & 20 \end{bmatrix} \times 10^3 \text{ kg} \cdot \text{m}^2$$

$$I_{sl} = \begin{bmatrix} 90 & 0 & 0 \\ 0 & 50 & 0 \\ 0 & 0 & 50 \end{bmatrix} \times 10^3 \text{ kg} \cdot \text{m}^2$$

in the **b** frame.

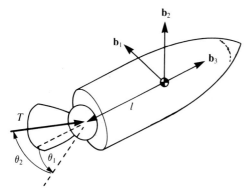

FIGURE 4.13
Thrust vector control.

6. A *thrust vector control* system swivels the rocket engine to maintain correct orientation of the booster. As shown in Figure 4.13, the small engine gimbal angles are θ_1 and θ_2. Show that the torques are

$$M_1 = lT\theta_1 \qquad M_2 = lT\theta_2 \qquad M_3 = 0$$

where T is the thrust and l is the distance from the center of mass to the engine. Also, if the control angles are made proportional to the angular rates,

$$\theta_1 = K\omega_1 \qquad \theta_2 = K\omega_2$$

we have a feedback control system, where K is a gain constant. The principal moments of inertia are A, A, and C about the $\mathbf{b}_1$, $\mathbf{b}_2$, and $\mathbf{b}_3$ axes. Write Euler's equations for a nonspinning missile ($\omega_3 = 0$), and show the system is stable if $K < 0$.

7. Show that the system in problem 6 is still stable if the missile spins about its long axis at constant rate $\omega_3 = \omega_{3o}$.

CHAPTER

5

SATELLITE ATTITUDE DYNAMICS

5.1 INTRODUCTION

A rigid body possesses six degrees of freedom. Three of these represent the motion of the center of mass, and this is the problem which is addressed by orbital mechanics. The other three degrees of freedom represent the rotational motion of the spacecraft about its center of mass. Usually in satellite attitude dynamics these two problems are almost perfectly decoupled: The orbit does not influence the rotational motion of the spacecraft, and the rotational motion does not alter the orbit. We will consider one case where this is not true, but very often the two problems are separate.

The problem of spacecraft attitude dynamics is just as important as the orbital problem. Almost all satellites have some attitude requirements, either explicit pointing requirements for antennas or cameras, requirements for solar panel orientation, or simply a requirement for a given spin-axis direction. The technique chosen to control the spacecraft attitude can limit the useful lifetime of the vehicle, as when an all-thruster control system depletes its propellant supply. Most spacecraft are nearly perfect torque-free systems. These are not familiar from our earthbound experience, where gravitational and aerodynamic torques are common. So, we will begin with a discussion of the dynamics of torque-free rigid bodies. We will then turn to available methods for satellite attitude control and will finally discuss several special cases of interest to satellite designers.

5.2 THE TORQUE-FREE AXISYMMETRIC RIGID BODY

Spin stabilization of satellites is an important attitude-control technique. An actively controlled satellite will have a "dead band" within its control loop and will slowly drift across this band, to be impulsively shoved back when it reaches the attitude error limit built into the system. Thus, the motion of a thruster-stabilized satellite is a continual nodding motion, with attitude-control fuel being expended twice per cycle. A spin-stabilized vehicle, on the other hand, can be inherently stable and will maintain a given attitude for a long period without the expenditure of attitude-control gas. The price paid for this economy, of course, is that the satellite's instruments must work from a rotating reference frame. Most spacecraft designed to be spun are *axisymmetric*; that is, two of the principal moments of inertia are the same. (The completely symmetric rigid body, with all three principal moments equal, is uninteresting.) Also, a satellite in orbit is an almost perfectly torque-free body, considerably simplifying the analysis.

If the principal moments of inertia are A, A, and C, then Euler's equations (section 4.6) become

$$M_1 = 0 = A\frac{d\omega_1}{dt} + (C - A)\omega_2\omega_3$$

$$M_2 = 0 = A\frac{d\omega_2}{dt} + (A - C)\omega_1\omega_3 \qquad (5.1)$$

$$M_3 = 0 = C\frac{d\omega_3}{dt} + (A - A)\omega_1\omega_2$$

The third equation above simplifies to the statement that $d\omega_3/dt = 0$, implying that $\omega_3 = \Omega$, a constant. So, the spin rate about the satellite's symmetry axis is fixed. This is not, however, the complete solution. The first and second Euler's equations now become

$$\frac{d\omega_1}{dt} = +\frac{A - C}{A}\Omega\omega_2 \qquad (5.2)$$

$$\frac{d\omega_2}{dt} = -\frac{A - C}{A}\Omega\omega_1 \qquad (5.3)$$

These are a pair of constant-coefficient, linear differential equations.

Perhaps the simplest technique for solving these equations is to differentiate (5.2) and substitute for $d\omega_2/dt$ from (5.3). The result is

$$\frac{d^2\omega_1}{dt^2} = -\alpha^2\omega_1 \qquad (5.4)$$

where α is the constant

$$\alpha = \frac{C-A}{A}\Omega \tag{5.5}$$

Equation (5.4) is a simple harmonic oscillator. Its solution is

$$\omega_1(t) = \omega_0 \sin\left[\alpha(t-t_0)\right] \tag{5.6}$$

where ω_0 is the amplitude of the oscillation and the constant t_0 allows an arbitrary initial phase. A parallel development will yield the solution for ω_2, but it is easier to simply solve for ω_2 from (5.2). The result is

$$\omega_2(t) = -\omega_0 \cos\left[\alpha(t-t_0)\right] \tag{5.7}$$

Equations (5.6) and (5.7), combined with the statement that $\omega_3 = \Omega$, constitute the complete solution of Euler's equations for the axisymmetric rigid body.

Figure 5.1 shows the behavior of the $\boldsymbol{\omega}$ vector in the body frame. The $\boldsymbol{\omega}$ vector has a constant projection on the satellite's symmetry axis, while the projection of the $\boldsymbol{\omega}$ vector on the satellite's plane of symmetry executes a circular motion with angular frequency α. The $\boldsymbol{\omega}$ vector thus describes a cone about the satellite's axis of symmetry, which explains the term *coning motion* applied to this solution.

Now, all that we have at this point is the description of the angular-velocity vector with respect to the body frame. We are ultimately interested in a description of the motion of the satellite seen from the inertial frame. As a first step in this direction, let us calculate the angular-momentum vector. Since

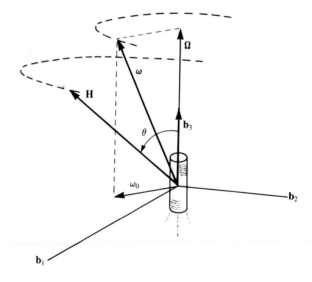

FIGURE 5.1
Angular velocity and angular momentum in the body frame.

$\mathbf{H} = I\boldsymbol{\omega}$, the components of the angular-momentum vector in the body frame are

$$\mathbf{H} = \begin{bmatrix} A\omega_0 \sin\left[\alpha(t - t_0)\right] \\ -A\omega_0 \cos\left[\alpha(t - t_0)\right] \\ C\Omega \end{bmatrix} \tag{5.8}$$

Now, the satellite has no applied torque, so the angular-momentum vector is constant *with respect to the inertial frame*. This is not true when viewed from the body frame. Since the components of the $\boldsymbol{\omega}$ vector have simply been multiplied by factors of A, A, and C to produce $\mathbf{H}$, the $\mathbf{H}$ vector also cones about the satellite's symmetry axis, again with angular frequency α. The angular-velocity vector $\boldsymbol{\omega}$ and the angular-momentum vector $\mathbf{H}$ do not coincide. The angle θ between the angular-momentum vector and the symmetry axis of the satellite is given by

$$\cos\theta = \frac{\mathbf{H} \cdot \mathbf{b}_3}{|\mathbf{H}|} = \frac{C\Omega}{\sqrt{A^2\omega_0^2 + C^2\Omega^2}} \tag{5.9}$$

which relates θ to ω_0 and Ω.

To complete the solution, we introduce the Euler angles ϕ, θ and ψ and align the angular-momentum vector $\mathbf{H}$ with the z inertial axis. The Euler angle θ is then just the angle given in (5.9), and it is a constant. The inertial-frame geometry is shown in Figure 5.2. The second set of equations of motion are obtained by equating the body-frame components of the angular-velocity vector to their expressions in terms of the Euler-angle rates $d\phi/dt$ and $d\psi/dt$.

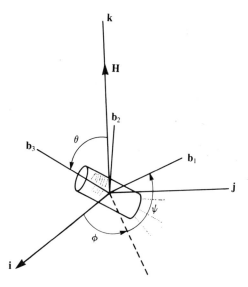

FIGURE 5.2
Satellite motion with respect to the inertial frame.

The result is

$$\omega_1 = \omega_0 \sin \alpha t = \dot{\phi} \sin \theta \sin \psi \qquad (5.10)$$

$$\omega_2 = -\omega_0 \cos \alpha t = \dot{\phi} \sin \theta \cos \psi \qquad (5.11)$$

$$\omega_3 = \Omega = \dot{\psi} + \dot{\phi} \cos \theta \qquad (5.12)$$

where we have suppressed the t_0 dependence. These three equations will give the behavior of the Euler angles with time. If we square and add equations (5.10) and (5.11), the result is

$$\dot{\phi}^2 \sin^2 \theta = \omega_0^2 \qquad (5.13)$$

showing that the precession rate $d\phi/dt$ is a constant. However, (5.13) leaves the sign of the precession ambiguous, so we need to be somewhat more careful. Knowing that $d\phi/dt$ is constant, direct comparison of (5.10) and (5.11) shows that

$$\dot{\phi} \sin \theta = \omega_0 \qquad (5.14)$$

and

$$\psi = 180° - \alpha t \qquad (5.15)$$

These furnish the behavior of both inertial position angles as a function of time.

One final result is of interest, however. If we solve for $d\phi/dt$ from (5.12) and substitute for Ω from the statement that $d\psi/dt = -\alpha$ implied by (5.15) and use (5.5), there results

$$\dot{\phi} = \frac{C\dot{\psi}}{(A - C) \cos \theta} \qquad (5.16)$$

This expresses the *precession rate* $d\phi/dt$ in terms of the *inertial spin rate* $d\psi/dt$. The term precession rate for the quantity $d\phi/dt$ is astronomical terminology. It is perhaps more common for astronautical engineers to refer to this component of the motion as the *nutation*. However, this conflicts with the accepted usage of nutation for the nodding motion of the axis of a top or torque-free rigid body, and can cause much confusion. If $A > C$, the satellite is a *prolate* rigid body, like a slim cylinder, and the precession and spin will occur in the same direction. If $A < C$, the satellite is an *oblate* rigid body, like a squat cylinder, and the precession is in the opposite sense from the spin. In this latter case the precession is *retrograde,* while in the former case the precession is termed *prograde.* Prograde and retrograde precession may be easily observed by tossing a spinning chalk stick and a spinning coin into the air. Note that the "free" precession is fast, in contrast to the "forced" precession of the earth's axis. Equation (5.15) shows that the spin rate of the body frame is constant, while (5.16) shows that the precession rate can be large, especially for a nearly symmetric body. This occurs because the angular-momentum vector is not aligned with the spin axis in free precession. In the forced precession of the earth's axis, it is aligned with the spin axis at all times.

The main features of the coning motion of an axisymmetric rigid body are familiar from observing the motion of a poorly thrown forward pass in football. It should be obvious that spinning a football for "stability" does not ensure that a state of pure spin about the symmetry axis is achieved. To achieve a state of pure spin, the initial conditions must also be correct: ω_0 should be zero. If this is not true, the object will cone. Even if the initial conditions *are* correct, a state of pure spin will not necessarily persist forever. Spinning the object merely gives it a large angular momentum, and this makes the object relatively immune to the influence of small external torques once a state of pure spin is achieved.

5.3 THE GENERAL TORQUE-FREE RIGID BODY

The general case of a torque-free rigid body is also very important for spacecraft. In the general case, the principal moments of inertia are all different, say, A, B, and C. For definiteness, let us assume that $A > B > C$, in which case the axis with the largest moment of inertia, say, the $\mathbf{b}_1$ axis, will be termed the *major axis* of inertia, the $\mathbf{b}_2$ axis will be aligned with B and called the *intermediate axis*, and the third axis $\mathbf{b}_3$ is the *minor axis* of inertia. The complete analytic solution (as we found for the axisymmetric case) is available for the general case as well, but this solution involves the elliptic integral functions. A geometric solution was discovered by the French mathematician L. Poinsoit in 1834, and permits us to obtain characteristics of general torque-free rigid-body motion without the use of unfamiliar functions.

We begin with the realization that a torque-free body conserves angular momentum $\mathbf{H}$, and without any external forces it will also conserve kinetic energy T. However, as we saw in the axisymmetric case, the angular-momentum expression $\mathbf{H} = I\boldsymbol{\omega}$ produces $\mathbf{H}$ in its body-frame components, and the angular momentum need not appear to be constant in the body frame. Only the magnitude of the angular momentum will be constant in the body frame. So, the statement of conservation of the magnitude of the angular momentum can be written

$$H^2 = |\mathbf{H}|^2 = A^2\omega_1^2 + B^2\omega_2^2 + C^2\omega_3^2 \tag{5.17}$$

The conservation of energy implies that

$$T = \tfrac{1}{2}A\omega_1^2 + \tfrac{1}{2}B\omega_2^2 + \tfrac{1}{2}C\omega_3^2 \tag{5.18}$$

is a constant. These two expressions suffice to characterize many features of the motion of the $\boldsymbol{\omega}$ vector as seen from the body frame.

If we divide each of (5.17) and (5.18) by its left side, we find

$$\frac{\omega_1^2}{(H/A)^2} + \frac{\omega_2^2}{(H/B)^2} + \frac{\omega_3^2}{(H/C)^2} = 1 \qquad (5.19)$$

$$\frac{\omega_1^2}{2T/A} + \frac{\omega_2^2}{2T/B} + \frac{\omega_3^2}{2T/C} = 1 \qquad (5.20)$$

Geometrically, each of these formulas describes three-dimensional ellipsoids in the body frame. The ω vector must lie on the surface of the *angular-momentum ellipsoid*, (5.19), and at the same time it must lie on the surface of the *kinetic-energy ellipsoid*, (5.20). These ellipsoids are not identical, since their axis lengths (the square roots of the denominators of the above expressions) are not the same. The axis lengths of the angular-momentum ellipsoid are

$$\frac{H}{A} \qquad \frac{H}{B} \qquad \frac{H}{C}$$

while those of the kinetic-energy ellipsoid are

$$\sqrt{\frac{2T}{A}} \qquad \sqrt{\frac{2T}{B}} \qquad \sqrt{\frac{2T}{C}}$$

The first set of axis lengths depend on the inverse of the principal moments of inertia, while the second set are functions of the square root of the principal moments, so the shapes of the ellipsoids are not the same. The values of H and T depend on the initial conditions for a particular case. Since the ω vector must lie on the surface of these two ellipsoids simultaneously, the possible path of the angular-velocity vector seen from the body frame will be given by the curve where the ellipsoids intersect. Only the information on the speed of movement of the ω vector has been sacrificed in this description.

Figure 5.3 shows the intersection of the kinetic-energy and angular-momentum ellipsoids for a particular set of values of H and T. Note that since the axis lengths are inversely proportional to the moments of inertia, the major axis of these two ellipsoids is the *minor* inertia axis and the minor axis of these ellipsoids is the *major* axis of inertia. The curve where the ellipsoids intersect is the path of the ω vector, seen, of course, from the body frame. It is termed a *polhode*.

To obtain a general feeling for the possible motions of a general torque-free rigid body, it is necessary to construct a series of polhodes for different sets of initial conditions. Let us hold the angular momentum H constant and consider different values for the kinetic energy. Begin with the rigid body in a state of pure spin about the minor axis of inertia, the C axis. As shown in Figure 5.4a, the kinetic-energy ellipsoid lies entirely outside the angular-momentum ellipsoid but tangent at the $\mathbf{b}_3$ axis. The polhode for this case consists of one point at the top and one point at the bottom of the angular-momentum ellipsoid, corresponding to the two possible spin direc-

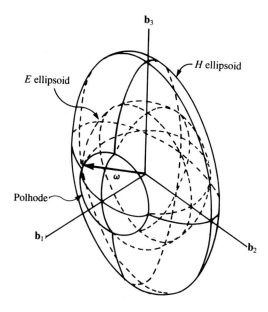

FIGURE 5.3
Intersection of the kinetic-energy and angular-momentum ellipsoids.

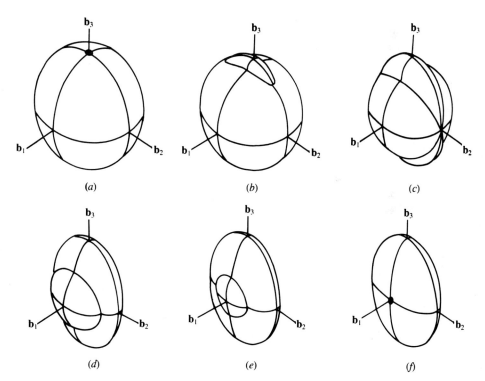

(a)   (b)   (c)

(d)   (e)   (f)

FIGURE 5.4
Polhodes for different kinetic energies.

tions. Now, hold the angular momentum constant and decrease the energy. This is a completely different possible state for this system, since a rigid body has no means to dissipate energy. The kinetic-energy ellipsoid will now lie partly inside the angular-momentum ellipsoid, and their intersection will be a pair of closed curves around the $\mathbf{b}_3$ axis, as in Figure 5.4b. As we continue to decrease the kinetic energy, more and more of the energy ellipsoid lies within the angular-momentum ellipsoid, and the polhodes grow in size around the $\mathbf{b}_3$ axis. The two branches of the polhodes first touch each other at the intermediate axis $\mathbf{b}_2$, as shown in Figure 5.4c. At this point, the kinetic-energy ellipsoid lies within the angular-momentum ellipsoid along the $\mathbf{b}_3$ direction but lies outside the angular-momentum ellipsoid in the $\mathbf{b}_1$ direction. As we continue to decrease the energy, the polhodes again form two branches, this time encircling the $\mathbf{b}_1$ axis (Figure 5.4d, e). Eventually we reach a point where the energy ellipsoid is entirely within the angular-momentum ellipsoid, just touching it at the $\mathbf{b}_1$ axis, shown in Figure 5.4f. This is the minimum energy configuration for the given value of the angular momentum. Once again, the polhode has become a single point.

Figure 5.5 shows the entire sequence of polhodes, all plotted on the same angular-momentum ellipsoid. Arrows have been added to show the direction of movement of the $\boldsymbol{\omega}$ vector along these curves, although we do not have in our possession any information on how fast $\boldsymbol{\omega}$ travels. The first point to be noticed from these curves is that rotation about either the major or minor inertia axes is *stable*, while rotation about the intermediate axis is *unstable*. A small displacement from pure rotation about either the major or minor axes

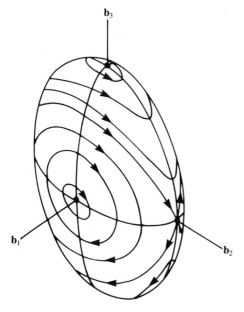

FIGURE 5.5
The complete set of polhodes.

puts the ω vector on a small closed polhode about that axis. A small initial displacement from pure spin thus stays small. This is not true at the intermediate axis of inertia, where a small displacement will place us on polhodes departing very far from the intermediate axis. An initially small displacement will thus grow to be large, and seen from the inertial frame the body will appear to tumble wildly. This polhode will, however, lead the ω vector to the vicinity of the intermediate axis on the other side of the figure (the opposite spin direction), only to depart again. This behavior can be readily seen in a textbook spun and tossed into the air about each of its principal axes. The book will be difficult to catch again when initially spun about the intermediate axis of inertia.

The second step of Poinsoit's solution is to carry this description of the motion into a description of what happens with respect to the inertial frame. Figure 5.6 shows the inertial frame, with the angular-momentum vector oriented along the negative z axis. The projection of the ω vector upon the angular-momentum vector $\mathbf{H}$ is given by

$$\frac{\boldsymbol{\omega} \cdot \mathbf{H}}{|\mathbf{H}|} = \frac{1}{H}(A\omega_1^2 + B\omega_2^2 + C\omega_3^2) = \frac{2T}{H} \tag{5.21}$$

This is a constant, since T and H are constant. So, the tip of the ω vector always lies within a plane perpendicular to the angular-momentum vector $\mathbf{H}$. This plane is termed the *invariable plane*. Now, draw the kinetic-energy

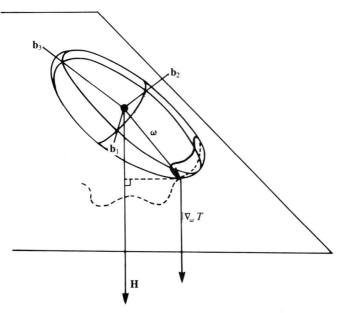

FIGURE 5.6
Motion seen from the inertial frame.

ellipsoid about the base of the **H** vector (the center of mass of the rigid body). The normal to the kinetic-energy ellipsoid is given by the gradient of (5.18), or

$$\nabla_\omega T = \left(\mathbf{b}_1 \frac{\partial}{\partial \omega_1} + \mathbf{b}_2 \frac{\partial}{\partial \omega_2} + \mathbf{b}_3 \frac{\partial}{\partial \omega_3}\right)$$
$$\times (\tfrac{1}{2}A\omega_1^2 + \tfrac{1}{2}B\omega_2^2 + \tfrac{1}{2}C\omega_3^2)$$
$$= A\omega_1\mathbf{b}_1 + B\omega_2\mathbf{b}_2 + C\omega_3\mathbf{b}_3 = \mathbf{H} \qquad (5.22)$$

Now, the tip of the **ω** vector lies on the surface of the kinetic-energy ellipsoid. The normal vector to the kinetic-energy ellipsoid is the angular-momentum vector, which is perpendicular to the invariable plane. Hence, the kinetic-energy ellipsoid is *tangent* to the invariable plane at the point of contact. Furthermore, the point of contact is at the tip of the **ω** vector, which is the instantaneous axis of rotation. This axis in the body is momentarily at rest, so there is no slippage between the kinetic-energy ellipsoid and the invariable plane. Since the spacecraft body frame is embedded within the kinetic-energy ellipsoid and moves with it, the motion of the body in the inertial frame is described by *the kinetic-energy ellipsoid pivoted at the center of mass and rolling without slipping on the invariable plane.*

As a special case, let us return for a moment to the axisymmetric rigid body. If the spacecraft is axisymmetric, then the kinetic-energy ellipsoid is an ellipsoid of revolution. When this rolls without slipping on the invariable plane, the resulting motion is once again the coning motion. The symmetry axis describes a cone about the angular-momentum vector, as shown in Figure 5.7. The general case of a triaxial kinetic-energy ellipsoid moving across the invariable plane is admittedly difficult to visualize, especially when motion is occurring near the unstable intermediate axis. However, when motion occurs reasonably near the major or minor axes, the motion is not too different from that of the axisymmetric case. One body axis will generally describe a cone

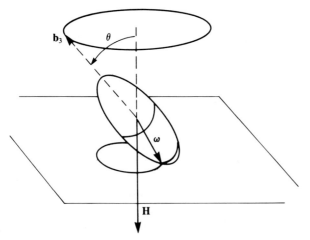

FIGURE 5.7
Poinsoit's construction for the axisymmetric case.

about the angular-momentum vector as the ellipsoid rolls on the plane. However, the ellipsoid now has two "flat spots" on it. So in addition to the coning (the free precession), the body axis will also nod up and down, twice per revolution of the object. This nodding motion is termed the *nutation* by astronomers, although we have already seen that this term is sometimes used for the precession by astronautical engineers.

5.4 THE SEMIRIGID SPACECRAFT

No spacecraft is ever a perfectly rigid body. The vehicle may have slightly flexible appendages, long whip antennas, for example, or it may have fuel tanks with fluids able to slosh about. Internal motions of spacecraft components are totally unable to alter the angular momentum of the vehicle, since internal torques always cancel. However, internal motions may be able to convert kinetic energy of rotation into *heat* by energy dissipation, in which case the energy is lost forever. Most semirigid bodies are, by intention at least, designed as rigid bodies. Thus, the rate of energy dissipation is likely to be small, and variations of the moment of inertia matrix can be neglected. Thus the *semirigid* body, conserving angular momentum but slowly dissipating energy, is important in spacecraft dynamics.

When energy is lost to the rotational motion, the ω vector will drift from polhode to polhode, moving toward ever-lower energy levels. If the rate of energy dissipation is low, the motion of the spacecraft can be well described by the current polhode at any given moment, but the long-term motion is toward lower energy levels. When the spacecraft is in a state of pure spin about a principal axis, the rate of energy dissipation will be zero. It is the time-varying accelerations experienced during precession and nutation that will drive oscillations in flexible and fluid components of the spacecraft, and these will be absent in a state of pure spin.

Figure 5.8 shows the polhodes of a general rigid body, modified by energy dissipation into a slow spiral outward from the minor inertia axis (the major axis of the angular-momentum ellipsoid). As the ω vector moves further from the minor axis, the amplitudes of the precession and nutation will increase, with a resulting increase in the rate of energy dissipation. Eventually the critical polhode is approached, the boundary between rotation about the minor inertia axis and rotation about the major inertia axis. When this polhode is crossed, the ω vector converges to a state of pure spin about the major inertia axis. Once this state is reached, the precession and nutational motions exciting the energy dissipation vanish, and a stable state is attained. So, for a general semirigid spacecraft, *in the long run only spin about the major axis of inertia is stable.* In the axisymmetric case, pure spin about the symmetry axis is stable *only if the symmetry axis is the major inertia axis.* Prolate semirigid bodies will eventually achieve an end-over-end tumble state.

This effect has been observed several times in the history of the space program, beginning with Explorer I. This satellite was a prolate axisymmetric

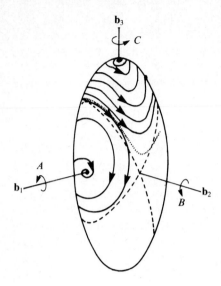

FIGURE 5.8
Polhodes for a semirigid body.

object, designed to spin about its symmetry axis for stability during the boost to orbit. Once in orbit, four whip antennas were deployed. Within only a few orbits the energy dissipated in the flexing of these antennas converted the motion of the satellite into a stable state of tumbling about the major axis. This effect however, once understood, can be made to work for the spacecraft designer. An oblate spinning spacecraft can be made more stable by including a *nutation damper,* a deliberate attempt to increase energy dissipation. This usually takes the form of a spring-mass system in a fluid-filled tube, as sketched in Figure 5.9. If the satellite departs from a state of pure spin, the nutation damper will automatically return it to this state without complicated control

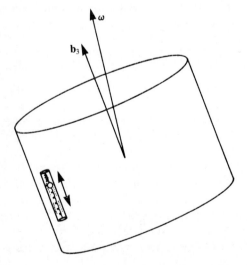

FIGURE 5.9
An oblate satellite with a nutation damper.

systems or the expenditure of control fuel. A nutation damper would, of course, actually drive a prolate spacecraft unstable all the faster. One practical difficulty with oblate spacecraft is that they do not fit well into the prolate payload shrouds of expendable launch vehicles. This problem should disappear with the space shuttle's 15-ft-wide payload bay available.

In the natural realm, planets and moons themselves are at best semirigid bodies, either with fluid cores or not quite rigid mantles. With the possible exception of Saturn's moon Hyperion, every object in the solar system for which we have data is in a nearly perfect state of pure spin about its major axis of inertia. While planets naturally form themselves into stable oblate shapes because of their own rotation, smaller objects need not do this. Mars' moon Phobos is shaped like a triaxial ellipsoid with axis lengths of 12, 18, and 24 km, spinning, of course, about the shortest geometric axis.

Finally, in the earthly realm, the transition from spin about the symmetry axis to end-over-end tumble can be observed by tossing prolate cans of your favorite variety of soup into the air. The moment of inertia tensor of a soup can is constant, while the state of the contents governs the rate of internal energy dissipation. Tuna fish cans, on the other hand, are quite stable.

For the triaxial rigid body, note that it is not possible to predict which branch of the critical polhode will be crossed first. In either case the object approaches a state of pure spin about the major axis, but the *direction* of this spin is not predictable in advance. The ATS-3 satellite was spun for stability during its Hohmann transfer to geosynchronous orbit and was carefully designed to be a major-axis spinner during this period. After the orbit insertion burn, the empty rocket casing was to be ejected, and the satellite would be again in a state of major-axis spin. However, with the empty motor casing in place the satellite was in a state of minor-axis spin. After the injection motor burn, anomalous modulations in the telemetry signal were noticed (only with hindsight), indicating that the transition to major-axis spin had begun. In attempting to find the source for this problem, the command to eject the motor case was not transmitted, allowing enough time for the transition to major-axis spin to be completed. Of course, if the case was ejected the satellite would transition back to its original spin axis, with a 50:50 chance that the spin sense would be reversed.

Unfortunately, ATS-3 was to have been completely despun by a yo-yo mechanism (see section 5.7). When the motor case was finally ejected, the spacecraft, of course, crossed the wrong critical polhode (as mandated by Murphy's Law), and the yo-yo device was useless. Without thrusters, ATS-3 still spins to this day, unable to perform its mission.

There is one case where it is common to spin a *prolate* configuration. A spin-stabilized satellite with its two kick motors will often have a prolate shape before the first Hohmann transfer burn. Such a stack may be spin-stabilized so long as the rate at which energy is internally dissipated is small. Such a configuration is, of course, actually unstable. The time required for the instability to develop is simply made longer than the time interval required for the mission. This time interval is usually from injection into parking orbit to the first Hohmann maneuver, a matter of a few hours at most. The longer coast out to apogee is best done with an inherently stable configuration. This, of course, requires a dynamically oblate shape once the first motor case is discarded.

5.5 ATTITUDE CONTROL: SPINNING SPACECRAFT

One of the simplest attitude maneuvers is the reorientation of the spin axis of an axisymmetric spacecraft. When spinning satellites execute a Hohmann transfer to geosynchronous orbit, an inclination change is required. This necessitates changing the spin axis of the stack from the original orientation to the orientation necessary for the second maneuver. Also, deep-space probes of the Pioneer and Voyager families are spun to eliminate the constant drain on attitude-control fuel experienced by a three-axis stabilized vehicle. These vehicles have the large dish antenna pointing out the spin axis, which must therefore be aligned with the direction to the earth. As the spacecraft sweeps through the solar system, it is necessary to periodically realign the spin axis with the current direction to the earth lest communications be interrupted.

As shown in Figure 5.10, assume the satellite is initially in a state of pure spin about the symmetry axis and that it is to be returned to a state of pure spin about this axis. However, the inertial direction of the symmetry axis is to be changed by an angle 2θ. This reorientation maneuver is performed by deliberately exciting the natural coning motion of the vehicle and allowing it to precess through one half cycle. A pair of attitude thrusters in the orientation shown in the figure deliver an impulsive torque **M** during a time period Δt, which changes the angular-momentum vector by the amount $\Delta \mathbf{H} = \mathbf{M} \, \Delta t$. During this impulsive thruster burn, then, the spacecraft angular-momentum

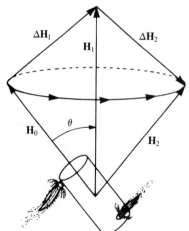

FIGURE 5.10
Attitude maneuver for a spinning axisymmetric satellite.

vector becomes

$$\mathbf{H}_1 = \mathbf{H}_0 + \Delta \mathbf{H} \tag{5.23}$$

but the spacecraft itself does not immediately move. In particular, since the torque is perpendicular to the spin axis, the satellite's spin rate $d\psi/dt$ remains unaltered. Once the thruster placement is known, the torque supplied by the engines can be calculated. Since we also know the required change in the angular momentum, the thrusters must fire for a time interval of $\Delta t = |\Delta \mathbf{H}| / |\mathbf{M}|$.

Borrowing a result from our discussion of the axisymmetric spacecraft, (5.16), the satellite now begins to precess at a rate given by

$$\dot{\phi} = \frac{C\dot{\psi}}{(A - C)\cos \theta} \tag{5.24}$$

In one half precession cycle the spacecraft spin axis will reach the desired new position. This will occur when ϕ changes by π rad, or it will occur a time interval

$$t = \frac{\pi(A - C)\cos \theta}{C\dot{\psi}} \tag{5.25}$$

after the original impulse. If the original spin rate is small, this will be a long time interval. At this point, a second impulse, identical to the first in magnitude and with the same orientation with respect to the spacecraft, will cause the angular-momentum vector to change by the amount $\Delta \mathbf{H}_2$. The new angular-momentum vector $\mathbf{H}_2$ is identical in magnitude to the original angular momentum $\mathbf{H}_0$ and is once again aligned with the spacecraft spin axis. So, after this second maneuver the satellite is once again in a state of pure spin and no longer coning.

The total cost of the two impulses is given by

$$\Delta H_{\text{tot}} = 2|\Delta \mathbf{H}| = 2|\mathbf{H}_0| \tan \theta \tag{5.26}$$

showing that reorientations by $\theta = 90°$ are infinitely expensive. This case corresponds to an end-for-end reversal of the spacecraft. Even moderately large values of θ can be quite costly. However, many attitude-control thrusters have a minimum impulse, and larger impulses are generated by commanding repeated firings of this size. A less expensive method to perform a large reorientation would then be to execute a series of small reorientation maneuvers, as shown in Figure 5.11. In the limit where a series of vanishingly small coning maneuvers are performed, the angular-momentum vector $\mathbf{H}$ is brought to its new position by moving it around the arc of a circle. The total change to the angular momentum is then given by

$$\Delta H_{\text{tot}} = 2|\mathbf{H}_0|\theta \tag{5.27}$$

Equation (5.27) may be directly compared with (5.26). Since $|\theta| < |\tan \theta|$, it is always cheaper to move a spinning spacecraft to a new orientation by a series

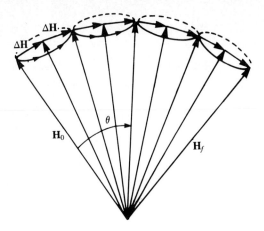

FIGURE 5.11
Large-angle reorientation by repeated impulses.

of small coning maneuvers rather than by one large maneuver. However, (5.25) shows that a small-angle coning maneuver will take the longest possible time interval. Thus, a compromise may need to be struck between saving attitude-control fuel and saving time.

5.6 ATTITUDE CONTROL: NONSPINNING SPACECRAFT

Most nonrotating spacecraft have some form of pointing requirement. The attitude of the spacecraft must be controlled to maintain the required inertial orientation for instruments, to align solar panels, or to point antennas. There are three main devices for performing this function: attitude-control thrusters, momentum wheels, and control-moment gyroscopes. In this section we will consider all three methods in turn.

Consider the three-axis thruster-controlled spacecraft of Figure 5.12. Assume that the principal moments of inertia about the $\mathbf{b}_1$, $\mathbf{b}_2$, and $\mathbf{b}_3$ axes are A, B, and C, respectively. If we fire the two small rocket engines shown mounted along the $\mathbf{b}_2$ direction, then the resulting torque on the satellite is

$$\mathbf{M} = 2\mathbf{r} \times \mathbf{F} \tag{5.28}$$

where $\mathbf{r}$ locates the thruster in the spacecraft reference frame and $\mathbf{F}$ is the thrust of the engine. Typical attitude-control thrusters fire for only a brief period of time, say, Δt. During this time the spacecraft does not rotate appreciably. So, if the initial angular momentum of the vehicle was zero, it now has angular momentum given by

$$\mathbf{H} = \mathbf{M}\,\Delta t = 2\mathbf{r} \times \mathbf{F}\,\Delta t \tag{5.29}$$

which will be in the $\mathbf{b}_1$ direction.

Now, since $\mathbf{H} = I\boldsymbol{\omega}$, the satellite has acquired an angular velocity in the $\mathbf{b}_1$

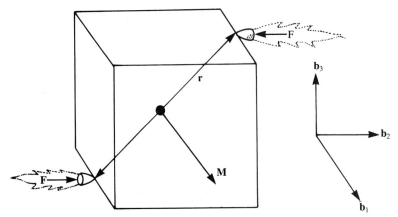

FIGURE 5.12
Thruster-controlled spacecraft.

direction given by

$$\omega_1 = \frac{|\mathbf{H}|}{A} \tag{5.30}$$

and will rotate about this principal axis. It is desirable that the thruster system be aligned with the principal-axis frame, since attempting to rotate about all three principal axes at once will produce the complicated motion characteristic of the torque-free general rigid body. If the thrusters are aligned with the spacecraft principal axes, then any general rotation can be performed by sequentially rotating the vehicle about each axis in turn. Of course, rotation about the intermediate axis is unstable, but this will not necessarily cause great difficulty. If the time required for the instability to develop is longer than the time needed to perform the rotation, then the instability will not be a problem. Also, the angular-velocity components about the major and minor inertia axes can be kept zero by firing the appropriate thrusters as necessary while rotating about the intermediate inertia axis. The rotation is stopped, of course, by firing an opposing pair of thrusters to once again zero the spacecraft angular momentum. Another possible way to avoid rotations about the intermediate inertia axis (say, $\mathbf{b}_2$) is to perform a general reorientation by using the classical Euler angles with three rotations: first about $\mathbf{b}_1$, then about $\mathbf{b}_3$, and the third about $\mathbf{b}_1$ again.

Attitude-control thrusters are usually low-performance devices. The two types most often used are cold-gas jets, in which a pressurized gas is vented through a rocketlike nozzle, or hydrazine monopropellant devices, in which hydrazine chemically decomposes in a catalyst bed and the hot exhaust is vented. A newer type increases the exhaust temperature (and thus the exhaust speed) by electrically heating the by-products of the hydrazine-breakdown reactions. The advantage of these systems is that they require only one set of plumbing and valves, which greatly increases reliability. A spacecraft with a thruster valve stuck open (or closed) is in severe straits, and monopropellant systems halve the possibility of this occurring.

Also, it is never possible to exactly zero the spacecraft angular momentum. With a small residual angular momentum, the vehicle slowly and inevitably drifts away from the desired orientation, and after some period of time it must be maneuvered back. This results in a steady drain on the supply of attitude-control fuel, and once it is depleted there is no more available. This was the cause of death of every NASA Mariner and Viking deep-space probe. Now, it is not possible to control the spacecraft trajectory (e.g., to maneuver) without throwing something overboard. However, there are two devices able to control the *attitude* of a spacecraft without using rocket fuel: the momentum wheel and the control-moment gyroscope.

The momentum wheel is shown in Figure 5.13. It consists of a flywheel, moment of inertia I, mounted rigidly in the satellite. We begin with the satellite stationary with respect to the inertial frame and the flywheel not rotating. The total angular momentum $\mathbf{H}_{\text{tot}}$ of this system is then initially zero. Now, use an electric motor to rotate the flywheel at angular rate ω_f compared to the satellite. The equal and opposite reaction torque of the electric motor on the satellite will cause it to rotate at rate ω_1 with respect to inertial space. The total angular momentum of the system will still be zero. Since the flywheel will be rotating at an angular rate of $\omega_f - \omega_1$ in inertial space, conservation of angular momentum requires

$$H_{\text{tot}} = 0 = I(\omega_f - \omega_1) - A\omega_1 \qquad (5.31)$$

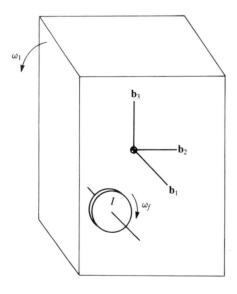

FIGURE 5.13
Spacecraft with a momentum wheel.

This can be solved to yield the spacecraft rate as

$$\omega_1 = \frac{I\omega_f}{A + I} \tag{5.32}$$

Typically, $A \gg I$, so it requires a large flywheel rate ω_f to produce a moderate spacecraft rate ω_1. This, however, makes the system sensitive, and it is able to null small spacecraft rates with ease. Of course, three such momentum wheels are required to obtain complete control over the spacecraft attitude.

The second system for attitude control without fuel expenditure is the control-moment gyroscope. The flywheel is now mounted in a gimbal pivoted about the $\mathbf{b}_2$ axis, as shown in Figure 5.14. Unlike the momentum wheel, the rotor of the control-moment gyroscope is always spinning, say, at rate ω_f, even when the spacecraft is stationary in inertial space. This gives the system a total angular momentum

$$\mathbf{H}_{\text{tot}} = I\omega_f \mathbf{b}_1 \tag{5.33}$$

in the initial state. Now, suppose we use an electric motor on the gimbal axis to rotate the gimbal downward by an angle θ compared to the spacecraft. The angular momentum of the rotor changes by the amount

$$\Delta \mathbf{H}_{\text{CMG}} = -I\omega_f \sin \theta \mathbf{b}_3 \tag{5.34}$$

Now, the total angular momentum of the complete system cannot change. So, in the process of turning the gimbal, the satellite acquires an angular

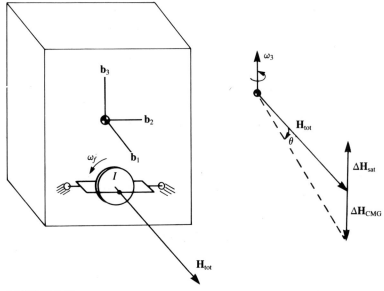

FIGURE 5.14
The control-moment gyroscope.

momentum opposite to the change in that of the rotor, $\Delta \mathbf{H}_{sat} = -\Delta \mathbf{H}_{CMG}$. Since the spacecraft has a moment of inertia of C about the $\mathbf{b}_3$ axis,

$$\omega_3 = \frac{I\omega_f \sin \theta}{C} \tag{5.35}$$

Note that the momentum wheel directly affects the spacecraft around the flywheel rotation axis, while the control-moment gyroscope is rotated about a second axis and produces a spacecraft rate about the third axis. Typically the rotor spin ω_f is large, and only small control angles θ are used. If $\theta = 90°$, the gyroscope saturates and is no longer able to control rotations about the $\mathbf{b}_3$ axis. The control-moment gyroscope is, in reality, a rate gyroscope run in reverse.

Now suppose we add a second gimbal to the system shown in Figure 5.14, able to rotate about the $\mathbf{b}_3$ axis. The single control-moment gyroscope will then be able to produce spacecraft rates about either the $\mathbf{b}_3$ axis or the $\mathbf{b}_2$ axis. (In fact, it is necessary to have a second gimbal. Otherwise a spacecraft rate ω_3 will cause a change in the angular momentum $\mathbf{H}$. The second gimbal must be allowed to drift about the $\mathbf{b}_3$ axis while maintaining the $\mathbf{b}_2$ displacement necessary to control the spacecraft ω_3 rate.) Two such dual units could control all three spacecraft axes, with one axis redundantly controlled; or three such dual units could control all axes with full redundancy. This latter method is normally chosen. So, both the momentum wheel and the control-moment gyroscope are able to control the attitude of a torque-free spacecraft without expending attitude-control fuel. Perhaps preference should be given to the control-moment gyroscope, since with the same three rotors it is able to provide redundant coverage of the three spacecraft axes.

Now, both these systems can have difficulties if the spacecraft is subject to small external torques. If these act predominantly in the same direction over a long period of time, there will be a gradual buildup of angular momentum within the spacecraft system. In a reaction-wheel system, this shows up as the flywheels having to spin faster and faster to maintain a *stationary* spacecraft inertial attitude. In a control-moment gyroscope system the control angles θ_i become progressively larger to maintain a stationary attitude. The new angular momentum generated by the external torque is being absorbed by the control system, which has only a finite capacity for retaining angular momentum. Eventually the spin rate of the momentum wheel will reach dangerous levels, or the control angles of the control-moment gyroscope will approach the 90° saturation point. When this occurs, a *momentum dump maneuver* must be performed. Attitude-control thrusters are used to hold the spacecraft stationary while the momentum wheels are braked or while the control-moment gyroscopes are returned to their null positions. So, in the presence of external torques on the spacecraft *acting predominantly in one sense,* it is still necessary to have a rocket-thruster system on the spacecraft. However, these two technologies still offer considerable fuel savings over an all-thruster system, since the continual "nodding" motion of the all-thruster system is handled by the gyroscopes, not by expenditure of fuel.

The Skylab space station was equipped with a set of dual-axis control-moment gyroscopes. These functioned acceptably for several years in orbit. It was the combination of gravity-gradient and aerodynamic torques on the Skylab just before decay which finally saturated these devices.

5.7 THE YO-YO MECHANISM

Satellites spun during launch and initial orbit transfer maneuvers may not require a high spin rate when in their final orbit. The spin rate may need to be greatly reduced or brought to zero before the satellite can become operational. One dependable device for performing this function is the so-called yo-yo mechanism. This consists of one or more cords with weights wrapped around the midplane of the satellite and released to despin the vehicle. As the cords unwrap, Figure 5.15, angular momentum is transferred from the satellite to the masses. When the cords reach full extension, they are released by a "split-hinge" device, which allows the masses to depart tangentially. This device uses no attitude-control fuel. Since the weights are held in position by explosive bolts, a very reliable release mechanism, the possibility of the device failing to operate is remote. And, as we will see, if the satellite is to be totally despun the initial satellite spin need not be known in advance.

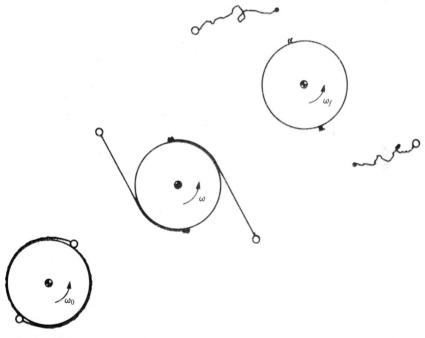

FIGURE 5.15
A two-mass yo-yo mechanism.

Figure 5.15 shows a two-mass version of this device. Since the two point masses and the rigid satellite form a closed system free of external torques, the system conserves angular momentum. Also, there is no provision for the dissipation of energy within this system, so the total mechanical energy is also constant. Assume that the satellite has a radius R and a moment of inertia C about the spin axis. It is spinning at angular rate ω_0 initially. The two masses m are identical and are attached by cords of length l. Before release, the system rotates as a unit, so the initial energy and angular momentum are given by

$$E = (\tfrac{1}{2}C + mR^2)\omega_0^2 \qquad (5.36)$$

and

$$H = (C + 2mR^2)\omega_0 \qquad (5.37)$$

As the cords unwind, the masses cease to rotate rigidly with the satellite. To calculate their inertial velocity, let us place a reference frame on the satellite with its y axis through the point where the cord is tangent to the satellite, as shown in Figure 5.16. The satellite is now rotating at angular rate ω compared to an inertial frame, while the cord xy frame has rotated by an angle ψ since the cord was released. The angular velocity of the cord frame is then $(\omega + d\psi/dt)\mathbf{k}$ in the inertial frame. Also, seen from the cord frame, the mass m is moving out along the x axis at a rate $dl/dt = R\,d\psi/dt$. This gives an

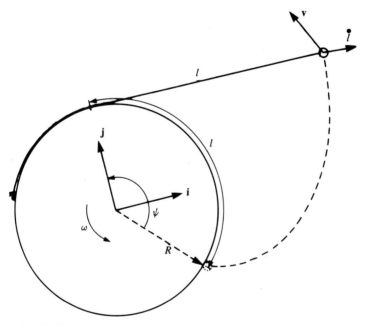

FIGURE 5.16
Velocity calculation for the masses.

alternate expression for the angular rate of the cord frame with respect to the satellite, $d\psi/dt = (dl/dt)/R$. The length of cord unwound since release is simply $l = R\psi$. The position vector of the mass is $\mathbf{r} = R\psi\mathbf{i} + R\mathbf{j}$, so the inertial velocity of a mass is given by

$$\mathbf{v} = \dot{l}\mathbf{i} + \left(\omega + \frac{\dot{l}}{R}\right)\mathbf{k} \times (l\mathbf{i} + R\mathbf{j})$$

$$= -R\omega\mathbf{i} + l\left(\omega + \frac{\dot{l}}{R}\right)\mathbf{j} \tag{5.38}$$

A simple calculation then gives the angular momentum of the two masses as

$$\mathbf{H}_m = 2m\mathbf{r} \times \mathbf{v}$$

$$= 2m\left[\omega(R^2 + l^2) + l^2\frac{\dot{l}}{R}\right]\mathbf{k} \tag{5.39}$$

The kinetic energy of one mass is $T = \frac{1}{2}m\mathbf{v} \cdot \mathbf{v}$, so the energy of both masses is given by

$$T_m = m\left[\omega^2 R^2 + l^2\left(\omega + \frac{\dot{l}}{R}\right)^2\right] \tag{5.40}$$

Then, since the angular momentum of the rigid satellite is $\mathbf{H}_{\mathrm{sat}} = C\omega\mathbf{k}$ and its energy is $T_{\mathrm{sat}} = \frac{1}{2}C\omega^2$, the total angular momentum of the system can be written

$$H = [C + 2mR^2]\omega_0$$

$$= C\omega + 2m\left[\omega(R^2 + l^2) + l^2\frac{\dot{l}}{R}\right] \tag{5.41}$$

while the total system kinetic energy is given by

$$E = \frac{1}{2}(C + 2mR^2)\omega_0^2$$

$$= \frac{1}{2}C\omega^2 + m\left[\omega^2 R^2 + l^2\left(\omega + \frac{\dot{l}}{R}\right)^2\right] \tag{5.42}$$

There are two variables of interest in this problem: the angular velocity of the satellite ω and the length of the cord l. The two available conservation laws should completely specify the behavior of these variables. They involve at worst first derivatives of ω and l and are to be preferred over the equations of motion for this system involving second derivatives of these quantities. Extracting the desired information, of course, involves some algebraic manipulation. Begin by dividing both equations by $2mR^2$. The angular-momentum equation (5.41) becomes

$$K(\omega_0 - \omega) = \frac{l^2}{R^2}(\omega + \dot{l}/R) \tag{5.43}$$

while the energy equation (5.42) becomes

$$K(\omega_0^2 - \omega^2) = \frac{l^2}{R^2}(\omega + \dot{l}/R)^2 \tag{5.44}$$

The constant K is given by

$$K = 1 + \frac{C}{2mR^2} \tag{5.45}$$

and groups most of the parameters of the system into one quantity.

In the last form above the two conservation laws have a symmetric form. Remembering that the difference of two squares is the product of their sum and difference, we can divide the energy equation (5.44) by the angular-momentum equation (5.43) to find

$$\omega_0 + \omega = \omega + \frac{\dot{l}}{R} \tag{5.46}$$

After the obvious simplification, this gives $dl/dt = R\omega_0$, a constant. So, the cord unwraps from the satellite at a fixed rate. Using this result, we can eliminate the quantity dl/dt from (5.44) and solve for the angular rate of the satellite as function of the cord length:

$$\omega = \omega_0 \frac{KR^2 - l^2}{KR^2 + l^2} \tag{5.47}$$

Alternately, if we wish the satellite to have an angular velocity ω_f when the cord reaches its end at length l_f, then (5.47) can be solved for the required cord length as

$$l_f = R\left(K\frac{\omega_0 - \omega_f}{\omega_0 + \omega_f}\right)^{1/2} \tag{5.48}$$

If the initial and final angular velocities are specified, then (5.48) gives us the required cord length to despin the satellite. A major advantage of the yo-yo mechanism is its simplicity. However, another major advantage of this device can be seen by calculating the cord length required to completely despin the satellite. If $\omega_f = 0$, then (5.48) becomes

$$l_f = R\sqrt{K} = \sqrt{R^2 + \frac{C}{2m}} \tag{5.49}$$

The cord length required to strip all angular momentum from the satellite *does not depend on the initial spin of the satellite*. Thus, if the ω_0 planned before launch is not the actual spin rate, a yo-yo mechanism will still be able to completely despin the satellite. When the weights are released, they fly off tangentially to their current path, and all the angular momentum goes with them. Of course, the string must be wound around the satellite in the correct direction, or the yo-yo mechanism will not function.

5.8 THE GRAVITY-GRADIENT SATELLITE

A small satellite in high orbit is the most nearly torque-free system ever created by human agency. However, there is one case of importance where a satellite may have to deal with external torques. A large object in orbit close to a planet may experience significant gravity-gradient torques. These torques arise from the slightly different attraction of gravity across the satellite. Figure 5.17 shows a large satellite in orbit close to the earth. The gravitational torque on the satellite is given by

$$\mathbf{M} = \int \mathbf{r} \times \mathbf{a}_g \, dm \qquad (5.50)$$

where the gravitational acceleration is

$$\mathbf{a}_g = -GM_e \frac{\mathbf{R} + \mathbf{r}}{|\mathbf{R} + \mathbf{r}|^3} \qquad (5.51)$$

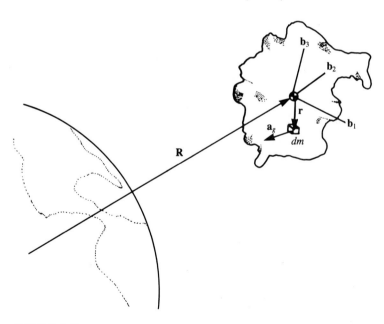

FIGURE 5.17
Gravity-gradient torques on a close-earth satellite.

Here G is the gravitational constant, M_e is the mass of the earth, and $\mathbf{R}$ locates the center of mass of the satellite seen from the earth. If we ignore the variation in gravitational force over the satellite, (5.50) will produce zero for the torque. In our normal daily environment this gravity torque is so small that it is never noticed. However, in the otherwise disturbance-free environment of an earth satellite, the gravity-gradient torque may be important.

In preparation for using the torque $\mathbf{M}$ in Euler's equations, we will evaluate the integrals (5.50) in the principal-axis body frame of the satellite. Introduce the components X, Y, and Z of the vector $\mathbf{R}$ in the body frame and the components x, y, and z of the vector $\mathbf{r}$. Then

$$\mathbf{r} \times (\mathbf{R} + \mathbf{r}) = \mathbf{r} \times \mathbf{R}$$

$$= \mathbf{b}_1(yZ - zY) + \mathbf{b}_2(zX - xZ) + \mathbf{b}_3(xY - yX) \tag{5.52}$$

Also, assuming the size of the satellite is small, we can write the denominator of (5.51) as

$$|\mathbf{R} + \mathbf{r}|^{-3} = (R^2 + 2\mathbf{R} \cdot \mathbf{r} + r^2)^{-3/2}$$

$$\approx R^{-3}\left(1 + \frac{2\mathbf{R} \cdot \mathbf{r}}{R^2} + \cdots\right)^{-3/2}$$

$$\approx R^{-3}\left(1 - \frac{3(xX + yY + zZ)}{R^2} + \cdots\right) \tag{5.53}$$

using the binomial theorem in the last step above and ignoring second-order terms in x, y, and z.

Now, inserting the intermediate results (5.51) to (5.53) into (5.50), the $\mathbf{b}_1$ torque component can be expanded as

$$M_1 = -\frac{GM_e}{R^3}\left[Z\int y\, dm - Y\int z\, dm\right.$$

$$-\frac{3XZ}{R^2}\int xy\, dm - \frac{3YZ}{R^2}\int (y^2 - z^2)\, dm$$

$$\left.-\frac{3Z^2}{R^2}\int yz\, dm + \frac{3XY}{R^2}\int xz\, dm + \frac{3Y^2}{R^2}\int zy\, dm\right] \tag{5.54}$$

The first two integrals are zero because the origin of the body frame is at the center of mass. All but two of the other integrals can be recognized as the off-diagonal products of inertia, which are zero since we are using the principal-axis body frame. The $\mathbf{b}_1$ torque component thus simplifies to

$$M_1 = \frac{3GM_e YZ}{R^5}\int [x^2 + y^2 - (x^2 + z^2)]\, dm$$

$$= \frac{3GM_e}{R^5} YZ(C - B) \tag{5.55}$$

The other two torque components are similarly found to be

$$M_2 = \frac{3GM_e}{R^5} XZ(A - C) \tag{5.56}$$

$$M_3 = \frac{3GM_e}{R^5} XY(B - A) \tag{5.57}$$

Note that a fully symmetric body will not experience any gravitational torque.

Now, consider the satellite of Figure 5.18, in a circular orbit about the earth. Then, Euler's equations for this object become

$$A\dot{\omega}_1 + (C - B)\omega_2\omega_3 = \frac{3GM_e}{R^5} YZ(C - B) \tag{5.58}$$

$$B\dot{\omega}_2 + (A - C)\omega_1\omega_3 = \frac{3GM_e}{R^5} XZ(A - C) \tag{5.59}$$

$$C\dot{\omega}_3 + (B - A)\omega_1\omega_2 = \frac{3GM_e}{R^5} XY(B - A) \tag{5.60}$$

Before these equations can be integrated, it will be necessary to introduce three orientation angles and to write the position components X, Y, and Z in these angles. However, instead of attempting to find the general solution, let us examine (5.58) to (5.60) for equilibrium points. Assume that the angular velocity of the satellite's orbit is Ω. Align one principal axis with the orbit normal and spin the satellite about this axis with rate $\omega_i = \Omega$. If the other two components ω_j are zero, then seen from the satellite body frame the earth is stationary. Since two ω_j are zero, the nonlinear terms on the left sides of (5.58) to (5.60) are zero. We can also zero the torque components on the right of (5.58) to (5.60) if two of the position components X, Y, or Z are zero. This corresponds to a second principal axis pointing toward the earth. When these choices are made, (5.58) to (5.60) reduce to the statements that $d\omega_i/dt = 0$, which are just what is needed to maintain these special configurations.

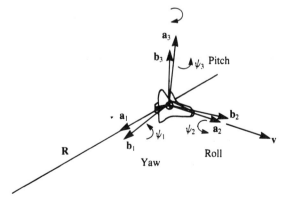

FIGURE 5.18
Gravity-gradient satellite and orbital reference frame.

So, there are three equilibrium configurations of the gravity-gradient satellite in a circular orbit. These correspond to the three possible choices of which principal axis to align with the orbit normal, times the two choices of axis to align with the direction of the earth, when the three mirror-image situations are eliminated. Of course, not all these configurations are stable. To study the stability of these equilibria, introduce the **a** frame aligned with the direction to the earth, the velocity vector, and the orbit normal, as shown in Figure 5.18. We may treat all three configurations at once if we do not make any assumptions about the relative magnitudes of the principal moments A, B, and C. Now, the **a** frame has an inertial angular-velocity vector $\Omega\mathbf{a}_3$. The body frame is not aligned with the **a** frame but differs by the three small angles ψ_1 (yaw), ψ_2 (roll), and ψ_3 (pitch). (Note that the classical Euler angles cannot be used since they are singular in this case.) Relative to the **a** frame, the body rates are simply $d\psi_1/dt\mathbf{b}_1 + d\psi_2/dt\mathbf{b}_2 + d\psi_3/dt\mathbf{b}_3$, while the angular velocity of the **a** frame itself is $-\psi_2\Omega\mathbf{b}_1 + \psi_1\Omega\mathbf{b}_2 + \Omega\mathbf{b}_3$ when resolved along the body axes. The total inertial angular velocity of the body, resolved along the body axes, is then

$$\boldsymbol{\omega} = (\dot{\psi}_1 - \psi_2\Omega)\mathbf{b}_1 + (\dot{\psi}_2 + \psi_1\Omega)\mathbf{b}_2 + (\dot{\psi}_3 + \Omega)\mathbf{b}_3 \qquad (5.61)$$

Similarly, the position of the earth, seen from the body frame, is

$$X \approx R \qquad Y \approx -\psi_3 R \qquad Z \approx \psi_2 R \qquad (5.62)$$

We now are now ready to substitute in (5.58) to (5.60).

Since we are interested in the stability of small oscillations, any term in the equations of motion involving two small factors (two of any ψ_i) can be ignored. The equations of motion then become

$$A\ddot{\psi}_1 + (C - B - A)\Omega\dot{\psi}_2 + (C - B)\Omega^2\psi_1 = 0 \qquad (5.63)$$

$$B\ddot{\psi}_2 + (B + A - C)\Omega\dot{\psi}_1 - 4(A - C)\Omega^2\psi_2 = 0 \qquad (5.64)$$

$$C\ddot{\psi}_3 + 3(B - A)\Omega^2\psi_3 = 0 \qquad (5.65)$$

We have made use of Kepler's third law

$$\Omega^2 = \frac{GM_e}{R^3} \qquad (5.66)$$

from section 2.5.

The pitch equation of motion, (5.65), is a simple harmonic oscillator so long as $B > A$. If this is not true, the pitch equation is unstable and the satellite will swing away from the equilibrium configuration when disturbed. If the pitch motion is stable, then the pitch solution is

$$\psi_3 = \psi_{30}\cos\left[n_p(t - t_0)\right] \qquad (5.67)$$

where the pitch frequency n_p is given by

$$n_p = \Omega\left[\frac{3(B - A)}{C}\right]^{1/2} \qquad (5.68)$$

The yaw and roll equations (5.63) and (5.64) are coupled. Introduce the vector $\Psi_{YR} = (\psi_1, \psi_2)$, the yaw-roll state, and the conventional definitions

$$k_Y = \frac{C-B}{A} \qquad k_R = \frac{C-A}{B} \tag{5.69}$$

Then, the roll-yaw equations can be put into the form

$$\begin{bmatrix} 1 & 0 \\ 0 & 1 \end{bmatrix} \ddot{\Psi}_{YR} + \begin{bmatrix} 0 & \Omega(k_Y - 1) \\ \Omega(1 - k_R) & 0 \end{bmatrix} \dot{\Psi}_{YR}$$

$$+ \begin{bmatrix} \Omega^2 k_Y & 0 \\ 0 & \Omega^2 k_R \end{bmatrix} \Psi_{YR} = 0 \quad (5.70)$$

Assume a solution of the form $\Psi_{YR} = \Psi_{YR0} e^{\lambda \Omega t}$, so that the roll-yaw frequencies are $\lambda \Omega$. The characteristic equation of the system (5.70) becomes

$$\begin{vmatrix} \lambda^2 + k_Y & \lambda(k_Y - 1) \\ \lambda(1 - k_R) & \lambda^2 + 4k_R \end{vmatrix} = 0 \tag{5.71}$$

Or,

$$\lambda^4 + \lambda^2(1 + 3k_R + k_Y k_R) + 4k_Y k_R = 0 \tag{5.72}$$

This is a quadratic equation in the quantity λ^2.

For stability we would like all four λ values to be imaginary. (In a dissipation-free system, it is not possible for all roots λ to all have negative real parts.) This means that the two λ^2 values from (5.72) should both be real and negative. A quadratic equation has real roots if the discriminant $b^2 - 4ac > 0$. In our case, this implies

$$\boxed{1 + 3k_R + k_Y k_R > 4\sqrt{k_Y k_R}} \tag{5.73}$$

Also, to ensure that both real roots are negative, we need $b^2 > b^2 - 4ac$, which becomes

$$\boxed{k_Y k_R > 0} \tag{5.74}$$

These two requirements, combined with the earlier requirement that $B > A$ for pitch stability, govern the complete stability of the gravity-gradient spacecraft equilibria.

It is convenient to plot stability regions on the $k_R k_Y$ plane. Each of these variables lies between -1 and 1, and between them they specify the inertial shape of the object. Now, the pitch-stability criterion $B > A$ is equivalent to $C - A > C - B$, or, dividing by A, pitch stability requires

$$k_Y = \frac{C-B}{A} < \frac{C-A}{A} < \frac{C-A}{B} = k_R \tag{5.75}$$

since $B > A$. So, the pitch-stability criterion $k_R > k_Y$ eliminates the upper left half of the $k_R k_Y$ plane shown in Figure 5.19. Equation (5.74) then eliminates the lower right quadrant of this plane. Condition (5.73) has nothing to say

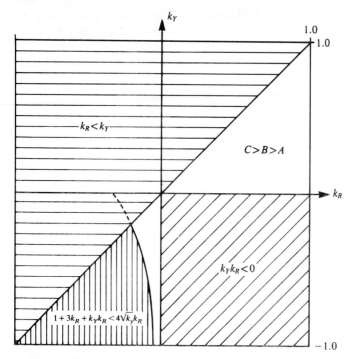

FIGURE 5.19
Gravity-gradient spacecraft stability regions.

about the upper triangular region in quadrant I but eliminates most of the rest of quadrant III in Figure 5.19, leaving two stability regions. Of these two, the large triangular region, where $C > B > A$, is preferable, since this orientation corresponds to the minimum total energy configuration for a gravity-gradient satellite. In the presence of energy dissipation, this is the only stable region. Almost every moon in the solar system has a captured rotation rate, and they are always found to have their major inertia axis along the orbit normal, their intermediate inertia axis along the velocity vector, and their minor inertia axis pointing at their mother planet. However, the spacecraft designer may not be able to wait the geologically long timescales required for energy dissipation to produce this stable configuration.

A small satellite may be modified for gravity-gradient stabilization by adding a small mass on a long boom. This will increase the moment of inertia in the directions transverse to the boom, and the spacecraft will be stable with the mass pointed toward or away from the earth. Yaw stability will be poor if the configuration is nearly axisymmetric. However, gravity-gradient stabilization normally must be aided by some active system. Although the preferred orientation may be a stable equilibrium, the gravity-gradient torques are small, and there is nothing forbidding oscillations about the equilibrium. It is the job of the active system to suppress these oscillations when they occur, leaving the satellite in the stable position. Alternately, ignoring the gravity-gradient torques for a large satellite can in the long run impose a burden on an active control system. The Skylab space station was a large object designed to be sun-pointing, keeping the solar panels perpendicular to the sun. Just before its demise, the gravity-gradient torques overwhelmed the control-moment gyroscope system and brought the station to the stable position. Unfortunately, this meant that the long axis of the Skylab pointed earthward, with the large solar panels broadside to the direction of motion: the worst position possible for air drag. Gravity-gradient stabilization has also proven important for the space shuttle. The stable attitude here would again be long axis toward the earth, with the wings along the orbit normal. Small aerodynamic torques will roll the shuttle slightly until some of the vertical stabilizer is also within the airflow. In this attitude, the shuttle's attitude-control thrusters may be turned off, and the vehicle will remain in this position. (The noise of occasional thruster firings, likened by one astronaut to having a cannon in the next room, otherwise interferes with sleeping!) The U.S. space station is also being designed with gravity-gradient stabilization.

[handwritten marginal note: wings in orbit plane (cf. Chobotov p. 99)]

5.9 THE DUAL-SPIN SPACECRAFT

A problem occurs when one attempts to mate a large, spin-stabilized spacecraft to an expendable booster. For spin stability the spacecraft should be dynamically oblate: It should have the shape of a short, squat cylinder. However, upper stages of boosters usually have a small diameter, and aerodynamic efficiency dictates that the payload be a long, slim cylinder. The advantages of spin stabilization in terms of reduced use of attitude-control fuel are so great, however, that they are not lightly given up. The solution, found in 1964 by V. D. Landon and B. Stuart, is the *dual-spin spacecraft,* which combines the stability of the oblate simple spinner with the geometric shape of the usual expendable booster payload shroud. In this section we offer an heuristic treatment of this problem, referred to as the *energy-sink* approach. Dual-spin configurations are now standard for most communications satellites, and were used for the first time in a deep-space mission on the Voyager program.

Consider the dual-spin spacecraft of Figure 5.20, consisting of a platform p and a rotor r. In normal operation, the platform will spin at rate ω_{p0}, while the rotor will have an angular rate of ω_{r0}. There is no coupling of these rates between the platform and rotor. However, consider what happens when the system is disturbed and has a transverse angular-velocity component of ω_t. Assume that the moment of inertia of the platform about its symmetry axis is C_p, the corresponding moment of inertia of the rotor is C_r, and the transverse moment of inertia of the entire assembly is given by A_t. Then, since the system is isolated, it must conserve angular momentum. This implies that

$$H^2 = (C_p\omega_p + C_r\omega_r)^2 + A_t^2\omega_t^2 \tag{5.76}$$

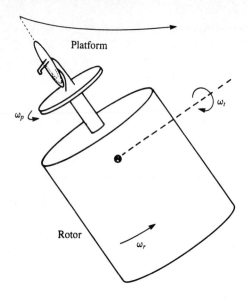

Platform

ω_p

ω_t

Rotor

ω_r

FIGURE 5.20
The dual-spin spacecraft.

is a constant. The kinetic energy of the system will be given by

$$T = \tfrac{1}{2}C_p\omega_p^2 + \tfrac{1}{2}C_r\omega_r^2 + \tfrac{1}{2}A_t\omega_t^2 \tag{5.77}$$

This is not a constant of the motion. Since there will be internal energy dissipation in the system,

$$\dot{T} = C_p\omega_p\dot{\omega}_p + C_r\omega_r\dot{\omega}_r + A_t\omega_t\dot{\omega}_t \tag{5.78}$$

will be less than zero. Differentiating (5.76), and remembering that angular momentum will be constant, we have

$$0 = (C_r\omega_r + C_p\omega_p)(C_r\dot{\omega}_r + C_p\dot{\omega}_p) + A_t^2\omega_t\dot{\omega}_t \tag{5.79}$$

Now, assume that the platform and rotor have separate rates of energy dissipation given by dT_p/dt and dT_r/dt, both of which will be negative. Solving (5.78) and (5.79) for the quantity $A_t\omega_t\dot{\omega}_t$, we have

$$A_t\omega_t\dot{\omega}_t = (\dot{T}_r - C_r\omega_r\dot{\omega}_r) + (\dot{T}_p - C_p\omega_p\dot{\omega}_p)$$

$$= -\frac{1}{A_t}(C_r\omega_r + C_p\omega_p)(C_r\dot{\omega}_r + C_p\dot{\omega}_p) \tag{5.80}$$

Identifying dT_r/dt with terms in (5.80) involving the rate of change of the rotor spin, $\dot{\omega}_r$, we find

$$\dot{T}_r = \left[\left(1 - \frac{C_r}{A_t}\right)C_r\omega_r - \frac{C_rC_p}{A_t}\omega_p\right]\dot{\omega}_r \tag{5.81}$$

For the platform the corresponding result is

$$\dot{T}_p = \left[\left(1 - \frac{C_p}{A_t}\right)C_p\omega_p - \frac{C_rC_p}{A_t}\omega_r\right]\dot{\omega}_p \tag{5.82}$$

Now, the platform usually carries instruments which must be despun in order to perform their function. This implies that ω_p will be almost perfectly zero for inertially fixed instruments or very close to zero for communications satellites, where the platform rotates once per day to follow the earth. Equations (5.81) and (5.82) are then approximated for small ω_p by

$$\dot{T}_r \approx \left(1 - \frac{C_r}{A_t}\right) C_r \omega_r \dot{\omega}_r \tag{5.83}$$

$$\dot{T}_p \approx -\frac{C_r C_p}{A_t} \omega_r \dot{\omega}_p \tag{5.84}$$

Now, both the platform and rotor should dissipate energy, implying that both dT/dt's should be negative. The most usual design for a dual-spin spacecraft is a prolate configuration, which implies that $C_r < A_t$ and $C_p < A_t$. For a prolate configuration (5.83) implies that $\dot{\omega}_r < 0$ if $\omega_r > 0$, while (5.84) says that $\dot{\omega}_p > 0$ if $\omega_r > 0$. Thus, the rotor spin will decrease, while the platform spin will increase.

Of course, the whole point to providing energy dissipation is to make the dual-spin configuration stable, meaning that any tendency of the vehicle to cone will vanish. For this to be true, we need

$$\omega_t \dot{\omega}_t < 0 \tag{5.85}$$

as the stability criterion. This ensures that coning motion should damp out, no matter the algebraic sign of ω_t. Using (5.83) and (5.84) to eliminate $\dot{\omega}_p$ and $\dot{\omega}_r$ from the energy part of (5.80), we have

$$A_t \omega_t \dot{\omega}_t = \frac{-C_r/A_t}{1 - C_r/A_t} \dot{T}_r + \frac{C_r \omega_r + A_t \omega_p}{C_r \omega_r} \dot{T}_p \tag{5.86}$$

The coefficient of dT_p/dt is nearly 1 for a slowly rotating platform, while the coefficient of dT_r/dt is positive for an oblate rotor ($C_r > A_t$) and negative for a prolate rotor ($C_r < A_t$).

So, as might have been expected, an oblate dual-spin spacecraft is inherently stable, since $\omega_t \dot{\omega}_t$ will always be negative. The case of a prolate rotor is of much greater practical interest. In this instance, energy dissipation on the platform must dominate the positive term from the rotor. So, for a prolate dual-spin satellite with a slow platform, the stability criterion is

$$|\dot{T}_p| \gg \left| \dot{T}_r \frac{C_r/A_t}{1 - C_r/A_t} \right| \tag{5.87}$$

The platform must dissipate energy at a rate far greater than the rotor to ensure that any disturbance will die out.

Even when not disturbed, a dual-spin spacecraft will still dissipate energy while conserving angular momentum. This will take the form of a frictional torque at the bearing between the platform and rotor, tending to spin up the platform and spin down the rotor. If allowed to proceed unhindered, this frictional torque would quickly bring the two angular velocities ω_r and ω_p to equality. To prevent this, the spin bearing must incorporate an electric motor, whose function is to replace energy lost to friction between the platform and rotor and (usually) to provide precise pointing for instruments or antennas on the platform. This latter function is carried out by linking the motor to a control loop and some form of platform attitude sensor. For example, a communications satellite would need to point antennas on the platform at the earth. An earth sensor (probably located on the spinning rotor, where it can see both edges of the earth once per rotor revolution) would then provide control for keeping the platform pointed at the earth.

5.10 REFERENCES AND FURTHER READING

Spacecraft attitude stability and control is the subject of excellent books by Kane, Likens, and Levinson and by Hughes. The Poinsoit construction of section 5.3 is geometric in nature and does not lend itself to actually calculating the motion of the spacecraft. The solution for this case, in terms of the elliptic integral functions, can be found in Thomson. Kaplan's book also goes further into the applications of control theory to attitude dynamics than we have space for in this book.

Kane, Thomas R., Peter W. Likens, and D. A. Levinson, *Spacecraft Dynamics,* McGraw-Hill, New York, 1983.

Hughes, Peter C., *Spacecraft Attitude Dynamics,* Wiley, New York, 1986.

Thomson, W. T., *Introduction to Space Dynamics,* Wiley, New York, 1961.

Kaplan, M. H., *Modern Spacecraft Dynamics and Control,* Wiley, New York, 1976.

5.11 PROBLEMS

1. Attitude-control thrusters are rated according to their "minimum impulse bit" $p(N \cdot s)$, the minimum amount of linear momentum that the thruster can provide in one pulse. Four thrusters are mounted on a satellite with moment of inertia I a distance R from the center of mass, as shown in Figure 5.21. The thrusters fire

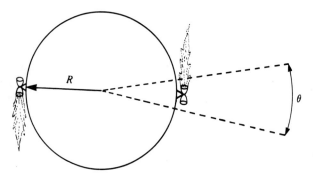

FIGURE 5.21
"Bang-bang" attitude control.

whenever the satellite strays from within an angular window of width θ, a technique referred to as "bang-bang" control. Argue that the satellite usually has angular momentum

$$|H| \approx 2pR$$

and an average angular velocity

$$|\omega| \approx \frac{2pR}{I}$$

This causes the time interval between thruster firings to be about

$$\Delta t \approx \frac{I\theta}{2pR}$$

so the satellite eats through its fuel supply at a rate proportional to

$$\frac{p^2R}{I\theta}$$

Why are small minimum-impulse-bit (MIB) thrusters desirable?

2. A spacecraft must sometimes be rotated about the intermediate inertia axis, although this is unstable. Write Euler's equations for a torque-free rigid body. For initial conditions $\omega_2 = \omega_{2o}$ (not small) and ω_1 and ω_3 very small, show that Euler's equations imply $\omega_2 \approx \omega_{2o}$, a constant, and

$$\begin{bmatrix} A & 0 \\ 0 & C \end{bmatrix}\begin{bmatrix} \dot{\omega}_1 \\ \dot{\omega}_3 \end{bmatrix} + \begin{bmatrix} 0 & (C-B)\omega_{2o} \\ (B-A)\omega_{2o} & 0 \end{bmatrix}\begin{bmatrix} \omega_1 \\ \omega_3 \end{bmatrix} = 0$$

These are a pair of linear, constant-coefficient equations. Substitute an assumed solution of the form

$$\begin{bmatrix} \omega_1 \\ \omega_3 \end{bmatrix} = \mathbf{A}e^{\lambda t}$$

and show that, as expected, rotation about the intermediate inertia axis is unstable.

Now, assume that the spacecraft has rate gyroscopes, capable of supplying measured values of the spacecraft rates ω_1 and ω_3. Use these measured rates to produce control torques according to the feedback laws

$$M_1 = k\omega_3 \qquad M_3 = k\omega_1$$

where k is a gain constant. Show that the augmented system is stable when rotating about the intermediate axis if k/ω_{2o} is between $|C-B|$ and $|B-A|$.

3. A dual-spin satellite shown in Figure 5.22, sometimes referred to as a *gyrostat*, has principal moments of inertia of A, B, and C about the axes $\mathbf{b}_1$, $\mathbf{b}_2$, and $\mathbf{b}_3$. It contains a rotor with moment of inertia I_r which spins on frictionless bearings with constant angular rate ω_r. If the angular-velocity vector of the satellite is $\boldsymbol{\omega} = (\omega_1, \omega_2, \omega_3)^T$, show that the total angular momentum is

$$H^2 = A^2\omega_1^2 + B^2\omega_2^2 + (C\omega_3 + I_r\omega_r)^2$$

while the total kinetic energy is

$$T = \tfrac{1}{2}(A\omega_1^2 + B\omega_2^2 + C\omega_3^2 + I_r\omega_r^2)$$

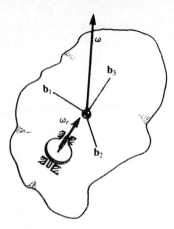

FIGURE 5.22
Dual-spin satellite.

and that both are constant. Show that both these expressions describe ellipsoids but that one ellipsoid is no longer centered at the satellite's center of mass. Argue that it is still possible to define polhodes for a gyrostat as the intersection curves of the two ellipsoids.

4. A dual-spin satellite, as in problem 3, is symmetric $(A = B)$ and normally rotates about the $\mathbf{b}_3$ axis at angular rate

$$\omega_{3o} = -\frac{I_r \omega_r}{C}$$

and thus has zero net angular momentum if ω_1 and ω_2 are zero. Write $\omega_3 = \omega_{3o} + \delta\omega_3$, and write the expressions for the kinetic-energy and angular-momentum ellipsoids. Argue that such a satellite can be controlled as if it were not spinning, although it does rotate.

5. The space-shuttle orbiter has a moment of inertia matrix

$$I = \begin{Bmatrix} 1.29 & 0 & 0 \\ 0 & 9.68 & 0 \\ 0 & 0 & 10.1 \end{Bmatrix} \times 10^6 \, \text{kg} \cdot \text{m}^2$$

in the principal-axis frame shown in Figure 5.23. Describe the stable gravity-gradient attitudes for the shuttle, and calculate the three oscillation frequencies for this vehicle in a 90-min-period orbit.

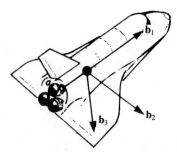

FIGURE 5.23
Space-shuttle principal axes.

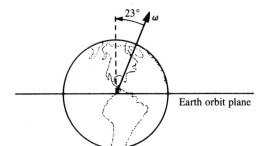

FIGURE 5.24
Reorientation maneuver for the earth.

6. Show that a single-mass yo-yo mechanism needs to have a string length of

$$l_f = \sqrt{R^2 + \frac{C}{m}}$$

to completely despin a satellite. However, unlike the two-mass yo-yo device, the satellite will not remain in its previous orbit. Calculate the velocity of the mass at release, and if the mass of the satellite is M, find the speed of the satellite with respect to the center of mass of the system after release.

7. To eliminate the rigors of the seasons, the government is studying a proposal to reorient the earth's spin axis normal to its orbital plane, eliminating the 23° tilt of the planet's axis shown in Figure 5.24. After the first attitude maneuver, how long will it take to precess the earth's axis to its new position? For the earth, $\dot{\psi} = 2\pi$ rad/day, and

$$\frac{A - C}{C} = 0.003279$$

8. The satellite in Figure 5.25 initially rotates at $\omega = 0.2\mathbf{b}_3$ rad/s and has a moment of inertia tensor

$$I = \begin{bmatrix} 500 & 0 & 0 \\ 0 & 600 & 0 \\ 0 & 0 & 300 \end{bmatrix} \text{kg} \cdot \text{m}^2$$

Internal energy dissipation is negligible within the body of the satellite. Two 1-kg

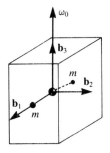

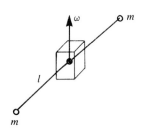

FIGURE 5.25
Antenna deployment from a spinning spacecraft.

masses m, not included in the moments of inertia above, will slowly pull out two 50-m wire antennas. When the wires have a length l, show that the moment of inertia becomes

$$I_{\text{tot}} = \begin{bmatrix} 500 & 0 & 0 \\ 0 & 600 + 2ml^2 & 0 \\ 0 & 0 & 300 + 2ml^2 \end{bmatrix}$$

What is the final spin rate when $l = 50$ m? Will this work as designed, or is there a fatal flaw in this scheme? If so, can it be redesigned so that it will work?

CHAPTER
6

GYROSCOPIC INSTRUMENTS

6.1 INTRODUCTION

The theory of gyroscopic instruments is implicit in the work done on rigid-body dynamics in the nineteenth century. However, the development of practical gyroscopic instruments occurred in the first part of the twentieth century. The first impetus for these developments was naval architecture, which was moving from the wooden ship to the metal hull. This rendered the traditional magnetic compass unreliable. Even if the magnetic compass was corrected for the mass of the ship, a vessel carrying a cargo of metal, or a vessel with very large moving metal parts (e.g., the turrets of a battleship) could not use a magnetic compass. The *gyrocompass* was already a small, dependable unit in 1929, eliminating the magnetic interference problems caused by a metal hull and also eliminating the magnetic compass' declination correction: the difference between true and magnetic north. The second gyroscopic instrument developed for shipboard use was the gyroscopic *stabilizer*; originally intended to enable long-range guns to be fired from a rolling ship, it was also applied to increase the comfort of passengers on liners.

Gyroscopic instruments were also applied to rocket flight at an early date. Goddard's experiments in New Mexico included the first complete gyroscopic stabilization system for a rocket. A large rocket cannot depend on aerodynamic fins for stabilization, since at lift-off it is moving too slowly for the fins to operate, and later in the trajectory it may be above the sensible atmosphere. Goddard's gyroscopes controlled the action of vanes in the exhaust stream, a method independently adopted by the Germans in the development of the V-2. The V-2 control system was a further advance, since this vehicle was not intended to fly a simple vertical trajectory but to fly a carefully controlled curve to a given burnout point. The V-2 gyroscopic control system thus represents the first inertial-rocket control system: A combination of gyroscopes and clockwork mechanism caused the vehicle to fly a predetermined powered-flight trajectory.

Gyroscopic instrument development received further impetus after World War II, spurred on by the demands of aircraft development and the infant ICBM program. This led to the first true inertial-navigation units, whose development was led by Charles Stark Draper at the MIT laboratory now bearing his name. The ballistic-missile program demanded a higher order of accuracy in guidance, since an error of even a few feet per second at burnout will cause a miss of many miles over intercontinental ranges. These guidance systems then needed further improvement when ICBM's were converted to satellite boosters, since the accuracy requirements for obtaining low earth orbit are even tighter.

In this chapter we will consider the theory of basic gyroscopic instruments, from the simple three-axis gyro to the complete stable platform. Finally, we will see how these devices are used in inertial navigation.

6.2 THE FULLY GIMBALED GYROSCOPE

Flight in space requires some attitude reference system, since the familiar clues of gravity and horizon are not available. It would seem that a spinning gyroscope, conserving angular momentum, would maintain a constant direction with respect to inertial space. Such a fully gimbaled gyroscope, shown in Figure 6.1, could then serve as an attitude reference system for space navigation. This is quite false.

To understand the problem with this system, let us calculate the total angular momentum. It is critical to realize that we are not dealing with one rigid body (the rotor) here, but *three*. Assume that the gyroscope is mounted in an inertial frame. The total angular momentum of the system consists of the angular momentum of the rotor plus the angular momentum of the outer and inner gimbals. Write the moment of inertia matrices for the rotor, inner gimbal, and outer gimbal as I_r, I_{ig}, and I_{og}, respectively. Now, the outer gimbal can only rotate about the vertical axis. Write its angular-velocity vector as $\boldsymbol{\omega}_{og}$. It will then possess angular momentum $\mathbf{H}_{og} = I_{og}\boldsymbol{\omega}_{og}$. The inner gimbal can rotate about its pivot axis with the outer gimbal, with relative angular-velocity vector $\boldsymbol{\omega}_{ig}$ with respect to the outer gimbal. Since angular velocities add, the inertial angular velocity of the inner gimbal is $\boldsymbol{\omega}_{og} + \boldsymbol{\omega}_{ig}$, and the inner gimbal will possess angular momentum equal to $\mathbf{H}_{ig} = I_{ig}(\boldsymbol{\omega}_{og} + \boldsymbol{\omega}_{ig})$. Finally, the rotor

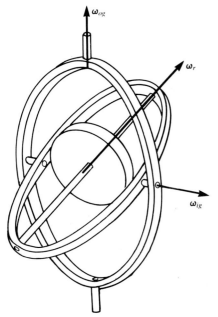

FIGURE 6.1
The fully gimbaled gyroscope.

spins with the large angular-velocity vector $\boldsymbol{\omega}_r$ with respect to the inner gimbal. It, of course, is carried along with any motions of the outer and inner gimbals. Adding all three angular-momentum expressions, the total angular momentum is given by

$$\mathbf{H}_{\text{tot}} = I_{og}\boldsymbol{\omega}_{og} + I_{ig}(\boldsymbol{\omega}_{og} + \boldsymbol{\omega}_{ig}) + I_r(\boldsymbol{\omega}_{og} + \boldsymbol{\omega}_{ig} + \boldsymbol{\omega}_r) \qquad (6.1)$$

Equation (6.1) is in a set of mixed basis vectors, but it is not worth the effort to express this in one consistent set of unit vectors. Since the gimbal system is considered perfect, the total angular momentum is a constant. The problem is that *part of the angular momentum resides in the gimbals.*

Since the spin rate of the rotor is large, the angular-momentum expression (6.1) is dominated by the last term. However, the rotor-spin axis and the total angular momentum do not line up unless the initial angular velocities of the gimbals are zero. Great care must be taken when starting such a system to ensure that the angular momentum in the gimbals is small. The gimbals are "caged" while the rotor is first spun up, which is best done while the vehicle is at rest. Once the rotor is spinning at the correct rate, the gimbals are very carefully released, and the system runs freely. However, in uncaging the gimbals, it is inevitable that they will be nudged slightly. This puts some angular momentum into the gimbals, and they begin to move with respect to inertial space. The gimbals are said to "drift." So, the three-axis gimbaled gyroscope is not used in space navigation.

One possible solution to this problem is to totally eliminate the gimbals.

If a spinning sphere is magnetically supported within a spherical cavity, the gimbals are completely missing, and the spin axis of the sphere will remain constant with respect to inertial space. Such devices have been built but suffer from the difficult problem of sensing the spin axis of the sphere without touching it. The classical solution to the problem of providing an inertial reference frame has been the *stable platform*. However, before we can study this system we will need to discuss several types of single-axis gyroscopes.

6.3 THE RATE GYROSCOPE

One of the most important gyroscopic instruments is the rate gyroscope. This instrument, shown in Figure 6.2, is a rotor with moment of inertia C_r spinning at angular rate ω_r. This angular rate is maintained fixed by an electric motor, so the angular momentum of the rotor $H_r = C_r\omega_r$ may be considered a constant with respect to the gimbal. The one gimbal has a rotary spring and damper mechanism attached to its axis, supplying restoring torque $-k\theta$ from the spring and damping torque $-c\,d\theta/dt$ from the damper.

Now, suppose that the vehicle to which this device is attached begins to rotate about the vertical axis in Figure 6.2 with angular rate $d\psi/dt$. This angular rate will be taken as a given quantity, since we assume that the rate gyroscope is too small to influence the motion of the vehicle. The vertical axis is termed the *input axis* of the instrument. Then, if the gimbal angle θ remains small, the rotor angular momentum will have a rate of change given by

$$\dot{\mathbf{H}}_r = \dot{\boldsymbol{\psi}} \times \mathbf{H}_r$$
$$\approx C_r\omega_r\dot{\psi}\mathbf{b}_2 \tag{6.2}$$

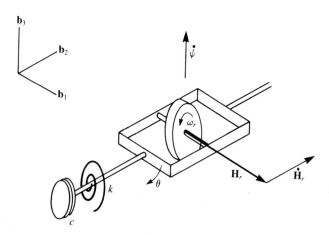

FIGURE 6.2
The rate gyroscope.

assuming θ remains small. This rate of change of the rotor angular momentum is actually caused by a torque from the gimbal, since the gimbal is the only object in contact with the rotor. There must then be an equal and opposite torque $-C_r\omega_r \, d\psi/dt \, \mathbf{b}_2$ exerted by the rotor *on the gimbal*. It is this torque that we must use to formulate the equations of motion for the gimbal. Since the gimbal is free to rotate in only one direction, its equation of motion is given by

$$B_g\ddot{\theta} = -k\theta - c\dot{\theta} - C_r\omega_r\dot{\psi} \qquad (6.3)$$

about the *output axis* $\mathbf{b}_2$. Here, B_g is the moment of inertia of the gimbal and rotor about the gimbal axis. This is the equation of motion of a forced, damped harmonic oscillator.

Of particular interest here is that (6.3) possesses the equilibrium solution

$$\theta_e = -\frac{C_r\omega_r}{k}\dot{\psi} \qquad (6.4)$$

found by zeroing the rates $\dot{\theta}$ and $\ddot{\theta}$ in (6.3) and solving for θ. This is the value of θ the instrument will give after any transients are allowed to damp out. Note that this value is linearly proportional to the rate $d\psi/dt$. In the steady state, the instrument will indicate the vehicle rate around the vertical axis (the input axis) by a steady displacement of the gimbal around the output axis. Hence the name given this instrument.

Of course, should the vehicle make a sudden change in its angular rate $d\psi/dt$, the instrument will need a certain amount of time to assume the correct reading. Since (6.3) is a damped oscillator, there are three possible forms that the solution can take. If the damping is low, the gimbal will be free to oscillate many times before it comes to the equilibrium position (6.4). This is the underdamped case. On the other hand, if the damping is large, the damping term in (6.3) will not permit a large angular rate $d\theta/dt$ to develop, and the gimbal will move sluggishly toward the equilibrium position (6.4). Finally, if the unit is *critically damped*, the harmonic oscillator will assume the equilibrium position (6.4) in the minimum time, and without overshooting. This is the preferred choice for the damping constant c.

However, we still have at our disposal the spring constant k. If we allow the gimbal displacement angle θ to become large, the rate of change of the rotor angular momentum is given by $\dot{H}_r = C_r\omega_r\dot{\psi}\cos\theta$, and the equation of motion for the rate gyroscope becomes

$$B_g\ddot{\theta} = -k\theta - c\dot{\theta} - C_r\omega_r\dot{\psi}\cos\theta \qquad (6.5)$$

when θ does not remain small. As the angular rate $d\psi/dt$ becomes large, the gimbal moves toward a position given by $\theta = \pm90°$, and the instrument is said to be *saturated*. The larger the spring constant k, the larger the input rate $d\psi/dt$ the instrument can tolerate before it saturates. However, large spring constants imply that small vehicle rates produce small output angles θ_e, which is not desirable either. Given the characteristics of a vehicle, a compromise could be made between the ability to sense high angular rates and sensitivity to low angular rates. The most common use of the rate gyroscope is in the stable platform, in which case the instrument can be made very sensitive without fear of saturation effects. In fact, in the stable platform the restoring spring is omitted entirely.

6.4   THE INTEGRATING GYROSCOPE

The integrating gyroscope shown in Figure 6.3 is similar to the rate gyroscope, but the rotary spring has been removed. The rotary damper now plays the key role in the response of the gyroscope to an input vehicle rate $d\psi/dt$ along the $\mathbf{b}_3$ axis. With a large amount of damping on the gimbal axis, the acceleration of the gimbal, $d^2\theta/dt^2$, may be ignored. In this case, the system will rapidly approach an equilibrium, where the torque on the gimbal from the rate of change of angular momentum, $\mathbf{M} = -C_r\omega_r\, d\psi/dt\, \mathbf{b}_2$, is balanced by the torque developed by the damper, $-c\, d\theta/dt$. We may then write

$$c\dot{\theta} = -C_r\omega_r\dot{\psi} \tag{6.6}$$

This equation may be immediately integrated (hence the name) to yield

$$\theta = -\frac{C_r\omega_r}{c}(\psi - \psi_0) \tag{6.7}$$

This arrangement thus integrates the vehicle angular rate, and its output is proportional to the inertial rotation the vehicle has performed.

However, two factors limit the usefulness of the integrating gyroscope. First, if the vehicle insists on performing a large rotation around the $\mathbf{b}_3$ axis, the output θ rapidly approaches $\pm 90°$ and the unit saturates. For the rate

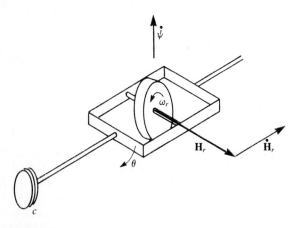

FIGURE 6.3
The integrating gyroscope.

gyroscope we could at least ensure that saturation occurred at a completely absurd vehicle *rate*. It is not possible, however, to ensure that a vehicle will never rotate, say, 17 complete revolutions about the $\mathbf{b}_3$ axis. Saturation is thus almost inevitable if we cannot control the possible motions of the vehicle to prevent it. Second, a vehicle input rate around the $\mathbf{b}_2$ axis will cause the gimbal to turn with respect to the craft, since the rotor's angular momentum will tend to hold the spin axis constant with respect to inertial space. Thus, the integrating gyroscope can mistake a vehicle rate about the $\mathbf{b}_2$ axis for a vehicle rotation about the $\mathbf{b}_3$ direction. This also occurs for the rate gyroscope, but in that case a unit sensitive to the $\mathbf{b}_2$ direction would clear up the difficulty. To remove the ambiguity for the integrating gyroscope, we would need a rate gyroscope sensitive to the $\mathbf{b}_3$ axis as well.

There is one case where vehicle rotations can be carefully controlled, and in which integrating gyroscopes are sometimes used. The guidance systems of most rockets are designed to begin the pitchover maneuver into inclined flight (see section 7.9) with respect to a preprogrammed axis *fixed in the vehicle*. In the early German V-2, for example, the pitchover occurred in the direction of a given fin, which usually was painted a different color. The launch crew simply pointed this fin in the direction of London by rotating the entire vehicle on its launch turntable. As rockets became larger, launchpads became huge structures absolutely fixed in place. It is still necessary, however, to launch vehicles toward different azimuths. Since rotating the launchpad is out of the question, the vehicle itself is rotated to the proper azimuth before the pitch program is begun. This is done *in flight,* and since there is every assurance that the initial roll maneuver will not exceed 180°, an integrating gyroscope could be used to control this maneuver.

6.5 THE LASER GYROSCOPE

One of the most recent developments in the area of "gyroscopic" instruments is the laser gyroscope. In this device, Figure 6.4, two beams of laser light are made to follow a triangular path in opposite senses. The laser gyroscope is a rate instrument. It depends for its operation on the statement from special relativity that light will *always* propagate at speed c with respect to an inertial frame, *absolutely independent of the motion of its source and receiver.* If this device is not rotating, the path lengths around the triangle will be equal. If the device is rotating about an axis perpendicular to the light path, the inertial path followed by the light beam moving against the rotation will be slightly shorter than the path length followed by the light beam moving in the same sense as the rotation. The first beam finds the mirrors moving toward it, while the second beam has to catch up with each mirror in turn.

Now, the difference in the path lengths is very small, since light moves at $c = 2.998 \times 10^{10}$ cm/s and the velocity of the mirrors is likely to be at least 9 to 10 orders of magnitude slower than this. The path lengths each way will thus differ from each other by about 1 part in 10^9, a very small quantity to measure.

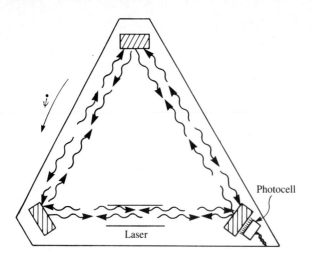

FIGURE 6.4
The laser gyroscope.

This is why a laser is employed. With monochromatic laser light, the two light beams combine with each other to produce interference patterns on the surface of each mirror. These are patterns of light and dark bands across the mirror surface, marking the places where the two different path lengths differ by one-half wavelength of light. When the peak of the light wave moving in one direction combines with a peak from the wave propagating in the opposite direction, their intensities add. Where a peak combines with a trough of the other wave, darkness results. When the assembly rotates, the points where constructive and destructive interference occur shift across the mirror surface, by one bandwidth every time the path length changes by one-half wavelength. The band pattern can be seen if one mirror is made partially transparent, and can be automatically sensed by a photocell looking through a grid pattern with the expected band spacing. This measures the difference in the two path lengths with an optical interferometer.

Even using the property of interference to measure the difference in the two path lengths, early laser gyroscopes lacked sensitivity to low angular rates. This has now been largely overcome. To avoid thermal expansion problems (which could also change the path lengths), laser gyroscopes are generally fabricated within a single block of fused quartz, a material with a very small coefficient of thermal expansion. They are available in single-axis units, or in quadruple units built into the faces of a tetrahedron of fused quartz. This gives one spare unit should one fail. Also under development for NASA's next generation of planetary spacecraft is a ring-laser gyroscope in which the light path is over 4 km of optical quartz fiber wound on a spool. The great path length gives this instrument very great sensitivity indeed, without the saturation problems at high angular rate which arise in the mechanical rate gyroscope. Another advantage over the mechanical rate gyroscope is, of course, the total lack of moving parts. Other than the need for a zero-point adjustment, there is no need for calibration; that is built into the geometry and size of the instrument.

6.6 THE SINGLE-AXIS STABLE PLATFORM

Attitude control in orbit, inertial navigation, and satellite-booster control systems all require access to an inertial reference frame to perform their functions. We have seen that a fully gimbaled gyroscope will not maintain a constant orientation with respect to inertial space. The solution to this problem is the *inertial platform,* in which three rate gyroscopes sense the inertial rates of a platform, and three electric motors remove these rates by torquing the gimbal system supporting the platform. Now, the platform can never be perfectly torque-free. If it is not perfectly balanced, it will experience torques from gravity and vehicle accelerations, and would try to rotate like a pendulum to the position of lowest energy. Also, there will be frictional torques on the platform through the gimbal axes, and these will attempt to drag the platform along as the vehicle changes its attitude. The rate gyroscopes sense these disturbances and command the electric motors to counteract these torques, keeping the platform stationary with respect to inertial space. In the process, this also solves the saturation problem with the mechanical rate gyroscope. Since the rate gyroscopes are now mounted on a nearly stationary platform, they can be made sensitive without fear of driving the gyroscope gimbal beyond the limits of the small-angle approximation.

The complete three-axis platform is a complex device, with considerable coupling between the three axes. Before we discuss this system, then, it is advantageous to consider the single-axis platform shown in Figure 6.5. A rate gyroscope is mounted so that it is sensitive to rotation rates about the x axis, the only axis about which the platform is free to rotate. The gimbal of the rate gyroscope can rotate about the y axis with an angle from the vertical of θ. The orientation of the platform is specified by the angle ϕ, and it is our desire to keep this angle zero in the presence of disturbing torques on the platform. Note that this system consists of two rigid bodies: the gyroscope rotor and gimbal and the platform itself. It will be necessary to obtain equations of motion for each of these units.

Now, the rotor of the gyroscope has angular momentum **H**. If the platform and gimbal develop angular rates $d\phi/dt$ and $d\theta/dt$, then the inertial rate of change of the rotor angular momentum is

$$\dot{\mathbf{H}} = -H\dot{\theta}\mathbf{i} + H\dot{\phi}\mathbf{j} \tag{6.8}$$

We have assumed small angles θ and ϕ in writing the above expression. Note that the rate gyroscope in this application has no spring attached to the output axis. Assume that the moment of inertia of the rotor and gimbal about the y axis is I. The negative of the y axis torque on the rotor, $-H\dot{\phi}$, is one torque on the gimbal, and in addition there are frictional torques in the gimbal axle.

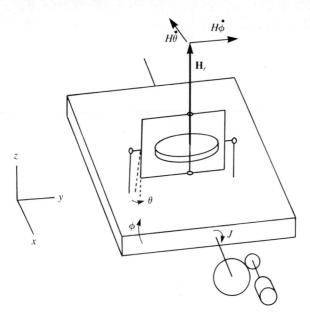

FIGURE 6.5
A single-axis stable platform.

Thus, the equation of motion of the gimbal is

$$I\ddot{\theta} = -H\dot{\phi} - f_g\dot{\theta} \qquad (6.9)$$

where f_g is the gimbal-axle friction constant. Now, the gimbal is not free to rotate about the x axis, so the other component of $d\mathbf{H}/dt$ is caused by a torque on the gimbal due to the platform. Its negative, $+H\,d\theta/dt$, is then the torque exerted by the gimbal *on the platform*. The platform will also be subject to the motor torque M_m, the frictional torque in the platform axle, and any disturbing torque M_d about the x axis. If the moment of inertia of the platform about the x axis is J, then the equation of motion of the platform is given by

$$J\ddot{\phi} = M_m + M_d + H\dot{\theta} - f_p\dot{\phi} \qquad (6.10)$$

These two equations of motion are not complete in themselves, since the output of the rate gyroscope, θ, is linked to the motor via a *control loop*. We need to design this control loop (or feedback circuit) to keep the platform stable with respect to inertial space.

Now, a simple direct-current electric motor develops torque in proportion to the current i flowing through its coils. Thus, $M_m = ci$, where c is a constant of the motor. Assume that the current is made proportional to the rate-gyroscope angle θ, giving

$$M_m = ck\theta \qquad (6.11)$$

where k is the control-loop *gain constant*. We will be free to change k to obtain acceptable platform behavior. Now, introduce the vector $\boldsymbol{\Psi}^T = (\theta, \phi)$. Then,

(6.9) to (6.11) can be put into the form

$$\begin{bmatrix} I & 0 \\ 0 & J \end{bmatrix} \ddot{\mathbf{\Psi}} + \begin{bmatrix} f_g & H \\ -H & f_p \end{bmatrix} \dot{\mathbf{\Psi}} + \begin{bmatrix} 0 & 0 \\ -ck & 0 \end{bmatrix} \mathbf{\Psi} = \begin{bmatrix} 0 \\ M_d \end{bmatrix} \tag{6.12}$$

These represent two coupled, constant-coefficient linear differential equations, with a forcing term from the external disturbances on the platform.

Consider first the unforced behavior of this system. Setting M_d to zero and assuming a solution to (6.12) of the form

$$\mathbf{\Psi}(t) = \mathbf{\Psi}_0 e^{\lambda t} \tag{6.13}$$

we obtain the characteristic equation

$$\begin{vmatrix} I\lambda^2 + f_g\lambda & H\lambda \\ -H\lambda - ck & J\lambda^2 + f_p\lambda \end{vmatrix} = 0 \tag{6.14}$$

Expanding the determinant, this gives the quartic equation

$$IJ\lambda^4 + (Jf_g + If_p)\lambda^3 + (H^2 + f_g f_p)\lambda^2 + Hck\lambda = 0 \tag{6.15}$$

One root is

$$\lambda_1 = 0 \tag{6.16}$$

while the other three roots are given by

$$IJ\lambda^3 + (Jf_g + If_p)\lambda^2 + (H^2 + f_g f_p)\lambda + Hck = 0 \tag{6.17}$$

The single-axis platform is stable without friction. By Descarte's rule of signs, either $k > 0$ and (6.17) gives a negative real root or $k < 0$ and (6.17) gives two conjugate roots with positive real parts. The first corresponds to wiring the motor correctly, so that it opposes the disturbing torques on the platform. The second case is a mechanical oscillation, the nutational motion of the platform and gyroscope, in which inertial lag in the platform gives the gyroscope opportunity to order too large a corrective torque. The platform then begins to oscillate the other way, with the gyroscope again ordering too large a correction. The result is an exponentially growing oscillation. However, with reasonable amounts of friction in the system, positive k values can be found which give all three roots of (6.17) negative real parts. The unforced response of the system is then stable.

Until now we have discussed only the unforced response of the one-axis platform. Of course, the rate gyroscope and feedback loop are only present because outside disturbances M_d are inevitable. Since the dynamical system (6.12) is a set of linear equations, the complete response will be the sum of the unforced solution and the response of the system in the presence of disturbances. Consider first the response of the system to a constant disturbing torque M_d. The particular solution can be found by assuming that the angles θ and ϕ are constant, in which case (6.12) becomes

$$\begin{bmatrix} 0 & 0 \\ -ck & 0 \end{bmatrix} \begin{bmatrix} \theta \\ \phi \end{bmatrix} = \begin{bmatrix} 0 \\ M_d \end{bmatrix} \tag{6.18}$$

These equations have the solution

$$\theta = -\frac{M_d}{ck} \qquad \phi = \phi_0 \tag{6.19}$$

where ϕ_0 is the original ϕ value. So, in response to a constant disturbing torque the rate gyroscope displaces, while *the platform itself does not move with respect to inertial space*. In this state the torque developed by the motor exactly balances the disturbing torque. The reason the rate gyroscope has no spring can now be seen. A spring would exert equal and opposite torques on the gimbal and the platform and would force a nonzero solution for ϕ in (6.18). Thus, the platform *would* move when subjected to an external torque. If the disturbing torque suddenly appears (for example, M_d is a step function) or changes with time, the system will oscillate about the equilibrium given by (6.19). Frictional torques in the gyroscope gimbal and platform axle will damp these oscillations, bringing the system back to the steady state.

6.7 THE STABLE PLATFORM

The single-axis platform discussed in the previous section has, of course, no practical use. A complete inertial platform must include three rate gyroscopes and make provision for control of the platform torques in all three directions. This introduces considerable complexity into the dynamics of the system. However, the general procedure is exactly the same as in the single-axis case. We will calculate the rate of change of the angular-momentum vectors of the three rate gyroscopes. This will enable us to write the equations of motion for the gyroscope gimbal axes. Then, it will be necessary to find the equations of motion of the platform itself. Finally, the control loops must be designed to ensure that the system is stable when disturbing torques are present.

Figure 6.6 shows one possible arrangement of three rate gyroscopes on the platform. Other arrangements serving the same purpose are possible, and the dynamics will differ slightly depending on the actual assembly used. Assume that all three units are identical, with the same angular momentum H and gimbal moment of inertia I. Now, the platform orientation with respect to inertial space will be specified by the three angles ϕ_x, ϕ_y, and ϕ_z. By examination of Figure 6.6, the rate of change of the three gyroscope angular-momentum vectors are given by

$$\dot{\mathbf{H}}_x = -\dot{\phi}_x H \mathbf{j} + \dot{\theta}_x H \mathbf{i} \tag{6.20}$$

$$\dot{\mathbf{H}}_y = \dot{\phi}_y H \mathbf{i} + \dot{\theta}_y H \mathbf{j} \tag{6.21}$$

$$\dot{\mathbf{H}}_z = \dot{\phi}_z H \mathbf{i} + \dot{\theta}_z H \mathbf{k} \tag{6.22}$$

The first terms on the right in (6.20) to (6.22) are along the respective gimbal axes of each gyroscope and are the negatives of the torque terms which rotate the rate-gyroscope gimbals. The second terms on the right in (6.20) to (6.22)

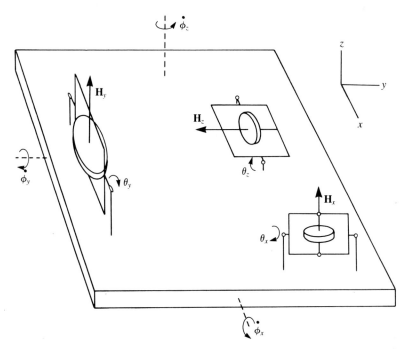

FIGURE 6.6
Rate gyroscopes on a stable platform.

appear in directions in which the respective gimbals are not free to move. These components are torque components on the gimbals due to the platform. Their negatives are therefore the equal and opposite torques exerted by the gimbals *on the platform.*

Now, we can obtain the equations of motion of the rate-gyroscope gimbals by equating the appropriate torque components to $I\,d^2\theta_j/dt^2$, where θ_j is the orientation angle of the gyroscope gimbal *with respect to inertial space.* The θ_j must be defined in this way to obtain valid equations of motion. The equations of motion for the three gimbals are then

$$I\ddot{\theta}_x - H\dot{\phi}_x = 0 \tag{6.23}$$

$$I\ddot{\theta}_y + H\dot{\phi}_y = 0 \tag{6.24}$$

$$I\ddot{\theta}_z + H\dot{\phi}_z = 0 \tag{6.25}$$

These three equations represent one-half the dynamics of the system.

Before we can discuss the platform dynamics, it is necessary to choose a means of support. The platform must be capable of rotating in all three directions and must be able to apply control torques to cancel disturbing torques inevitably arising in any imperfect system. In the past it was common to support the platform at the center of a complete mechanical gimbal system.

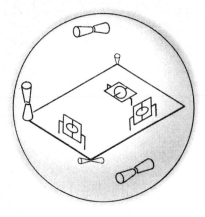

FIGURE 6.7
Inertial reference sphere.

However, simpler solutions are now available. The Apollo spacecraft and the space shuttle have spherical fluid-bearing platforms, shown schematically in Figure 6.7. In this concept, the platform is contained in a spherical housing and floats within this spherical cavity in a fluid. If the density of the platform is matched to that of the fluid, the platform will float with neutral buoyancy and will not contact the walls of the cavity when the unit is accelerated. Small fluid jets on the outside of the sphere supply the control torques necessary to stabilize the system. We are, of course, free to specify that all three principal moments of the platform will be the same, say, J. This eliminates any coupling terms from Euler's equations for the platform.

The platform experiences several different torques. First, there are the reaction torques given by the negatives of the last terms in (6.20) to (6.22). There are also the control torques $\mathbf{M}_c$, the disturbing torques $\mathbf{M}_d$, and viscous friction from the thin layer of fluid in which the spherical platform floats. These last torques can be written

$$\mathbf{M}_f = -f(\dot{\phi}_x \mathbf{i} + \dot{\phi}_y \mathbf{j} + \dot{\phi}_z \mathbf{k}) \tag{6.26}$$

where f is the coefficient of viscous friction. Actually, if the vehicle itself decides to rotate, there will be friction from the angular rate of the housing about the nearly stationary platform. However, we will agree to include these torques in $\mathbf{M}_d$. Euler's equations for the platform then become

$$J\ddot{\phi}_x = M_{dx} + M_{cx} - f\dot{\phi}_x - H\dot{\theta}_x \tag{6.27}$$

$$J\ddot{\phi}_y = M_{dy} + M_{cy} - f\dot{\phi}_y - H\dot{\theta}_y \tag{6.28}$$

$$J\ddot{\phi}_z = M_{dz} + M_{cz} - f\dot{\phi}_z - H\dot{\theta}_z \tag{6.29}$$

These represent the other three equations of motion for this system, but in their present form they are incomplete. The control torques must be written in terms of the rate-gyroscope output angles.

Unfortunately, the angles read out by the rate gyroscopes are not the θ_j, which measure the gimbal orientation with respect to inertial space, but the σ_j, the angles of the gimbals *with respect to the platform*. Consider the x-axis rate gyroscope. This unit will develop a nonzero angle if an x-axis rate $d\phi_x/dt$ forces a θ_x value through (6.23), or also if the platform rotates out from below the gimbal about the y axis. The x-axis gyroscope output will then be

$$\sigma_x = \theta_x - \phi_y \qquad (6.30)$$

Similarly, for the gyroscope arrangement in Figure 6.6,

$$\sigma_y = \theta_y + \phi_x \qquad (6.31)$$

$$\sigma_z = \theta_z + \phi_x \qquad (6.32)$$

It is the three σ_j which can be measured and fed back to control the restoring torques $\mathbf{M}_c$.

Now, let us group the six orientation angles θ_j and ϕ_j into the vector

$$\mathbf{\Psi} = (\theta_x, \theta_y, \theta_z, \phi_x, \phi_y, \phi_z)^T \qquad (6.33)$$

Then, define the matrices

$$A = \begin{bmatrix} I & 0 & 0 & 0 & 0 & 0 \\ 0 & I & 0 & 0 & 0 & 0 \\ 0 & 0 & I & 0 & 0 & 0 \\ 0 & 0 & 0 & J & 0 & 0 \\ 0 & 0 & 0 & 0 & J & 0 \\ 0 & 0 & 0 & 0 & 0 & J \end{bmatrix} \qquad (6.34)$$

$$D = \begin{bmatrix} 0 & 0 & 0 & -H & 0 & 0 \\ 0 & 0 & 0 & 0 & H & 0 \\ 0 & 0 & 0 & 0 & 0 & H \\ H & 0 & 0 & f & 0 & 0 \\ 0 & H & 0 & 0 & f & 0 \\ 0 & 0 & H & 0 & 0 & f \end{bmatrix} \qquad (6.35)$$

Also, write the control torques as

$$\begin{bmatrix} 0 \\ \mathbf{M}_c \end{bmatrix} = GB\mathbf{\Psi} \qquad (6.36)$$

where G is a 6-by-3 control-gain matrix

$$G = \begin{bmatrix} \{0\} \\ G_2 \end{bmatrix} \qquad (6.37)$$

which has an upper 3-by-3 block of zeros, since control torques are not applied to the gyroscope gimbals. Here, G_2 is a 3-by-3 matrix, and

$$B = \begin{bmatrix} 1 & 0 & 0 & 0 & -1 & 0 \\ 0 & 1 & 0 & 1 & 0 & 0 \\ 0 & 0 & 1 & 1 & 0 & 0 \end{bmatrix} \qquad (6.38)$$

is the equivalent of (6.30) to (6.32). Then, the complete system dynamics (6.23) to (6.25) and (6.27) to (6.29) and the control system (6.36) can be written as

$$A\ddot{\mathbf{\Psi}} + D\dot{\mathbf{\Psi}} + GB\mathbf{\Psi} = \begin{bmatrix} 0 \\ \mathbf{M}_d \end{bmatrix} \qquad (6.39)$$

The gain matrix G must be chosen so that this six-degree-of-freedom dynamical system is stable.

We will not discuss techniques for choosing the gain matrix G_2. That is the province of control theory. However, notice that if the system is subjected to a constant disturbing torque $\mathbf{M}_d$, then (6.39) admits the solution

$$\phi_x = \phi_{x0} \qquad \phi_y = \phi_{y0} \qquad \phi_z = \phi_{z0} \qquad (6.40)$$

where the zero subscripts indicate the original values and

$$\begin{bmatrix} \theta_x \\ \theta_y \\ \theta_z \end{bmatrix} = G_2^{-1}\mathbf{M}_d \qquad (6.41)$$

So, just as in the single-axis platform, the rate gyroscopes will displace enough to produce constant control torques, exactly canceling the disturbing torques. The only requirement for this to occur is that the control-gain matrix G_2 must not be singular. The individual elements of G_2 must still be chosen to ensure stability, in which case the complete dynamical system will exhibit stable behavior about the equilibrium solution given in (6.40) and (6.41). The stable platform will be stable.

When a mechanical gimbal system is used, three additional complexities appear. Such a system is shown in Figure 6.8, from which it may be noticed that the three torquing motors are not aligned with the axes of the three rate gyroscopes. The observed angular rates of the platform must be apportioned among the three torquing motors in such a way that the net torque on the platform is zero. This is the function of the *resolvers* shown on these axes. Second, a mechanical gimbal system introduces two additional rigid bodies into the analysis. Besides the three rate gyroscopes and the platform, a six-degree-of-freedom system, we now must include the rotational dynamics of the two gimbals. This increases the system order by another six degrees of freedom.

Finally, a mechanical gimbal system is subject to *gimbal lock*. If the outer and inner gimbal axes align, there is a direction in which the platform is not free to move. This is shown in Figure 6.9, where the platform is incapable of responding to an angular rate out of the page. In such a plight, the platform will attempt to force infinite rates in the outer and inner gimbals in response to a small disturbing torque out of the page. Of course, it will not succeed in forcing infinite rates in the gimbals but will only obtain very large rates in these axes. Once the gimbals have moved away from the locked orientation, the platform will eliminate the large angular rate it developed while the gimbals were locked. The platform will appear to move violently, quickly stopping in

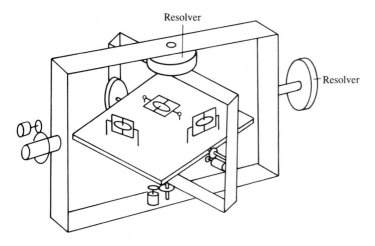

FIGURE 6.8
Stable platform with mechanical gimbals.

some new, random inertial orientation. The platform is said to have *tumbled*. If the platform is connected to the vehicle control system, platform tumbling is immediately followed by vehicle tumbling, usually with catastrophic results. Gimbal lock can be prevented by adding a third gimbal and arranging that no two gimbal axes will ever approach each other. This adds still further complexity to the mechanically gimbaled platform.

All these difficulties are missing in the floating-sphere device. However, it has new difficulties in their place. The orientation of the sphere must be obtained by a complicated sensing arrangement of coils. Also, a telemetry

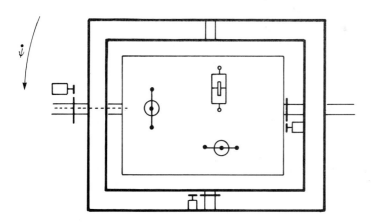

FIGURE 6.9
Gimbal lock.

system is required simply to talk to the accelerometers mounted inside the sphere. Finally, notice that without friction, and with a general moment of inertia matrix, the reference-sphere problem is identical to the attitude-stabilization problem for a torque-free satellite.

6.8 INERTIAL NAVIGATION

An inertial-navigation system calculates the vehicle position by actually solving the equations of motion. It does not do this in the sense we do it in this book. Since we are interested in studying the general characteristics of a dynamical system, we are free to make some simplifying assumptions. Also, some system forces, aerodynamic forces in particular, are notorious for their poor characterization. So, instead of working from an analytic model of the vehicle dynamics, an inertial-navigation unit simply integrates the three equations of motion

$$\ddot{X} = a_x \qquad \ddot{Y} = a_y \qquad \ddot{Z} = a_z \qquad (6.42)$$

where X, Y, and Z are the position components of the vehicle in some inertial reference frame and a_x, a_y, and a_z are the inertial accelerations of the vehicle along these axes. No information is assumed on the dynamics causing the accelerations; equations (6.42) are simply integrated once to supply the velocity and then integrated again to supply the vehicle position. A computer is thus an indispensable part of any inertial-navigation system.

To perform the integrations, the system must have two things available. It must have an inertial reference frame handy, and it must have the acceleration components with respect to this frame. There are two methods for obtaining a portable inertial reference frame. The first of these is the stable platform. The platform itself is an inertial reference frame, kept inertial by the rate gyroscopes and feedback control system within the platform. This is the analogue solution to this problem. A more recent alternative to the mechanical platform is the *strapdown* inertial-navigation system. In a strapdown system, the rate gyroscopes are rigidly attached to the vehicle, and they read the inertial rates ω_i in the unit vectors of the vehicle body frame. The kinematic equations for the roll, pitch, and yaw angles from section 4.7,

$$\begin{aligned}
\omega_1 &= \dot{\psi}_3 - \dot{\psi}_1 \sin \psi_2 \\
\omega_2 &= \dot{\psi}_2 \cos \psi_3 + \dot{\psi}_1 \cos \psi_2 \sin \psi_3 \\
\omega_3 &= -\dot{\psi}_2 \sin \psi_3 + \dot{\psi}_1 \cos \psi_2 \cos \psi_3
\end{aligned} \qquad (6.43)$$

must then be numerically integrated. Alternately, the unit vectors of the body frame, $\mathbf{b}_i$, obey the differential equations

$$\dot{\mathbf{b}}_i = \boldsymbol{\omega} \times \mathbf{b}_i \qquad (6.44)$$

When assembled by columns into a matrix, the $\mathbf{b}_i$ form the rotation matrix from the inertial to the body frame, without the possible singularities which appear when orientation angles are used. Either of these equations supplies the vehicle attitude with respect to the inertial frame and, by inference, locates the inertial frame itself. In the strapdown system the digital computer must be fast enough to integrate (6.43) or (6.44) as quickly as the vehicle can rotate, which may be quite rapidly. With a mechanical platform, the computer need only be fast enough to follow the motion of the vehicle across the earth's surface.

Once the inertial frame is available, the inertial-navigation system will still need the inertial-acceleration components. For the mechanical stable platform, these are directly obtainable from three accelerometers mounted on the platform itself. Since these accelerometers are maintained stationary while the vehicle rotates around them, their readings need not be corrected for spurious accelerations from vehicle rotation. In a strapdown system, the accelerometers are mounted directly on the vehicle. Unless they can be positioned at the center of mass of the vehicle, rotational accelerations must be subtracted to get the inertial acceleration of the vehicle. These must then be transformed to the inertial frame so that equations (6.42) may be integrated. In either method, the inertial navigation system can find the vehicle position without any assumptions about the vehicle dynamics: (6.42) to (6.44) represent only kinematical relations. Mechanical platforms, at least, need not know anything about the characteristics of the vehicle in which they are mounted; in fact, they are commonly "flight-tested" in the back of a truck.

However, there is one force on the vehicle which does cause problems: the force of gravity. The problem with this force is that the accelerometers actually read the *sum* of the vehicle inertial acceleration and the local acceleration of gravity. An aircraft sitting stationary on the runway is *not* accelerating upward at $9.8\,\mathrm{m/s^2}$. So, before the acceleration components may be used in (6.42), the local acceleration of gravity must be calculated and subtracted. The inertial-navigation system must keep track of which way is down and subtract the acceleration of gravity from the readings of the accelerometers before they are integrated in (6.42). Since the acceleration of gravity varies by a few percent at different points on the earth's surface, it is necessary to have a model of gravitational variations available. Finally, the apparent gravitational acceleration is reduced by the centrifugal "force" for a fast-moving vehicle. This, however, is an issue which we will discuss in the next section.

6.9 THE STABILITY OF INERTIAL NAVIGATION

The inertial-navigation system, comprising an inertial reference frame, a set of accelerometers, and a computer, can be thought of as a single instrument. For an instrument to have acceptable behavior, it must not only be capable of indicating the true value of what is being measured but it must also tend to return to this value if the reading is disturbed. The reading will inevitably be

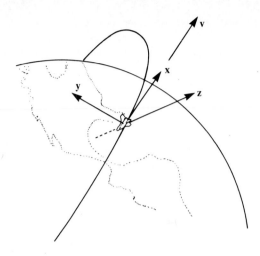

FIGURE 6.10
Vehicle path and local reference frame.

disturbed by measurement error, numerical error in the computer, vibration and calibration errors in the stable platform and accelerometers, and a host of other error sources. The requirement that the instrument return to the correct value when perturbed is a requirement for *stability* of the instrument. This is an issue that does not arise with simple instruments (e.g., a ruler), but must be considered when something as complex as an inertial-navigation unit is designed.

Consider Figure 6.10, which shows a vehicle following a great circle path at speed v, not necessarily orbital velocity. The vehicle is at true distance r_0 from the center of the earth, and the local rectangular reference frame moves with the vehicle. Let us assume for a moment that everything is working perfectly. Then, the accelerometers on the stable platform sense the true difference between the local acceleration of gravity and any accelerations caused by the noninertial nature of the local horizon frame **h**:

$$\mathbf{a}_0 = -\frac{GM}{r_0^3}\mathbf{r}_0 - \frac{{}^h d^2}{dt^2}\mathbf{r}_0 - 2\boldsymbol{\omega} \times \frac{{}^h d}{dt}\mathbf{r}_0 - \boldsymbol{\omega} \times (\boldsymbol{\omega} \times \mathbf{r}_0) \tag{6.45}$$

Time derivatives on the right are taken with respect to the horizon reference frame **h**. This frame has an inertial angular velocity of

$$\boldsymbol{\omega} = \frac{v_0}{r_0}\mathbf{j} \tag{6.46}$$

We are interested in the behavior of (6.45) when the inertial-navigation system commits an error in the position of the vehicle.

Now, assume that the true position vector of the vehicle is not given by $\mathbf{r}_0 = r_0\mathbf{k}$, as believed by the navigation system, but is really

$$\mathbf{r}_0 + \delta\mathbf{r} = \delta x\,\mathbf{i} + \delta y\,\mathbf{j} + (r_0 + \delta z)\mathbf{k} \tag{6.47}$$

Then, with respect to the horizon frame, the rotating-frame derivatives required for (6.45) will be given by

$$\frac{^h d}{dt}\mathbf{r} = \delta \dot{x}\,\mathbf{i} + \delta \dot{y}\,\mathbf{j} + \delta \dot{z}\,\mathbf{k} \tag{6.48}$$

$$\frac{^h d^2}{dt^2}\mathbf{r} = \delta \ddot{x}\,\mathbf{i} + \delta \ddot{y}\,\mathbf{j} + \delta \ddot{z}\,\mathbf{k} \tag{6.49}$$

The inertial-navigation unit, of course, has no information on its own errors, thinking that δx, δy, and δz are all zero. The calculation of the gravity acceleration in the computer proceeds under this assumption. The difference between the real acceleration of gravity and what the computer thinks it is is given by

$$\delta\mathbf{g} = -GM\frac{\delta x\,\mathbf{i} + \delta y\,\mathbf{j} + (r_0 + \delta z)\mathbf{k}}{[\delta x^2 + \delta y^2 + (r_0 + \delta z)^2]^{3/2}} + GM\frac{\mathbf{k}}{r_0^2} \tag{6.50}$$

Expanding the first term above by the binomial theorem, the gravity-calculation error caused by the incorrect position is given by

$$\delta\mathbf{g} \approx -\frac{GM}{r_0^3}(\delta x\,\mathbf{i} + \delta y\,\mathbf{j} - 2\delta z\,\mathbf{k}) \tag{6.51}$$

retaining only first-order terms in the result. Finally, the accelerometers themselves are likely to be a source of error, not indicating the true acceleration $\mathbf{a}_0$ but $\mathbf{a}_0 + \delta\mathbf{a}$ instead.

When the intermediate results (6.48), (6.49), and (6.51) are inserted in (6.45), we obtain the equation of motion of the error vector $\delta\mathbf{r}$ committed by the system. Broken into its three components, this becomes

$$\delta\ddot{x} + \left(\frac{GM}{r_0^3} - \frac{v_0^2}{r_0^2}\right)\delta x = -\frac{2v_0}{r_0}\delta\dot{z} - \delta a_x \tag{6.52}$$

$$\delta\ddot{y} + \frac{GM}{r_0^3}\delta y = -\delta a_y \tag{6.53}$$

$$\delta\ddot{z} - \left(\frac{2GM}{r_0^3} + \frac{v_0^2}{r_0^2}\right)\delta z = \frac{2v_0}{r_0}\delta\dot{x} - \delta a_z \tag{6.54}$$

These three equations govern the stability of the inertial-navigation system. For inertial navigation to be a trustworthy technique, small errors should preferably decay to zero, or at worst oscillate in magnitude. If they grow, then inertial navigation will have serious problems, since its predicted position cannot be trusted.

The lateral-error equation (6.53) is always well-behaved. This equation is a forced harmonic oscillator with fundamental frequency $\sqrt{GM/r_0^3}$. This is referred to as the *Schuler frequency* and near the earth's surface gives a period of about 84 min. Note that the Schuler frequency is also the angular rate of a satellite in orbit just above the earth's surface. The in-track error (6.52) will also be stable if the bracketed coefficient is positive, which implies

$$v_0 < \left(\frac{GM}{r_0}\right)^{1/2} \tag{6.55}$$

Since the right side is the expression for circular orbital velocity at radius r_0, this is an explicit requirement that the vehicle be moving at suborbital speed. The vertical-error equation (6.54) has a negative coefficient on its δz term for any value of the vehicle velocity, so the vertical-error channel is *always* unstable.

The in-track error (6.52) is forced by a Coriolis term involving the vertical-error rate. Since the vertical error will grow exponentially, this forcing term will drive the in-track error unstable as well. To avoid this, slow-moving craft (meaning anything moving at suborbital velocity) obtain information on their vertical position by other means and simply do not integrate the vertical equation of motion in the navigation computer. Aircraft have altimeters, submarines have depth gauges, and ships not at sea level generally have no further need for navigational data. With $\delta\dot{z}$ supplied from a trustworthy source, the in-track channel will be stable, and the error in this direction will oscillate with a frequency given by

$$f_{it} = \left(\frac{GM}{r_0^3} - \frac{v_0^2}{r_0^2}\right)^{1/2} \tag{6.56}$$

For almost all vehicles, this will be nearly equal to the Schuler frequency.

The case of a satellite in a circular orbit is quite different. To eliminate the vertical-channel instability, let us assume that the satellite is using a radar altimeter, although the power requirements for such a device are generally prohibitive. Errors in (6.53) will then oscillate once per orbit, duplicating the behavior of an error in the orbit plane. Equation (6.52) then becomes

$$\delta\ddot{x} = -\delta a_x \tag{6.57}$$

Now, since the spacecraft is in free fall, the x accelerometer *should* be reading zero acceleration. If it is not reading zero, the error δa_x will force the in-track error to grow unbounded. Alternately, we could simply eliminate all three accelerometers, since we know very well exactly what they should read in free fall. This, however, corresponds to the case of simply propagating the initial estimate of the orbit, without any new information. If the original value of the orbital period is slightly wrong, the in-track error will still grow unbounded. We are faced with the unpalatable decision either to integrate no information or to integrate wrong information. The conclusion is unavoidable: Inertial navigation does not work on an orbiting spacecraft. Some source of external information is required.

Inertial navigation is, however, used on satellite boosters. The instability in the x equation is approached from the stable side, and the in-track frequency does not go to zero until orbit insertion. As satellite velocity is approached, the period of error oscillation becomes long and the errors essentially become static. The entire period of boost for a satellite launch is short compared to the Schuler time. Even the errors in the unstable δz equation may not grow enough to be a cause of concern. Thus, inertial navigation can be trusted for the duration of powered flight. However, the same system then cannot be expected to correctly follow the orbit for any substantial time period. The space shuttle may be capable of getting itself into and out of orbit with an inertial-navigation unit, but it must depend on other systems (at the present time, ground-based tracking) to tell it where it is while on orbit.

6.10 REFERENCES AND FURTHER READING

Gyroscopic instruments are also covered by Thomson. *Inertial Guidance,* by Draper et al., is now mainly of historical interest, while the book by Britting is much more comprehensive and current. While treating the dynamics of gyroscopic devices is possible with classical newtonian mechanics, it becomes very difficult when the gimbals must also be included and equations of motion written for several coupled rigid bodies. The advanced techniques of lagrangian and hamiltonian dynamics are a great aid here. Meirovitch, in particular, has an excellent chapter on the dynamics of gyroscopic instruments using these methods.

With satellite navigation systems becoming more common and very accurate, it is not clear how much longer the techniques of inertial navigation will remain in use. It seems unlikely that they will be abandoned for military systems, since they are impervious to jamming. Also, even if inertial navigation gives way to satellite-based systems, the suite of gyroscopic instruments will still find use in the flight-control systems of most aerospace vehicles.

Thomson, W. T., *Introduction to Space Dynamics,* Wiley, New York, 1961.
Draper, C. S., W. Wrigley, and H. Hovorka, *Inertial Guidance,* Pergamon, New York, 1960.
Britting, K. R., *Inertial Navigation System Analysis,* Wiley, New York, 1971.
Meirovitch, L., *Methods of Analytical Dynamics,* McGraw-Hill, New York, 1970.

6.11 PROBLEMS

1. An *integrating gyroscopic accelerometer,* shown in Figure 6.11, has a rotor with angular momentum $\mathbf{H}$ along the $\mathbf{b}_1$ axis and is free to rotate in a gimbal system pivoted about both the $\mathbf{b}_2$ and the $\mathbf{b}_3$ axes. A mass m is located a distance d from the axis of the inner gimbal, producing a torque on the inner gimbal when the unit accelerates in the $\mathbf{b}_3$ direction (the input axis). Show that this torque is

$$\mathbf{M} = -mad\mathbf{b}_2$$

and that this produces a rate of change of the angular-momentum vector

$$\dot{\mathbf{H}} = H\dot{\phi}\mathbf{b}_2$$

Thus, demonstrate that the angular rate $\dot{\phi}$ is proportional to the acceleration a along the $\mathbf{b}_3$ axis and that the rotation of the outer gimbal is proportional to the velocity change

$$\phi - \phi_0 = -\frac{md}{H}(v - v_0)$$

along the $\mathbf{b}_3$ axis.

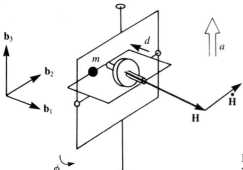

FIGURE 6.11
The integrating gyroscopic accelerometer.

2. In a strapdown inertial-navigation system, the accelerometers are also mounted rigidly in the vehicle body frame **b**. They are located at a fixed position **R** from the center of mass, as shown in Figure 6.12. If the accelerometers sense the acceleration vector **A**, give the expression for the inertial acceleration of the center of mass, $\mathbf{a}^i$. Where does the vehicle obtain measurements of $\boldsymbol{\omega}$ and $\dot{\boldsymbol{\omega}}$?

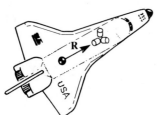

FIGURE 6.12
Strapdown accelerometers.

3. Without coupling to the δx error coordinate, show that the vertical error in an inertial-navigation unit grows like

$$\delta z = A_1 e^{\alpha t} + A_2 e^{-\alpha t}$$

where A_1 and A_2 are constants and

$$\alpha = \left(\frac{2GM}{r_0^3} + \frac{v_0^2}{r_0^2}\right)^{1/2}$$

A space-shuttle launch lasts 8 min. Use crude values of $r_0 \approx 6390$ km and $v_0 \approx 4$ km/s to calculate approximately how much the vertical error will grow during boost to orbit. If the error at insertion must be less than 100 m, how accurately must δz be known at lift-off? (The vertical error $\delta z \neq 0$ at launch, since the platform will drift even on the pad. This sets limits on the hold time once "guidance is internal.")

CHAPTER
7

ROCKET
PERFORMANCE

7.1 INTRODUCTION

The rocket is the symbol of space exploration. The sheer size and power of a large satellite booster place it in a category by itself in terms of human craft. The Saturn V, for example, was over eight times more massive at lift-off than the heaviest aircraft ever flown, or about the mass of a naval cruiser. Only a few fighter aircraft can fly straight up, and fewer still can accelerate while doing it. However, this is a necessary attribute of all rockets.

The two visionaries of rocket flight were Konstantin Tsiolkovsky in Russia and Robert Goddard in the United States. Born in 1857, the self-educated Tsiolkovsky spent his life as a schoolteacher. His contributions include demonstrating that liquid-fuel rockets would outperform solid-fuel rockets, since the entire structure did not have to withstand the combustion-chamber pressure. He first realized the advantages of staging a rocket and originated the concept of space stations, including artificial gravity produced by rotation. Goddard, born in 1882, began his career as a college teacher in Worcester, Mass. To him belongs the honor of constructing and flying the first liquid-fuel rocket in a cabbage patch in Auburn, Mass., in 1926. The site of this first flight is now a shopping-mall parking lot, ironically marked by a U.S.

Navy Polaris missile, a *solid-fuel* vehicle. Goddard moved his experiments to New Mexico for safety and privacy reasons and designed the first successful gyroscopic control mechanism for a rocket.

However, the modern rocket really took shape in Germany in the 1930s and early 1940s. Forbidden by the Treaty of Versailles from building long-range artillery, the German military undertook the development of the long-range rocket as an alternative. The result was the famous V-2. Weighing 12.5 tons at lift-off and carrying a payload of 1 ton over a range of 200 mi, it was far in advance of anything produced by the Allies. Developing this weapon required the obvious large engines, pumps, and structures but also required that automatic control and gyroscopic instruments be vastly improved. The V-2 had severe problems as a weapon. The explosion occurred below ground level due to the large impact velocity, and the alcohol for fuel consumed the entire potato crop of Germany in 1944. The V-2 was used only because the Allies' air superiority denied the use of the reusable bomber to the Germans. However, the development of the atomic bomb by the United States removed any doubt as to the future of the long-range rocket as a weapon.

At the close of the war, the leading German scientists of the V-2 program decided to surrender to the Allies. The Soviet Union acquired the test site at Peenemünde and most of the technicians and production facilities for the V-2. The development of long-range rockets became a matter of high priority. In 1954, Wernher von Braun, former head scientist at Peenemünde, proposed Project Orbiter to the U.S. Navy. It was rejected, since it used military rockets to do the same job as the civilian Vanguard program, and the United States wished to avoid any military taint on the purely peaceful infant space program. Nikita Khrushchev, however, apparently realized the psychological impact of being the first country to launch a satellite. If nothing else, it removed any doubt that Soviet ICBMs could reach the United States. In reply to Sputnik 1 and 2, the U.S. Vanguard crashed on the pad. Von Braun was given permission to proceed with Project Orbiter, and within three months Explorer 1 was in orbit.

Until the development of the Saturn I booster, *every* satellite-launch vehicle was in its inception a long-range delivery vehicle for weapons of mass destruction. While this is not necessarily pleasant to think about, the space program was helped mightily by the availability of large ICBMs. Many spacecraft have been carried into orbit by a recycled nuclear-weapons carrier. The space program has now reached the point where the need for satellite boosters is being met by the design of vehicles designed explicitly for that purpose. While the demand for the development of new boosters is admittedly small, the elements of rocket performance drive the design of spacecraft to a great extent. The engineer can be expected to integrate existing rocket components into new configurations and to ensure that they will perform the required mission. Finally, the elements of rocket performance are an inseparable part of the field of astronautical engineering.

7.2 THE ROCKET EQUATION

Care must be exercised in the derivation of the equation of motion of a rocket, because it is a variable-mass system. (By a "rocket" we will mean any vehicle producing thrust by ejecting mass, no matter what the propulsion technology.) One common fallacy is that the equation of motion for a rocket vehicle is obtained by modifying Newton's second law to read

$$\mathbf{F} = \frac{d}{dt}m\mathbf{v} = m\mathbf{a} + \frac{dm}{dt}\mathbf{v} \qquad (7.1)$$

where the term with dm/dt is interpreted as a thrust. This equation *is not correct*; note in particular that a rocket sitting stationary on the launchpad cannot lift off!

To derive the correct form of the equation of motion, it is necessary to follow the ejection of a blob of fuel during a short interval of time dt. This is a closed system of particles, shown in Figure 7.1, and it must conserve linear momentum. The momentum at the start is simply $m\mathbf{v}$. After time dt, the blob of fuel $-dm$ (we interpret dm as a negative quantity, since the booster is losing mass) has velocity $\mathbf{V}_e$ *with respect to the rocket* and velocity $\mathbf{v} + \mathbf{V}_e$ with respect to our inertial frame. The vehicle itself has lost mass dm and in return has gained speed $d\mathbf{v}$. The statement of momentum conservation becomes

$$(m + dm)(\mathbf{v} + d\mathbf{v}) - dm(\mathbf{v} + \mathbf{V}_e) = m\mathbf{v} \qquad (7.2)$$

Expanding the products and ignoring second-order terms, (7.2) becomes

$$m\,d\mathbf{v} - dm\,\mathbf{V}_e = 0 \qquad (7.3)$$

Or, dividing by dt, taking the limit, and realizing that there may be other

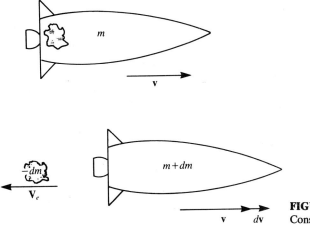

FIGURE 7.1
Conservation of momentum for a rocket.

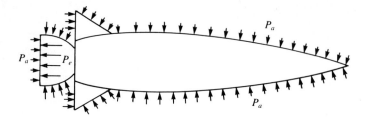

FIGURE 7.2
Atmospheric pressure effects on a rocket.

external forces $\mathbf{F}_{ext}$ acting on the rocket, its equation of motion is given by

$$\mathbf{F}_{ext} + \frac{dm}{dt}\mathbf{V}_e = m\mathbf{a} \qquad (7.4)$$

The quantity $\mathbf{V}_e$ is called the *exhaust velocity* of the rocket. The second term on the left above, which is positive since $\mathbf{V}_e$ points out of the tail of the vehicle and dm/dt is negative, is termed the *thrust*. Equation (7.4) is similar in form to the incorrect (7.1) but allows an initially stationary vehicle to accelerate.

There is an additional term added to (7.4) if the vehicle is operating in an atmosphere or if the exhaust does not exit with zero pressure. As shown in Figure 7.2, the ambient pressure P_a balances everywhere around the vehicle except across the bell of the engine. If the exhaust pressure is P_e and the engine nozzle has area A, then the complete equation of motion for a rocket is

$$\mathbf{F}_{ext} + \frac{dm}{dt}\mathbf{V}_e - \frac{\mathbf{V}_e}{V_e}A(P_e - P_a) = m\mathbf{a} \qquad (7.5)$$

So, if the vehicle is in vacuum, where P_a is zero, its performance will actually be better than when it is in an atmosphere. The additional term in (7.5) is negative if the ambient pressure is higher than the exhaust pressure. Since rockets leave the atmosphere so rapidly, the pressure correction is usually important only at the beginning of the trajectory. The pressure effect can be included by defining the *effective exit velocity*

$$V_{e,\text{eff}} = V_e - \frac{A}{dm/dt}(P_e - P_a) \qquad (7.6)$$

which is maximum in a vacuum.

In free space, with no external forces on the rocket, and assuming the exhaust gas is expanded to zero pressure, the equation of motion becomes

$$m\frac{dv}{dt} = -V_e\frac{dm}{dt} \qquad (7.7)$$

where we have dropped the vector notation of (7.4). The variables v and m are easily separated to yield the definite integrals

$$-V_e \int_{m_0}^{m} \frac{dm}{m} = \int_0^v dv \qquad (7.8)$$

assuming we start from zero velocity with initial mass m_0. The integrals give *the rocket equation*:

$$v = V_e \ln \frac{m_0}{m} \qquad (7.9)$$

The burnout velocity of the vehicle depends only on the exhaust velocity of the engine and how much of the vehicle is fuel. Equation (7.9) accounts for the great difference between aircraft and rockets: The former are mostly empty space, while the latter are almost all fuel tanks. A rocket will equal its own exhaust velocity at burnout if the mass ratio is $m_0/m = e$ and will exceed the exhaust velocity if the mass ratio is greater than this value.

The performance of a single-stage rocket is governed by the exhaust velocity V_e and the ratio of initial to final masses m_0/m_f. If we divide the initial mass m_0 into the propellant mass m_p, the structural mass m_s, and the payload mass m_*, then

$$m_0 = m_* + m_p + m_s \qquad (7.10)$$

and

$$m_f = m_* + m_s \qquad (7.11)$$

It is also customary to introduce the dimensionless *payload ratio*

$$\pi = \frac{m_*}{m_0} \qquad (7.12)$$

and the *structural ratio*

$$\varepsilon = \frac{m_s}{m_s + m_p} \qquad (7.13)$$

The payload ratio measures how much of the initial vehicle is payload. It is usually a small number. The structural ratio is a measure of how much of the vehicle is structure. This quantity is defined excluding the payload, lest ε change every time a new payload is launched. The burnout mass ratio is then given by

$$\frac{m_f}{m_0} = \frac{m_* + m_s}{m_* + m_p + m_s} = 1 - \frac{m_p}{m_0} \qquad (7.14)$$

when m_p is added and subtracted from the numerator above. With further manipulation, this becomes

$$\frac{m_f}{m_0} = 1 - \frac{m_s + m_p}{m_0} \frac{m_p}{m_s + m_p}$$

$$= 1 - (1 - \pi)(1 - \varepsilon)$$

$$= \varepsilon + (1 - \varepsilon)\pi \qquad (7.15)$$

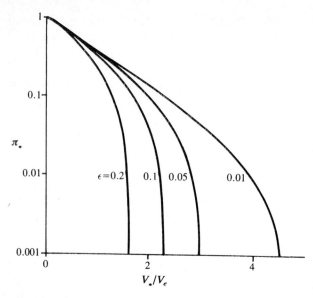

FIGURE 7.3
Payload ratio π_* versus burnout velocity.

The rocket equation now becomes

$$V_f = -V_e \ln \left[\varepsilon + (1 - \varepsilon)\pi \right] \tag{7.16}$$

From this it is obvious that there is an absolute limiting velocity for a given single-stage rocket. This occurs when the vehicle carries no payload and is given by

$$V_{f,\max} = -V_e \ln \varepsilon \tag{7.17}$$

Figure 7.3 shows the dimensionless payload ratio π for several values of the structural factor. A structural ratio of $\varepsilon = 0.05$ seems to be a reasonable current limit to what is possible, especially since the mass of engines and thrust structure must be counted into m_s. A rocket with this structural factor can barely achieve three times its own exhaust velocity.

7.3 PROPULSION TECHNOLOGY

Although we are principally interested in the dynamics of rocket flight, it is necessary to have some familiarity with the propulsion technologies now available or on the drawing board. Any propulsion scheme producing thrust by ejecting mass obeys the rocket law. However, different technologies have their strengths and weaknesses, and their performances differ considerably. The one conceptual technology which does not follow this fundamental relation is the photon rocket, and we will treat this in a later section.

It is common in the English engineering literature to cite the *specific impulse* I_{sp} instead of the exhaust velocity V_e. To a physicist, "specific" means per unit mass, and the amount of momentum gained by the rocket per mass of fuel expended is

$$\frac{dm\, V_e}{dm} = V_e \qquad (7.18)$$

So, with this definition, specific impulse is really just the exhaust velocity itself. However, the "specific impulse" cited in the engineering literature is the amount of momentum gained per *weight* of fuel consumed,

$$I_{sp} = \frac{dm\, V_e}{g\, dm} = \frac{V_e}{g} \qquad (7.19)$$

As defined in (7.19), I_{sp} has units "seconds." Extreme care must be exercised in using this definition. The exhaust velocity, which is the quantity needed for the dynamics, is obtained by multiplying I_{sp} by g. This g is *never* the local acceleration of gravity. The exhaust velocity of a rocket engine does not become smaller on a planet with lower gravity nor is it zero when the vehicle is in a state of free fall. The g to be used is the standard mean acceleration of gravity on the earth. The one advantage of this convention is that the exhaust velocity V_e may be obtained in English or metric units by simply using the appropriate value of g.

Table 7.1 lists some current and conceptual propulsion technologies with their exhaust velocities and specific impulses. The technologies with the lowest performance are chemical rocket engines. If E_{ch} is the amount of energy per molecule liberated by the chemical reaction, then

$$E_{ch} = \tfrac{1}{2}mV_e^2 \approx \tfrac{5}{2}kT \qquad (7.20)$$

where m is the mass of one molecule of the reaction product, T is the absolute combustion temperature, and k is Boltzmann's constant, 1.38×10^{-23} J/K. The first half of (7.20) comes from dynamics, while the second half comes from the kinetic theory of gases, assuming simple diatomic reaction products. The

TABLE 7.1
Representative propulsion technologies

V_e, km/s	I_{sp}, s	Technology
1.6–2.1	170–220	Solid fuel
1.9–3.4	200–350	Hydrocarbon liquid fuel
4.4	455	Liquid hydrogen and liquid oxygen
3.0–5.0	300–500	Nuclear and hydrogen
3.0–7.0	300–700	Plasma jet, arc jet
10–50+	1000–5000+	Mass driver
$10^2 \rightarrow$?	$10^4 \rightarrow$?	Ion, MHD thrusters

reaction products are of course created with random molecular velocities (they are hot), and it is the sole function of the rocket nozzle to straighten and align these random velocities. Solving the left half of (7.20) for the exhaust velocity, we find

$$V_e \approx \left(\frac{2E_{ch}}{m}\right)^{1/2} \tag{7.21}$$

To obtain the highest possible exhaust velocity, it is necessary to use energetic chemical reactions or to use a fuel with a low-mass exhaust product. However, the number of possible chemical reactions is limited. The best chemical rocket engine currently available is the space shuttle main engine (SSME), which uses liquid hydrogen and liquid oxygen and achieves an I_{sp} of 455 s. The theoretical I_{sp} is 457 s for this fuel combination, so the SSME operates at over 99.5% efficiency. A combination of liquid oxygen and liquid fluorine offers somewhat improved performance, but the exhaust product is hydrofluoric acid. At room temperature this substance eats its way out of glass bottles, and at high temperature it has a pronounced tendency to alter the geometry of the engine.

To achieve still higher performance it is necessary to obtain energy from some source other than a chemical reaction. This breaks the first bond in (7.20), since we will be able to insert as much energy per molecule as we wish. Equation (7.21) then implies that, all other things being equal, we should use the lowest-mass "working fluid" possible: liquid hydrogen. Into this category fall the nuclear rocket and the plasma jet and arc jet technologies. However, solving the second half of (7.20) for the exhaust temperature, we have

$$T \approx \frac{mV_e^2}{5k} \tag{7.22}$$

According to this relation, the SSME is already operating at a combustion temperature of 6000 K. The temperature goes up as the *square* of the exhaust velocity, so not too much progress should be expected of any propulsion technology which uses a *thermal* process. In fact, the improved performance cited in Table 7.1 for these technologies is largely due to the use of liquid hydrogen as the working fluid instead of the water-vapor exhaust of the SSME. Very high exhaust velocities imply extreme temperatures, and it is difficult to tolerate these temperatures without vaporizing the engine in the process.

Although these technologies offer improved performance over most chemical rockets, they do have problems. Nuclear reactors produce heat by fission, but over half the energy liberated in this process appears via the radioactive decay of secondary isotopes. This occurs over a period of several hours after the initial fission event. So, once the reactor has been used for several hours and is "scrammed," it will continue to produce heat at about the same rate for several more hours. If the nuclear engine is to be reused, considerable fuel must continue to flow through the engine to prevent core meltdown. Including cooling fuel, the effective specific impulse of a nuclear engine may be no higher than for a chemical rocket. The other two technologies require large-capacity electrical power supplies. Electrical power generation in space is difficult, and such power systems have a very considerable mass. This effectively restricts these technologies to "low-thrust" applications. They certainly cannot lift a vehicle off the earth's surface.

To make still further progress, it is necessary to eject mass by a *nonthermal* process. In this area are the ion engine, which electrostatically accelerates an ionized gas, the magnetohydrodynamic (MHD) engine, which electromagnetically accelerates an ionized gas, and the O'Neill "mass driver," which is an electromagnetic catapult. The ion and MHD engines use a hot, ionized gas, but only sufficient energy is supplied to ionize the gas, and the exhaust process is not thermal. To avoid thermal problems, the gas density within the engine is very low, which gives the device a small dm/dt and hence a low thrust. The mass driver accelerates slugs of matter along a linear electric motor and is a completely nonthermal process. The ion engine is selective about its fuels, with preference given to substances with low ionization energies. It can have such a high specific impulse that the mass of the fuel ceases to dominate the mass of the vehicle. However, when operated at high exhaust velocities the ion engine consumes large quantities of electrical power. Since space power supplies are massive, there is a temptation to feed ion engines less electricity and more fuel. This increases the thrust and can reduce the mass of the vehicle, but it also reduces the specific impulse. The mass driver shows the promise of being the equivalent of the prospector's mule in the old west in that it will eat anything.

Finally, we note that only chemical rocket engines have sufficient thrust to lift off from the earth's surface. To achieve orbit, one must supply about 7.7 km/s of orbital speed and pay approximately another 1.5 km/s to achieve orbital altitude and make up losses from air drag. The total cost of low earth orbit is thus almost 10 km/s. Using the rocket law (7.9) and performance figures for the SSME, a single-stage satellite booster would need a mass ratio $m/m_0 \approx 0.10$ to achieve orbit. With hydrocarbon liquid fuels, this drops to about 0.03. Of course, out of this fraction of the vehicle remaining at burnout we must subtract the mass of the engines, structure, guidance systems, and any other deadweight. What is left over, if anything, is the payload. There is a method by which these figures can be considerably improved, however. The solution is the multistage rocket.

7.4 THE MULTISTAGE ROCKET

Since the performance of a rocket depends sensitively on the structural mass of the vehicle, performance can be improved if some way can be found to dispose of useless structural mass whenever possible. The most common method for doing this is to stage the vehicle. Empty tanks and the large engines necessary to lift off the earth's surface are shed, and the smaller vehicle proceeds from that point with considerably less parasitic mass. The treatment of multistage

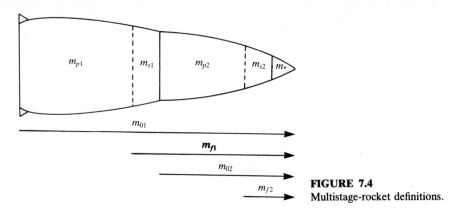

FIGURE 7.4
Multistage-rocket definitions.

rockets is not difficult, but there is a bookkeeping problem in keeping track of the different stages.

Figure 7.4 shows a two-stage rocket, with each stage broken down into its structural and propellant masses. For performance purposes, the initial mass m_{0k} of the kth stage is the mass of everything above the separation plane for that stage. The final mass of the kth stage, m_{fk}, is the structural mass of that stage plus the mass of stages still remaining. The $N+1$ stage in an N-stage rocket is the payload, mass m_*. We generalize the definition of the stage structural factor to be

$$\varepsilon_k = \frac{m_{sk}}{m_{sk} + m_{pk}} \tag{7.23}$$

a constant *of the stage*. The "payload" of the kth stage is the mass of everything above that stage, so the payload ratio of each stage is

$$\pi_k = \frac{m_{0,k+1}}{m_{0k}} \tag{7.24}$$

It is also useful to define the *overall payload ratio* as the fraction of the total vehicle which is payload:

$$\pi_* = \frac{m_*}{m_{01}} \tag{7.25}$$

With these definitions, each stage can be treated as if it were a single-stage rocket. In a vacuum, without gravity, the final burnout velocity of the vehicle V_* is the sum of the burnout velocities of the individual stages:

$$V_* = -\sum_{k=1}^{N} V_{ek} \ln\left[\varepsilon_k + (1 - \varepsilon_k)\pi_k\right] \tag{7.26}$$

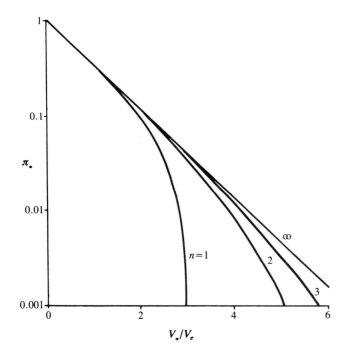

FIGURE 7.5
Payload ratio for a multistage rocket.

To see the advantage conferred by staging, expand the overall payload ratio as

$$\pi_* = \frac{m_*}{m_{01}} = \frac{m_*}{m_{0N}}\frac{m_{0N}}{m_{0,N-1}}\frac{m_{0,N-1}}{m_{0,N-2}}\cdots\frac{m_{02}}{m_{01}}$$

$$= \prod_{k=1}^{N} \pi_k \tag{7.27}$$

So a small overall payload ratio, essential for a large burnout velocity, can be generated as the product of several individual payload ratios, each of which need only be reasonably small.

Figure 7.5 shows the overall payload ratio π_* as a function of V_*/V_e, assuming the same exhaust velocities V_e, identical structural factors of $\varepsilon_k = 0.05$, and identical stage payload ratios π_k for each stage of a multistage rocket. We will see in a later section that equal π_k is actually the optimal arrangement for a multistage rocket. It is clearly evident that a two-stage rocket can achieve burnout speeds which are unattainable for a single-stage rocket with the same structural factor. A three-stage rocket can attain still higher burnout velocities. As the number of stages is increased, the curves rapidly approach a limiting case, that of the infinite-stage rocket. The use of

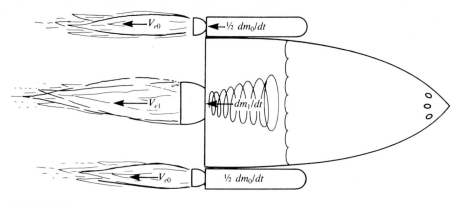

FIGURE 7.6
A parallel-staged vehicle.

many stages rapidly becomes inefficient and can also decrease the reliability of the vehicle.

A variation on simple staging is *parallel staging,* in which two stages are burning at the same time. The space shuttle is one obvious example, since it leaves the pad with both solid boosters and liquid-fuel engines running. But booster stages are often added to an existing vehicle to improve its performance. Figure 7.6 shows a parallel-staged vehicle, with the booster rockets and main engines burning together. The total thrust of both sets of engines is

$$T_{\text{tot}} = \frac{dm_0}{dt} V_{e0} + \frac{dm_1}{dt} V_{e1} = \frac{dm_{\text{tot}}}{dt} \bar{V}_{e0} \qquad (7.28)$$

The quantity $\bar{V}_{e0}$ is the averaged exit velocity of the combined engines in parallel burn:

$$\bar{V}_{e0} = \frac{dm_0/dt V_{e0} + dm_1/dt V_{e1}}{dm_0/dt + dm_1/dt} \qquad (7.29)$$

To treat this case, let us define the "zeroth" stage as the *combined booster rockets and first stage while they burn together,* and define the "first" stage as the burning of the *remaining* first-stage fuel after the booster rockets have been discarded. Assume that the first stage burns a mass m_{ip1} during parallel burn, and the remainder after the boosters are discarded. It is then appropriate to define the effective structural and payload factors for the zeroth stage as

$$\varepsilon_0 = \frac{m_{s0} + m_{s1}}{m_{s0} + m_{s1} + m_{p0} + m_{ip1}} \qquad (7.30)$$

$$\pi_0 = \frac{m_{01} - m_{ip1}}{m_{00}} \qquad (7.31)$$

since the remaining fuel in the first stage is payload for the zeroth stage. The corresponding ratios for the first stage are now

$$\varepsilon_1 = \frac{m_{s1}}{m_{s1} + (m_{p1} - m_{ip1})}$$ (7.32)

$$\pi_1 = \frac{m_{02}}{m_{02} + m_{s1} + (m_{p1} - m_{ip1})}$$ (7.33)

With these definitions, the parallel-staged vehicle can be treated as a simply staged rocket. Equation (7.26) generalizes to

$$V_* = -\bar{V}_{e0} \ln\left[\varepsilon_0 + (1 - \varepsilon_0)\pi_0\right] - \sum_{k=1}^{N} V_{ek} \ln\left[\varepsilon_k + (1 - \varepsilon_k)\pi_k\right]$$ (7.34)

where the only changes are the use of the average exit velocity for the zeroth stage and the redefinition of what constitutes the first stage.

The principle of staging is carried further than just discarding used rocket engines and fuel tanks. Because the performance of a booster depends critically on the masses involved, any opportunity to discard mass should be exercised. During the early part of the launch, the payload is usually covered by an aerodynamic fairing, or payload shroud. The satellite is designed for the space environment and cannot tolerate flight through the atmosphere at high Mach numbers. However, rockets leave the atmosphere extremely rapidly, and when the dynamic air pressure reaches tolerable levels the payload shroud is jettisoned. The shroud is usually heavy, since it must withstand supersonic flight loads, and it is positioned at the top of the vehicle, where it can greatly affect the overall performance of the system (see section 7.6). On more than one occasion, the failure of the shroud to jettison has resulted in the payload not achieving orbit. In this case, the shroud lives up to the other meaning of its name as the last stage and payload reenter the atmosphere together.

The vehicle guidance system is another candidate for waste mass under certain conditions. It would be very inefficient to have a separate guidance unit for each stage, so the guidance unit is usually placed in one of the upper stages. However, carrying this unit all the way to the final orbit would also be inefficient if its function can be replaced by a simpler technique. Once out of the atmosphere, complex inertial-navigation systems can be replaced by ground tracking and control, and the attitude of the vehicle can be maintained by spin stabilization. This technique was first used during the launch of Explorer I in 1958, where the massive guidance system of the early Jupiter C was shed with the first stage. It is still the preferred technique for executing Hohmann transfers, since the massive booster guidance system can be discarded in the parking orbit once it has performed its last function of aligning and spinning up the upper stages. A final possibility for excess mass are some of the engines in the first stage. A rocket needs a thrust-to-weight ratio greater than 1 to leave the ground, but once in nearly horizontal flight this is no longer true. Dropping extra engines not only sheds unnecessary mass but helps limit the acceleration buildup as the fuel is used. The Atlas leaves the pad with three engines but drops the outer two engines after a few minutes of flight.

7.5 TWO MULTISTAGE ROCKETS

As examples of multistage rockets, consider the two designs shown in Figure 7.7. The large vehicle on the left is a design for a three-stage space-station supply rocket. This design is by Wernher von Braun and first appeared in a famous series in *Colliers' Magazine* in 1951. This behemoth would have stood more than 80 m (265 ft) high and weighed about 6,500,000 kg (14,000,000 lb) on the pad. It is a classic design for a simple multistage rocket. For comparison, the current space shuttle system is shown on the right to the same scale. The shuttle is also in its inception a space-station supply vehicle. It is considerably shorter (about 50 m, or 165 ft) and much less massive (about 2,000,000 kg, or 4,400,000 lb) at lift-off. However, these two designs can be contrasted nicely, since their payloads to orbit are nearly equal!

The major difference in the two designs is the choice of propellant. The von Braun design uses nitric acid and hydrazine in all three stages, while the space shuttle uses solid rockets and liquid oxygen–liquid hydrogen in the main propulsion system. The details of the von Braun design are given in Table 7.2, while the space shuttle is detailed in Table 7.3. Note that the payload ratios for the first two stages of the von Braun design are nearly equal. This represents an optimal design, as we will see in a later section. The lower specific impulse for the first stage arises because this stage must operate in the atmosphere, where the rocket engines do not perform as well. The burning times for the three stages are also short compared to those of the shuttle. This means that

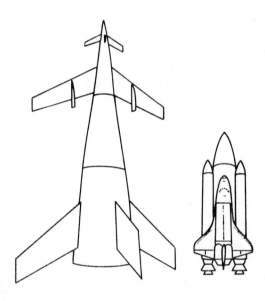

FIGURE 7.7
Two space shuttle designs.

TABLE 7.2
Von Braun three-stage rocket[†]

	Stage			
	1	**2**	**3**	*****
m_{0k}, kg	6,349,000	897,930	129,700	35,400
m_{fk}, kg	1,587,250	199,540	60,225	
m_{sk}, kg	698,390	69,840	21,950	
m_{pk}, kg	4,761,750	698,400	69,480	
dm/dt, kg/s	55,300	5,533	700	
T, N	124,544,000	15,568,000	1,957,120	
$T/m_0 g$	2.0	1.77	1.54	
I_{sp}, s	230	286	286	
t_{burn}, s	84	124	84	
ε_k	0.128	0.0909	0.240	
π_k	0.141	0.144	0.272	

[†] From Cornelius Ryan, *Across the Space Frontier,* Viking Penguin, New York, 1980. Reprinted by permission.

the vehicle accelerates quickly, in this case achieving $8g$ just at the burnout of the first and second stages. This also is a more efficient design compared to the space shuttle, since a quick ascent means that less fuel is used simply supporting the rocket against gravity. It does, however, subject the passengers to a more severe ride.

The structural ratio for the second stage is impressive, but the structural ratios for the first and third stages are considerably larger, indicating a large amount of structural mass. In the third stage this is excusable, since it is not a simple rocket. It has wings, a crew cabin, a payload bay, and other encumbrances not present on a pure rocket. In fact, von Braun also designed a cargo variant of this spaceship, in which the third stage simply consisted of engines, tanks, and payload. The structural ratio of the first stage suffers from the presence of the large wings. These are normally unnecessary in any large rocket. During the first critical seconds at lift-off, the wings would provide no stability whatever, so control must be maintained by swiveling the rocket engines. The first stage essentially leaves the atmosphere within the first minute of flight, and once again the wings are useless. So, modern designs simply omit them. However, in this case they are needed to keep the wings of the third stage at zero angle of attack during boost, a problem the modern shuttle also experiences.

However, perhaps the most impressive thing about this design is its sheer mass. It has a takeoff mass almost three times that of the Saturn V. This is due both to the choice of fuels and, more directly, to the rocket equation. Each stage must be smaller than the one below it, usually by a constant factor if the same fuels are being used in each stage. If we move upward this means that we will arrive at the top with a small payload compared to the vehicle. If we start

TABLE 7.3
The space shuttle[†]

	Stage			
	1	**2**	**3**	*****
m_{0k}, kg	2,015,600 (1,169,200)	670,500 (846,400)	108,300	29,500
m_{fk}, kg	834,300	143,300	104,700	
m_{sk}, kg	163,800	35,000	68,000	
m_{pk}, kg	1,181,300 (1,005,400)	527,200 (703,100)	3,600	
dm/dt, kg/s	9842 (8378)	1464	17.4	
T, N	29,900,000 (23,600,000)	6,300,000	53,400	
$T/m_0 g$	1.51	0.96	0.05	
I_{sp}, s	312 (287)	455	313	
t_{burn}, s	120	360	200	
ε_k	0.144 (0.140)	0.062 (0.047)		
π_k	0.332	0.161		

[†] Numbers in parentheses refer to components of the shuttle and not equivalent parallel stages.

with a given payload and move downward, we can arrive, as in this case, at a huge vehicle.

Table 7.3 gives the corresponding figures for the nominal space shuttle design. This is a parallel-staged vehicle, with the two solid boosters and the liquid-fuel engines burning together at lift-off. Two minutes into the flight, the solid boosters drop off, and the shuttle continues under the liquid-fuel engines. The third stage listed is really the orbital maneuvering system. The tank, containing all the fuel for the main engines, is jettisoned before reaching orbit (so it will burn up in the atmosphere), and final orbit is achieved with small maneuvering engines.

The thrust-to-weight ratio is less than 1 for the space shuttle just after the solid boosters are discarded. While it is traditional to design the first stage of a multistage rocket to have $T/mg > 1$, it is not necessary to do the same for upper stages of a satellite booster. Unlike a sounding rocket fighting gravity during its entire flight, a satellite booster is rapidly brought to a shallow climb angle. It will then build up horizontal speed quickly, even without the capability of lifting its own weight off the ground. The von Braun shuttle was to have been flight-tested in one-, two-, and finally three-stage versions, beginning with the top stage, so in this case each stage can lift its own mass. The structural ratio for the external tank is remarkable, especially since this is not a simple axially loaded structure. This least-massive element of the shuttle system must withstand the off-axis loads generated by two solid boosters on each side and an orbiter hung off its back. The structural factor for the orbital maneuvering system is too small to be worth citing.

The two designs share one problem in common: the presence of wings on the orbiter stage. It was found by German experimenters in the 1930s that rockets do not behave well with wings. In fact, the early attempts at flying winged rockets generally resulted only in a series of spectacular loops followed by a crash. For these two designs, it is absolutely imperative that the wings be kept at zero angle of attack during launch. While no figures are available for the von Braun design, the space shuttle would rip itself free from the external tank if the wings developed an angle of attack of $\frac{1}{2}°$ during boost. Such an unplanned separation event is obviously not desirable.

The rocket equation will yield V_* values of 9.49 km/s for either design. This is larger than the 7.7 to 8.0 km/s burnout velocity needed for low earth orbit. The difference lies in losses fighting gravity and air drag during the boost to orbit. Also, neither design ascends directly into a circular orbit. Rather, burnout occurs at about 80 km for either design at a velocity too large to permit a circular orbit. The vehicle's burnout conditions are correct, however, for a Hohmann-transfer ellipse to the correct orbital altitude. A small maneuver at apogee will then circularize the orbit. This process is generally more efficient than a direct ascent to a high-altitude orbit.

7.6 TRADE-OFF RATIOS

During the design of a vehicle it is often useful to know how a small change in one stage will affect the performance of the rest of the vehicle. Also, space hardware is still almost unique in that each vehicle is different from its predecessors. The tendency for payloads to "grow" during design forces changes in the booster vehicles. Also, there is an enormous pressure to improve the boosters to accommodate ever-larger payloads. Even "old" boosters continually change to incorporate small improvements. Trade-off ratios also furnish an easy method to assess performance changes when promised design values prove to be wrong. We will calculate the two most useful trade-off ratios for a multistage rocket. These are nothing more than the derivatives of the multistage-rocket equation with respect to structural or propellant masses within the vehicle.

The burnout velocity of a multistage rocket is given by

$$V_* = + \sum_{k=1}^{N} V_{ek} \ln \frac{m_{0k}}{m_{fk}} \qquad (7.35)$$

We cannot tolerate changes in this velocity, since the quantity V_* will be dictated by a given mission and this given velocity *must* be achieved. Since the payload velocity must not be allowed to vary, we are interested in how the mass of the payload changes with changes in the booster design *for a fixed payload velocity V_**. Now, the initial mass of the kth stage is

$$m_{0k} = m_* + m_{pk} + m_{sk} + \sum_{j=k+1}^{N} (m_{sj} + m_{pj}) \qquad (7.36)$$

and the final mass of this stage will be

$$m_{fk} = m_* + m_{sk} + \sum_{j=k+1}^{N} (m_{sj} + m_{pj}) \qquad (7.37)$$

When (7.36) and (7.37) are substituted into (7.35), we have an expression for the burnout velocity as a function of the component masses of the booster.

There are two trade-off ratios of interest. The first of these describes how the mass of the payload changes with changes in the structural mass of the kth stage, m_{sk}. For small changes in m_{sk}, this will be given by the quantity $\partial m_* / \partial m_{sk}$ calculated for no change in V_*. To impose this latter constraint, we take implicit partial derivatives of (7.35) and set the result to zero. Remembering that the structural mass m_{sk} is present in the first k stages, this gives

$$dV_* = 0 = \sum_{i=1}^{N} V_{ei} \frac{m_{fi}}{m_{0i}} \left(\frac{1}{m_{fi}} - \frac{m_{0i}}{m_{fi}^2} \right) \partial m_* + \sum_{j=1}^{k} V_{ej} \frac{m_{fj}}{m_{0j}} \left(\frac{1}{m_{fj}} - \frac{m_{0j}}{m_{fj}^2} \right) \partial m_{sk} \qquad (7.38)$$

Solving for the desired ratio, we have

$$\frac{\partial m_*}{\partial m_{sk}} \bigg|_{dV_*=0} = - \frac{\displaystyle\sum_{j=1}^{k} V_{ej} \left(\frac{1}{m_{0j}} - \frac{1}{m_{fj}} \right)}{\displaystyle\sum_{i=1}^{N} V_{ei} \left(\frac{1}{m_{0i}} - \frac{1}{m_{fi}} \right)} \qquad (7.39)$$

Note that, as expected,

$$\frac{\partial m_*}{\partial m_{sN}} = -1$$

That is, structural mass in the last stage trades kilogram for kilogram with payload mass.

The second trade-off ratio describes the change in the payload mass with changes in propellant mass of the booster vehicle. Again taking partial derivatives of (7.35), this time with respect to the propellant mass m_{pk}, we have

$$dV_* = 0 = \sum_{i=1}^{N} V_{ei} \left(\frac{1}{m_{0i}} - \frac{1}{m_{fi}} \right) \partial m_* + \sum_{j=1}^{k} V_{ej} \frac{1}{m_{0j}} \partial m_{pk} \qquad (7.40)$$

TABLE 7.4

Trade-off ratios for the space shuttle

k	V_{ek}, m/s	m_{0k} kg	m_{fk} kg	$\dfrac{\partial m_*}{\partial m_{sk}}$	$\dfrac{\partial m_*}{\partial m_{pk}}$
0	3060	2,015,000	834,000	−0.078	+0.055
1	4459	675,500	143,300	−0.964	+0.296
2	3067	108,300	104,700	−1.000	+0.055
*		29,500			

This gives the propellant trade-off ratio as

$$\frac{\partial m_*}{\partial m_{pk}}\bigg|_{dV_*=0} = \frac{-\displaystyle\sum_{j=1}^{k}\frac{V_{ej}}{m_{0j}}}{\displaystyle\sum_{i=1}^{N} V_{ei}\left(\frac{1}{m_{0i}}-\frac{1}{m_{fi}}\right)} \qquad (7.41)$$

This quantity will be positive, since adding fuel will increase the payload capability.

Table 7.4 shows these trade-off ratios for the space shuttle. This is a three-stage vehicle, where the zeroth stage is the parallel burn of the solid rocket boosters (SRBs) and the main engines, the second stage is the remainder of the burn of the main engines, and the third stage is the insertion burn in which the orbiter uses its own maneuvering engines. As discussed in the last section, the exit velocity listed for the zeroth stage is the averaged exit velocity of the two propulsion systems. This complication does not affect in any way the derivation above. Also, the entries in the table reflect the baseline space shuttle design for a payload of 29,500 kg (65,000 lb), a figure which was not achieved until the third orbiter became available. Like all space hardware, there are very significant differences among the existing shuttle vehicles. Trade-off ratios are one useful tool for assessing these differences.

As expected, the table shows that payload mass and orbiter mass trade one for one. However, since the external tank is carried almost all the way to orbit, the structural trade-off ratio here is also close to 1. To illustrate the use of these coefficients, a small change in the structural mass of the tank δm_{s1} is related to a change in the payload mass by

$$\delta m_* \cong \frac{\partial m_*}{\partial m_{s1}}\,\delta m_{s1} \qquad (7.42)$$

After the second shuttle launch, NASA decided not to paint future external tanks white. This reduced the mass of the tank by about 1000 kg, so $\delta m_{s1} = -1000$ kg. The payload capability was increased by 964 kg (about 1 ton) as a consequence of this change. The best point to add fuel would also be in the external tank, since the table shows that an increase of slightly more than 3 kg of liquid fuel will increase the payload by 1 kg. Care must be exercised here to ensure that the vehicle will still be able to leave the launchpad. Often tank stretching on a booster is accompanied by the addition of small solid fuel strap-on rockets for this reason.

However, the most tempting place to improve the shuttle as a cargo booster is at the orbiter itself. All the other trade-off ratios describe how *small* changes relate to each other, but the trade-off ratio for structural mass of the orbiter is correct for any size change. One proposal to upgrade the shuttle to a "heavy lift vehicle" is to discard *all* the orbiter except the engines and rear thrust structure. The wings, vertical tail, and cabin are not needed on a pure cargo vehicle. The payload bay would be replaced with a disposable payload shroud. The engines could be encased in a reusable heat shield and recovered by parachutes after reentry. These changes would make almost the entire mass of the orbiter available as payload, and more than doubles the payload which could be delivered to low earth orbit. This is essentially the path that the Soviet Union has followed with its new Energia booster, although a Soviet shuttle is also rumored to be under development.

7.7 THE OPTIMAL MULTISTAGE ROCKET

Staging brings obvious benefits over single-stage rockets, and it is all that makes some missions possible with chemical rockets. However, given that an N-stage rocket can achieve a given burnout velocity

$$V_* = \sum_{k=1}^{N} V_{ek} \ln \frac{m_{0k}}{m_{fk}} \tag{7.43}$$

there are still many ways to design each stage so that the complete vehicle can perform the mission. What is the optimal design for a multistage rocket? First, it is essential to carefully define what is meant by "optimal," or the result of the optimization can be absurd. In the nondimensional parameters, (7.43) is

$$V_* = - \sum_{k=1}^{N} V_{ek} \ln \left[\varepsilon_k + (1 - \varepsilon_k)\pi_k \right] \tag{7.44}$$

and, for each stage, we have three quantities at our disposal: π_k, ε_k, and V_{ek}. It does not make sense to maximize V_*, since the maximum burnout velocity will be obtained when the vehicle carries no payload. Nor does it make sense to attempt a mathematical optimization on either the stage exhaust velocity V_{ek} or the stage structural factor ε_k. The optimal values for these parameters are ∞ and 0, respectively, numbers which yield very good performance but which will cause problems in attempting to build the vehicle. So, we will accept V_{ek} and ε_k as values specified by the state of the art, and all we have left to change is π_k.

The π_k specify how the mass of the vehicle is partitioned among the various stages. A reasonable definition of "optimal" is to maximize the amount of payload carried by a vehicle of a given lift-off mass. Put another way, we would like to build a vehicle with the maximum value of the overall payload ratio π_*, since this produces the smallest booster for a given payload. The optimization problem is thus to maximize

$$\pi_* = \prod_{k=1}^{N} \pi_k \tag{7.45}$$

subject to the condition that the vehicle is capable of achieving the stated mission, equation (7.44). Instead of (7.45), we will find it easier to work with the logarithm of this quantity. This replaces the product with a summation:

$$\ln \pi_* = \sum_{k=1}^{N} \ln \pi_k \tag{7.46}$$

Since the logarithm is a strictly increasing function, maximizing $\ln \pi_*$ is fully equivalent to maximizing π_* itself.

If this was an unconstrained maximization problem, we would simply calculate the N partial derivatives of (7.46) and set them all to zero. However, with the constraint (7.44), one π_k is specified for us as a function of the other π's and the parameters V_*, V_{ek}, and ε_k. We could solve (7.44) for this π_k and substitute the result into (7.46). The remaining $(N-1)\pi_k$ would be independent, and calculating the $N-1$ partial derivatives would give us our result. However, a much nicer method for handling this type of problem was invented by Lagrange. Let us take the quantity we wish to optimize, $\ln \pi_*$, and add to it the constraint (7.44) after multiplying by an arbitrary constant λ, a *Lagrange multiplier*. The result is the modified performance index

$$\ln \pi_* = \sum_{k=1}^{N} \left(\ln \pi_k + \lambda \left\{ \frac{V_*}{N} + V_{ek} \ln \left[\varepsilon_k + (1 - \varepsilon_k)\pi_k \right] \right\} \right) \tag{7.47}$$

We have not changed the value of $\ln \pi_*$ in the slightest, since we have only added a complicated expression for zero to the original performance index.

Now, let us calculate the N partial derivatives of (7.47) as if the individual π_k were independent:

$$\frac{\partial \ln \pi_*}{\partial \pi_k} = \frac{1}{\pi_k} + \frac{\lambda V_{ek}(1 - \varepsilon_k)}{\varepsilon_k + (1 - \varepsilon_k)\pi_k} = 0 \tag{7.48}$$

We would normally set these equal to zero to find the optimal solution. One π_k is not independent from the others, but we still have at our disposal the unspecified quantity λ. Let us choose λ so that (7.48) for the nonindependent π_k is zero anyway. The remaining $N-1$ equations (7.48) can also be set to zero, since these π_k are now independent variables. This means that we can treat the N equations (7.48) on an equal basis, without having to specify the offending π_k. We can now solve for the desired optimal payload ratios

$$\pi_k = \frac{-\varepsilon_k}{(1 - \varepsilon_k)(1 + \lambda V_{ek})} \tag{7.49}$$

in terms of known parameters and the unknown value of λ.

Although we have appended the constraint (7.44) to the performance index, only partial derivatives of the constraint appear in (7.48) and (7.49). In particular, the value of V_* does not appear. If we insert the optimal payload ratios (7.49) into the constraint (7.44), we find

$$V_* = -\sum_{k=1}^{N} V_{ek} \ln \left(\varepsilon_k - \frac{\varepsilon_k}{1 + \lambda V_{ek}} \right) \tag{7.50}$$

This is a single, very nonlinear equation for the unknown constant λ. Once it is solved by numerical techniques, the λ value can be substituted into (7.48), and the optimal payload ratios are immediately found.

There is one special case in which we can obtain a closed-form result. If each stage uses the same propulsion and structures technology, then all V_{ek} are the same, V_e, and all ε_k are the same, ε. Each term in the sum (7.50) is then identical, and the value of λ is found to be

$$\lambda = \frac{e^{-\beta}}{V_e(\varepsilon - e^{-\beta})} \tag{7.51}$$

where $\beta = V_*/NV_e$. The optimal payload ratio is then given by

$$\pi_k = \frac{e^{-\beta} - \varepsilon}{1 - \varepsilon} \tag{7.52}$$

These are all equal, since k does not appear on the right side of (7.52). So, when the same engines and structures technology are used in each stage, each stage is a reduced "photocopy" of the stage below. The reduction factor is a constant.

After the initial design of a vehicle, it is very common for rockets to "grow" as modifications and improvements are made, and the initial optimal design is soon lost. For example, the Thor Delta has gone through at least 30 different incarnations since 1960. However, occasionally one can find examples of the optimal design we have been discussing. For example, the Athena sounding rocket, a two-stage solid-fuel vehicle, was designed to be an optimal booster with a payload of about 250 kg (554 lb). This does not mean that the Athena achieves its highest altitude with a payload of 554 lb. It achieves the highest possible altitude with *zero* payload. It is at its most efficient with this payload, however. Nor has the Athena proved immune to modifications. It was used as the third and fourth stages of the four-stage Scout satellite booster, and in this use it is not optimal.

7.8 THE SOUNDING ROCKET

The simplest rocket trajectory is that of the sounding rocket, which follows a vertical path. Figure 7.8 shows the free-body diagram of a such a vehicle, with gravity and thrust the only forces acting. The equations of motion for this system are

$$\frac{dH}{dt} = V \tag{7.53}$$

$$m\frac{dV}{dt} = -mg - \frac{dm}{dt}V_e \tag{7.54}$$

where H is the altitude and V is the vertical speed. After burnout, of course, the thrust term disappears from (7.54). We will assume that the thrust and hence the mass flow rate dm/dt are constant. (This is the *optimal* thrust profile for a sounding rocket, but we do not prove that here.) Assume that the constant mass flow rate is $dm/dt = \beta$.

The variables in (7.54) can be separated to yield the definite integrals

$$\int_0^V dV = -g\int_0^t dt - \int_{m_0}^m \frac{dm}{m} \tag{7.55}$$

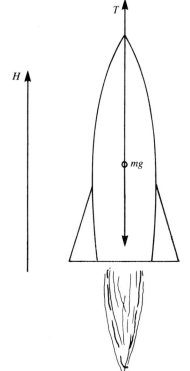

FIGURE 7.8
Forces on a sounding rocket.

When this is integrated, we have a modified form of the familiar rocket equation:

$$V = -gt - V_e \ln \frac{m}{m_0} \qquad (7.56)$$

The extra term in (7.56) represents the velocity losses due to gravity during powered flight. Now, since the mass flow rate is constant, the mass is given by

$$m = m_0 - \beta t \qquad (7.57)$$

during powered flight. Alternately, the time can be expressed as

$$t = \frac{m_0 - m}{\beta} \qquad (7.58)$$

while the vehicle is in boost. Now that the speed is known as a function of time, we can substitute (7.56) and (7.57) into the equation of motion for the altitude (7.53) to find

$$\frac{dH}{dt} = V = -gt - V_e \ln \frac{m_0 - \beta t}{m_0} \qquad (7.59)$$

Again the variables can be separated, giving

$$\int_0^H dH = -g \int_0^t t \, dt - V_e \int_0^t \ln \frac{m_0 - \beta t}{m_0} \, dt \qquad (7.60)$$

These integrals are also simple to perform. When the time is eliminated in favor of the mass via (7.58), we have the altitude as a function of vehicle mass as

$$H = -\frac{g}{\beta^2}(m - m_0)^2 - \frac{V_e}{\beta}\left(m \ln \frac{m}{m_0} + m_0 - m\right) \qquad (7.61)$$

In particular, if the mass at burnout is m_f, equations (7.56) and (7.60) give the burnout velocity V_{bo} and the burnout altitude H_{bo} as

$$V_{bo} = -\frac{g}{\beta}(m_f - m_0) - V_e \ln \frac{m_f}{m_0} \qquad (7.62)$$

and

$$H_{bo} = -\frac{g}{\beta^2}(m_f - m_0)^2 - \frac{V_e}{\beta}\left(m_f \ln \frac{m_f}{m_0} + m_0 - m_f\right) \qquad (7.63)$$

After burnout, the mass of the rocket is constant, and we must modify the original equations of motion (7.53) and (7.54) to follow the trajectory in the free-flight phase. Without the thrust term or air drag, these equations imply that energy is constant:

$$E = \frac{1}{2}m_f V^2 + m_f g H \qquad (7.64)$$

The total energy E can be evaluated at the burnout point using (7.62) and (7.63), with the result

$$E = -\frac{m_f g V_e}{\beta}\left(m_f \ln\frac{m_f}{m_0} + m_0 - m_f\right) + \frac{1}{2}m_f V_e^2 \ln^2\frac{m_f}{m_0} - \frac{1}{2}\frac{m_f g^2}{\beta^2}(m_f - m_0)^2$$

(7.65)

Equation (7.64) will then give the altitude as a function of velocity throughout the free flight of the vehicle. In particular, since the maximum altitude of the rocket occurs when the speed V goes to zero, the maximum altitude achieved is given by

$$H_{\max} = \frac{1}{2}\frac{V_e^2}{g}\ln^2\frac{m_f}{m_0} - \frac{1}{2}\frac{g}{\beta^2}(m_f - m_0)^2 - \frac{V_e}{\beta}\left(m_f \ln\frac{m_f}{m_0} + m_0 - m_f\right) \quad (7.66)$$

The last two terms in the above expression both depend on the engine mass flow rate β. Since $m_f < m_0$, the coefficients of both these terms are negative: They reduce the maximum achievable altitude. Both these terms can be eliminated if we make the mass flow rate β infinite. This corresponds to burning *all* the fuel while just leaving the launchpad, or an infinite thrust-to-weight ratio. This is the most efficient program for a sounding rocket in a vacuum, since no penalty is paid to transport unburned fuel to high altitude. However, when the effects of air drag are included, this ceases to be true. It is then desirable to postpone achieving very high velocities until some of the atmosphere is below the vehicle. So, in this case there will be an optimum thrust-to-weight ratio, striking a balance between the gravity losses due to lifting unburned fuel and the air drag losses due to high velocity at low altitude. It is common for sounding rockets to accelerate at 5 to 10 g when leaving the pad. In this instance, the "ideal" case is closely followed. Also, modern Fourth of July skyrockets are now actually mortar shells. Burning all the propellant charge at ground level increases the maximum altitude achieved.

7.9 GRAVITY-TURN TRAJECTORIES

Since the performance of a rocket is governed by the amount of fuel it carries, a premium is placed on reducing the structural mass of the vehicle. With structural mass reduced to the absolute minimum, it is not really possible to make the booster a strong, rigid structure in all directions. Since it must, of course, carry large loads in its axial direction, strength in the transverse direction is sacrificed when necessary to provide axial strength. The result is that most large rockets are incapable of flying through the atmosphere at an angle of attack. The aerodynamic loads generated by attempting this at several times the speed of sound could result in the catastrophic failure of the booster

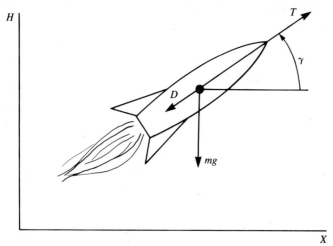

FIGURE 7.9
Forces acting on a satellite booster.

structure. It follows that the thrust vector must be kept aligned with the velocity vector of the vehicle at all times. However, during flight into orbit, the vehicle must obviously be rotated from its vertical position at launch to a horizontal position at burnout. This is done automatically by the dynamics in what is termed a *gravity-turn trajectory*.

Figure 7.9 shows the booster in flight over the earth, with the forces acting on it. Thrust and drag act along the vehicle axis, while the force of gravity is modified to include the apparent "centrifugal force." With this effect included, we may treat flight over a spherical earth as if the earth were flat. The inertial-acceleration components of the vehicle are dV/dt along the vehicle axis and $V\,d\gamma/dt$ transverse to this axis, where V is the vehicle speed and γ is the vehicle flight-path angle. The equations of motion then become

$$\frac{dX}{dt} = V \cos \gamma \tag{7.67}$$

$$\frac{dH}{dt} = V \sin \gamma \tag{7.68}$$

$$m\frac{dV}{dt} = T - D - \left(mg - \frac{m\dot{X}^2}{R+H}\right)\sin \gamma \tag{7.69}$$

$$mV\frac{d\gamma}{dt} = -\left(mg - \frac{m\dot{X}^2}{R+H}\right)\cos \gamma \tag{7.70}$$

where R is the radius of the earth. These equations must be supplemented with statements of how the mass of the vehicle changes with time, how the thrust changes with time, and the drag law

$$D = \tfrac{1}{2}C_d A \rho V^2 \qquad (7.71)$$

where ρ is the atmospheric density, A is the frontal area of the booster, and the coefficient of drag C_d will be a function of vehicle speed and configuration also.

   These equations have no known analytic solution when the mass varies with time and the air drag varies with speed. However, we can discuss the main features of the solution. Since we start in a vertical position, $\gamma = 90°$, equation (7.70) shows that γ does not change. The rocket will continue in vertical flight until it is nudged away from this attitude, either deliberately or accidentally. Once this happens, (7.70) shows that $d\gamma/dt$ is *negative*; that is, the vehicle will fall over. If this were done on the pad, only a few seconds would elapse until the booster was lying on the ground. However, at high speed it is not the vehicle's axis being turned by gravity, since the vehicle control system is attempting to maintain zero angle of attack, but its *momentum vector*. The effective force of gravity decreases rapidly with vehicle speed as the centripetal acceleration approaches g, while the momentum vector of the vehicle rapidly becomes large. So, it takes a long time to complete the rotation to a horizontal position. This rotation is performed by forces acting through the center of mass, so there is no torque on the booster. The booster's control system is then free to concentrate on maintaining a zero angle of attack. The trick in a gravity-turn trajectory is to nudge the vehicle by just the right amount so that it reaches the horizontal position just at burnout.

   This is a *boundary value problem* in the theory of differential equations. After a few seconds of vertical flight, when the vehicle is clear of the tower, the "pitch program" is begun. The vehicle is nudged away from the vertical by a small, and at this point unknown, amount. At the pitchover point, we know the small initial altitude H_0, the speed V_0, but we as yet do not know the initial flight-path angle γ_0. After a known interval of time, the booster has exhausted its fuel and we wish to be in a specified orbit. Assuming we want a circular orbit at altitude H_f, the burnout velocity must be given by

$$V_f = \left(\frac{GM}{R + H_f} \right)^{1/2} \qquad (7.72)$$

where M is the mass of the earth. The burnout-velocity condition would normally be met by adjusting the mass of the payload carried by the booster along this particular trajectory. Also, to ensure that the orbit is circular, we wish to achieve a final flight-path angle $\gamma = 0$ at burnout. However, we do not care about the downrange distance X_f at burnout. Thus, several conditions are specified at each end of the trajectory.

   Beginning shortly after lift-off, we may pick a reasonable altitude and speed to perform the pitch maneuver. It is best to do the initial pitchover at low speed, since the vehicle will briefly have an angle of attack during this time. However, it cannot be done at zero speed, since the vehicle will simply fall over. We can choose a trial value γ_0 for the initial flight-path angle at the pitchover point and integrate the equations of motion (7.67) to (7.70) through the various rocket stages until burnout. At the burnout point, the final booster conditions can be compared with the desired burnout conditions. The trial value of γ at the pitchover point can then be modified to more closely achieve the desired final conditions. Now, since we only have one quantity to modify at the start of the trajectory (γ_0), we only have at our disposal one condition at the end of the trajectory. Say we adjust γ_0 to achieve $\gamma_{bo} = 0$ at burnout. Then, the burnout velocity will be determined by the amount of payload and fuel carried by the booster. Modifying these last two quantities, we may achieve level flight, orbital speed, and burnout altitude simultaneously. Thus, the actual payload capability of the booster is determined by the trajectory boundary-value-problem calculation. Of course, the higher the desired orbit, the lower the payload.

Figure 7.10 shows three sample gravity-turn trajectories for the von Braun three-stage shuttle design discussed in an earlier section. Staging points are indicated by crosses. The lower plot uses the same vertical and horizontal scales, while the upper plot shows the same trajectories with the vertical axis

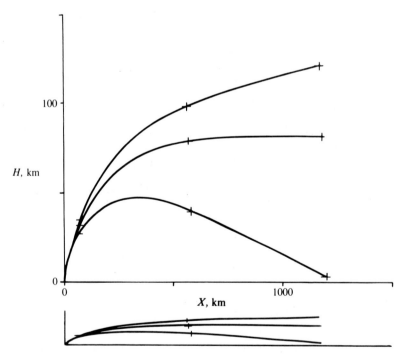

FIGURE 7.10
Gravity-turn trajectories.

enlarged. In the higher trajectory, the initial kick angle was too close to 90°, resulting in a lofted trajectory which does not level out. The lowest curve represents a trajectory achieving burnout just above ground level (moving at 8.5 km/s). In the middle trajectory, the flight-path angle goes to zero just at burnout. The three values of γ_0 differ by less than 1°. In this trajectory, most of the turning occurs in the earliest part of the first-stage burn. It is essential to get the vehicle over on its side early in the trajectory so that the vehicle is no longer fighting gravity and can build up horizontal speed rapidly. As $\dot{X}^2/(R+H) \to g$ and V becomes large, the rate of change of the flight-path angle becomes small. So, the upper stages are able to follow an almost horizontal path, the best possible conditions for building up orbital velocity. Note that all three paths become straight lines (over the "flat" earth) as the third stage approaches satellite velocities.

A gravity-turn trajectory directly into a high-altitude orbit is inefficient. In this case it is better to inject into the ascending branch of a Hohmann-transfer ellipse and circularize the orbit one-half revolution later. This, of course, requires some method for attitude control and either a separate kick stage or a restartable booster. Since the final burnout direction is directly opposite to the direction of the kick maneuver, spin stabilization is a possibility here. If the booster is really an ICBM, the only changes necessary are in the conditions at burnout. The booster must achieve some nonzero value of γ_{bo} at a given speed. Other than this, there is no difference in the ascent-trajectory calculation.

The booster does not have to fly the entire powered-flight trajectory without closed-loop control. After the initial kick maneuver, and while the vehicle is still within the atmosphere, the guidance system must concentrate on maintaining zero angle of attack. Any errors which appear in the actual path of the vehicle may be noted but cannot be corrected at this point. However, once out of the atmosphere, the booster can, if necessary, fly at an angle of attack, and the guidance unit can be allowed to correct position and velocity errors in the actual booster state. This is a closed-loop guidance problem. But as burnout is approached the structural problem reappears. If the vehicle were allowed to attempt to correct a 100-m position error when very close to the planned burnout position, spectacular maneuvering would be the result. The same effect occurs if errors in the direction of the velocity vector are still present just before burnout. So, as the moment of burnout nears, the guidance unit ignores position- and velocity-direction errors and turns its entire attention to obtaining the correct speed at burnout.

Finally, it should be mentioned that the gravity-turn trajectory is *not* the optimal trajectory for a satellite launch vehicle. The optimal trajectory requires flight at an angle of attack, and a nonvertical launch angle, both of which are impractical for launch vehicles starting from the earth's surface. Only the Apollo lunar module, able to perform its launch into lunar orbit in a pure vacuum, has been able to use the optimal (minimum fuel) launch trajectory.

7.10 THE AEROSPACE PLANE

Recently, interest in using airbreathing propulsion to reach orbital speeds has greatly increased. The United States has begun a research project aimed at producing a demonstration hypersonic aircraft, the X-30, with the hope that it could attain speeds which are a substantial fraction of orbital velocity. Although such a vehicle could not operate in space without auxiliary rockets, the hope is that it could substantially decrease the very high cost of delivering payloads to low earth orbit. Figure 7.11 shows such a vehicle moving at speed v through the atmosphere. Surround it with an imaginary surface, defining what is termed the *control volume*. Unlike a rocket which simply exhausts mass, the aerospace plane will both take mass in and exhaust mass. However, just as with the rocket, the equation of motion can be found by carefully following the momentum flow through the control volume.

It is simplest to think of the momentum transfer as a two-step process. Assume that the control volume moves at the speed v of the vehicle. In the first step, the engines take in a mass of air Δm_{in}, leading to a momentum transfer of

$$p_{in} = -\Delta m_{in} v \qquad (7.73)$$

This is negative, since the vehicle is assumed to be at rest (for the moment) and the air is moving to the left in Figure 7.11. In the second stage of the process, an amount of fuel Δm_f is added to the air and the resulting mixture is burned and exhausted out the rear of the vehicle at speed V_e with respect to the vehicle. The vehicle itself increases its speed from v to $v + \Delta v$ at this time. The momentum change in the second step is then

$$p_{out} = (m - \Delta m_f)\,\Delta v - (\Delta m_{in} + \Delta m_f)V_e \qquad (7.74)$$

However, this is a closed system, and it must conserve total linear momentum. If we equate (7.73) to (7.74) and divide by the short time interval Δt, we obtain

$$m\frac{\Delta v}{\Delta t} = \frac{\Delta m_{in}}{\Delta t}V_e + \frac{\Delta m_f}{\Delta t}V_e + \frac{\Delta m_f}{\Delta t}\frac{\Delta v}{\Delta t} - \frac{\Delta m_{in}}{\Delta t}v \qquad (7.75)$$

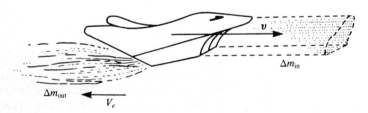

FIGURE 7.11
Momentum conservation for an aerospace plane.

Taking the limit, and allowing for the fact that there may be other forces F_{ext} on the vehicle, we have

$$m\frac{dv}{dt} = (\dot{m}_{in} + \dot{m}_f)V_e - \dot{m}_{in}v + F_{ext} \qquad (7.76)$$

In (7.76) we can recognize the usual rocket term $\dot{m}_f V_e$. However, this is usually the least important term for an airbreathing vehicle. If hydrogen is the fuel, then the combustion product is water vapor, and the fuel mass is only one-ninth the mass of oxygen the engine uses. But oxygen makes up only 21% of the atmosphere, so we would have $\dot{m}_f/\dot{m}_{in} \approx 0.023$ if the engine runs at the *stoichiometric ratio* of fuel to air. The rocketlike term in (7.76) is thus not the dominant contributor to the vehicle thrust.

Now, replace the quantity $\dot{m}_{in}$ with $\rho A v$, where ρ is the air density, A is the inlet area, and v is the speed. Expression (7.76) then becomes

$$m\frac{dv}{dt} = (\rho A v + \dot{m}_f)V_e - \rho A v^2 + F_{ext} \qquad (7.77)$$

To get a positive thrust from (7.77), it is necessary that $V_e > v$. If this is not true, the *ram drag* term $\rho A v^2$ exceeds the positive contribution from the exhaust, $\rho A v V_e$. As the worst possible case, assume that the incoming airflow is decelerated within the vehicle and then burned with the fuel. Then the engine could have an exhaust speed no higher than a hydrogen-oxygen rocket, and the ram drag and exhaust terms would balance when $v = V_e$. This is essentially what happens in a *turbojet* or a *ramjet*, where the incoming airflow is decelerated to subsonic speeds. If very high Mach numbers are to be obtained, the incoming airflow cannot be slowed down but must be allowed to proceed through the engine unhindered. This occurs in the *scramjet*, or supersonically combusting ramjet. However, this introduces new problems. Since the airflow through the engine is supersonic, it does not "know" where the engine is located, since it cannot communicate with the engine walls by pressure waves. The injected fuel might burn within the engine or several hundred meters behind the vehicle. In the latter case, of course, it would produce no thrust.

The dynamics of hypersonic flight are much different from those of rocket flight. A rocket attains high speeds by exiting the atmosphere quickly and rotating into a horizontal attitude. This completely eliminates drag after the first two or three minutes of flight. As an aerospace plane achieves high speeds, the net thrust (the difference between the thrust and ram drag terms) becomes relatively small, probably only about 10% at Mach 10. This does not leave much margin for external forces F_{ext}. In particular, there cannot be any extra drag in F_{ext}, so the aerospace plane must be essentially all inlet seen from its front. A sudden flameout of the engines would remove the thrust term, but not the ram drag, and the vehicle would undergo catastrophic deceleration. Combined with a very severe heating environment, these problems may make achieving a practical vehicle quite difficult.

7.11   THE RELATIVISTIC ROCKET

The rocket law is not obeyed when a vehicle approaches the speed of light. As first discovered by Albert Einstein in 1908, Newton's laws are incorrect when the velocity of a particle approaches the speed of light. We ignore the mythical "warp drive" as a means to approach the speed of light, since it clearly violates Newton's third law. One foreseeable technology is a matter-antimatter photon rocket. In such a vehicle, matter and antimatter would be allowed to annihilate each other, producing a shower of gamma rays. These would be focused (somehow) to produce thrust on the rest of the vehicle. The burnout velocity of this vehicle can be calculated by applying energy and momentum conservation laws in their relativistic form.

Assume that the fully fueled vehicle has rest mass m_0 and that a fraction π of this is payload. (No allowance is made for tankage. One advantage of the matter-antimatter drive is that the vehicle can consume its own tankage as fuel.) So, before ignition, the total linear momentum of the system p_{tot} is zero, while the total rest energy is given by the famous

$$E_{tot} = m_0 c^2 \qquad (7.78)$$

where c is the speed of light. After burnout, the system consists of two components. The vehicle, now with rest mass πm_0, is moving at velocity v with respect to its original rest frame, Figure 7.12. There is also a considerable quantity of radiation propagating in the opposite direction at the speed of light. Now, energy and momentum are related by $E_{rad} = p_{rad} c$ for radiation. Since c is large, a large amount of energy yields only a small amount of momentum. Also, kinetic energy and linear momentum are given by the expressions

$$E = \frac{mc^2}{\sqrt{1 - v^2/c^2}} \qquad (7.79)$$

$$p = \frac{mv}{\sqrt{1 - v^2/c^2}} \qquad (7.80)$$

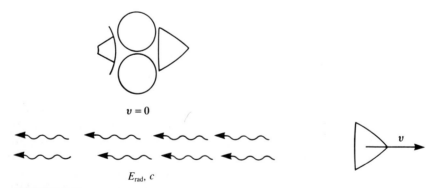

$v = 0$

E_{rad}, c

v

FIGURE 7.12
Photon rocket before and after burnout.

in special relativity. If the denominators of these expressions are expanded for small v, the usual newtonian formulas are recovered.

So, applying energy conservation before and after burnout, we obtain

$$E_{tot} = m_0 c^2 = \frac{\pi m_0 c^2}{\sqrt{1 - v^2/c^2}} + E_{rad} \tag{7.81}$$

The fact that the total linear momentum sums to zero gives

$$p_{tot} = 0 = \frac{\pi m_0 v}{\sqrt{1 - v^2/c^2}} - \frac{E_{rad}}{c} \tag{7.82}$$

It is customary to put $\beta = v/c$, making β the fraction of light speed, and to abbreviate $\gamma = (1 - v^2/c^2)^{-1/2}$. The momentum equation gives

$$E_{rad} = \pi m_0 \beta c^2 \gamma \tag{7.83}$$

When this is substituted into the energy equation (7.81) to eliminate E_{rad}, we find

$$m_0 c^2 = \pi m_0 c^2 + \pi m_0 \beta c^2 \gamma \tag{7.84}$$

This can be solved for the payload ratio to give

$$\pi = \frac{1}{\gamma(1 + \beta)} \tag{7.85}$$

Note that as $v \to c$, $\gamma \to \infty$ and the payload ratio $\pi \to 0$. The universe's speed limit of c is self-enforcing; there is no penalty for attempting to violate it except failure.

For example, suppose we put $\gamma = 10$, which implies that $\beta = 0.9949$, or over 99% of the speed of light. The payload ratio for such a vehicle is $\pi \cong 0.05$. This looks very promising until we realize that this is only sufficient to accelerate to this speed *once*. A round-trip would include two starts and two stops. Using staging, the overall payload ratio would then be

$$\pi_* = \pi^4 \cong 6 \times 10^{-6} \tag{7.86}$$

which is not as encouraging. If we are willing to accept lower burnout velocities, then the overall payload ratio can be made larger.

Finally, we note that γ is the time-dilatation factor. This is the factor by which a moving clock will appear to run slower and the amount by which a moving ruler will appear to contract. To observers left back on the earth, the clocks on the starship will appear to run slower, and the voyage will be completed in a shorter interval of "ship time" than will elapse on the earth. The occupants of the ship believe that nothing is wrong with their clocks, and they explain the time difference as the contraction of the distance to be traveled. Special relativity has been checked by many experiments and is the correct way to treat dynamics near the speed of light.

7.12 REFERENCES AND FURTHER READING

Rocket trajectories are not often discussed outside the propulsion literature. This is in contrast with aeronautical engineering, where aircraft performance is normally considered the province of the dynamicists. Exceptions to this are the books by Miele and by Thomson.

The field of propulsion is too large to cite here, and from our point of view, it does not really matter how the ejected mass is accelerated to exit velocity. However, Sutton is a classic introduction to chemical rockets, while the book by Stuhlinger holds the same place as a first look at more exotic propulsion schemes. After a long period of neglect, both chemical and nonchemical propulsion are undergoing a renaissance, and substantial advances can yet be made in this field.

Miele, A., *Flight Mechanics*, Addison-Wesley, Reading, Mass., 1962.
Thomson, W. T., *Introduction to Space Dynamics*, Wiley, New York, 1961.
Sutton, G. P., *Rocket Propulsion Elements*, Wiley, New York, 1963.
Stuhlinger, E., *Ion Propulsion for Space Flight*, McGraw-Hill, New York, 1964.

7.13 PROBLEMS

1. The total Δv required for transfer from a low earth orbit at $28°$ inclination to geosynchronous equatorial orbit is $V_* = 4.29$ km/s. A space tug with a restartable hydrogen-oxygen engine has an $I_{sp} = 453$ s, $m_p = 16,000$ kg, and $m_s = 1300$ kg. How much payload can the tug deliver to geosynchronous orbit? Can the tug make a round-trip without payload? If it can, how much payload could it carry to geosynchronous orbit and still return empty to low earth orbit?

2. Engine performance may sometimes be upgraded in a multistage rocket. Show that the trade-off ratio for engine improvement in stage k is given by

$$\frac{\partial m_*}{\partial I_{sp,k}} = \frac{-g_c \ln (m_{ok}/m_{fk})}{\sum_{j=1}^{N} V_{ej}(1/m_{oj} - 1/m_{fj})}$$

Evaluate $\partial m_*/\partial I_{sp,0}$ and $\partial m_*/\partial I_{sp,1}$ for the space shuttle. Use the quantities pertaining to individual components (in parentheses) in Table 7.3. Should effort be placed into improving the solid rocket boosters or the main engines first?

3. Calculate the trade-off ratios $\partial m_*/\partial m_{sk}$, $\partial m_*/\partial m_{pk}$, and $\partial m_*/\partial I_{sp,k}$ for the von Braun shuttle. If the I_{sp} could be increased to 295 s in the upper two stages at the expense of a 7% increase in the structural mass of these stages, is the payload increased or decreased? By how much?

4. For the sounding rocket in Table 7.5, calculate the approximate burnout speed (ignoring the effects of burn time) and maximum altitude achieved with a 250-kg payload. Suppose that instead of immediately igniting the second stage when the first burns out, the second stage is allowed to coast to its maximum altitude and is then ignited. Calculate the maximum altitude achieved by the payload for this burn-coast-burn-coast strategy. Why is there such a large difference?

TABLE 7.5
Sounding rocket for
Prob. 4

k	I_{sp}, s	m_p, kg	m_s, kg
1	282	1167	113
2	282	415	41
*			250

5. The solid rocket boosters of the space shuttle have an $I_{sp} = 290$ s, a dry mass of 81,900 kg each, and a full mass of 584,600 kg each. The external tank has a dry mass of 35,000 kg and contains 703,000 kg of propellant. Space shuttle main engines have an $I_{sp} = 455$ s.

Consider a vehicle consisting of *four* solid rocket boosters burning together as the first stage and an external tank with two main engines as the second stage, as shown in Figure 7.13. Approximately how much payload can this vehicle deliver to low earth orbit ($V_* = 31,000$ ft/s)?

(a) Set up the problem explicitly, and put in all numbers.

(b) The mass of the payload can be ignored in some places. Where and why?

(c) Calculate an approximate m_* to low earth orbit and compare to the shuttle payload (approximately 30,000 kg, or 65,000 lb).

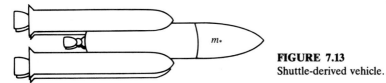

m_*

FIGURE 7.13
Shuttle-derived vehicle.

6. Using data on ε_k and π_k from the tables on the von Braun three-stage rocket and the space shuttle, calculate values of λ for each stage. Is either vehicle close to an optimal design (i.e., are the λ values nearly the same)?

7. Design a cargo third stage for the von Braun three-stage rocket. Use the structural factor ε_2 from the second stage, since this stage does not have wings. What total payload could it deliver to orbit?

CHAPTER

8

REENTRY
DYNAMICS

8.1 INTRODUCTION

The problem of surviving reentry does not arise in every space mission. Most earth satellites are given a one-way ride out of the earth's atmosphere. Their orbits are such that reentry will occur long after the useful lifetime of the satellite is exceeded, or in the case of very high orbits, reentry may never occur. In either case, the spacecraft itself is not designed to survive reentry forces and heat loading. However, there are three cases where surviving reentry is of paramount importance. The first of these is the ballistic missile. Problems with reentry first appeared in the German V-2 program, as rockets insisted upon exploding as they hit the atmosphere. The first reentry-vehicle designs to actually be tested were ballistic-missile warheads. The second case is the planetary-entry probe. These vehicles are similar in design to the ballistic-missile warhead but have more peaceful intent. Third, but perhaps first in the public mind, is the manned spacecraft. These latter vehicles must *always* make provision for surviving reentry.

The fundamental problem with reentry is the amount of kinetic energy per unit mass possessed by the spacecraft. This is simply $T = v^2/2$, and if we insert for v orbital velocity, then the energy content of the spacecraft is more than sufficient to melt and vaporize any known material. A significant fraction of the total energy of the launch vehicle has been concentrated in the spacecraft, and this energy must be dissipated during reentry. The spacecraft cannot be slowed to a stop with a rocket, since this would require a booster the size of the original launch vehicle, and then the booster required to orbit *that* assembly would be prohibitively large. Instead, atmospheric drag is used to slow the vehicle. The rate at which drag dissipates energy is simply $\dot{T} = \mathbf{D} \cdot \mathbf{v}$, from the work-energy relation. Since drag is proportional to the atmospheric density ρ times v^2, the rate of energy dissipation is proportional to ρv^3.

There are two approaches for enduring the high temperatures generated by friction during atmospheric entry. The first approach involves the vehicle aerodynamics and the technology of the heat shield: How does the vehicle endure and dispose of the frictional heating? This approach is the province of aerodynamics and heat transfer. The second approach involves the design of the reentry trajectory: By appropriate use of the dynamics the reentry can be stretched over a long period of time, under favorable conditions for disposing of waste heat. This last topic is the main concern of this chapter.

8.2 BALLISTIC REENTRY

In this section we will consider the reentry of a ballistic vehicle without lift for the case of steep entry angles. Figure 8.1 shows the forces of gravity and air drag acting on such a vehicle. The equations of motion are

$$\frac{dX}{dt} = V \cos \gamma$$

$$\frac{dH}{dt} = V \sin \gamma$$

$$m \frac{dV}{dt} = -D - mg \sin \gamma \tag{8.1}$$

$$mV \frac{d\gamma}{dt} = -mg \cos \gamma$$

where the flight-path angle γ is negative. Here, X is the downrange distance, H is the altitude, and the in-track and cross-track acceleration components are dV/dt and $V \, d\gamma/dt$. For high reentry speeds, the quantity g should be the apparent gravitational acceleration $g = g_c - V^2(\cos^2 \gamma)/r$, including the centripetal acceleration.

These equations of motion become more tractable if we switch to the altitude H as the independent variable. This will enable us to solve for the shape of the trajectory as a function of altitude. To change to altitude H as the new independent variable, divide each of equations (8.1) by $dH/dt = V \sin \gamma$.

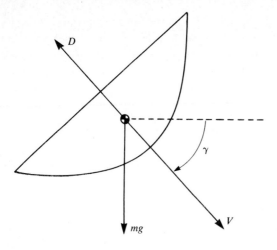

FIGURE 8.1
Forces on a ballistic-reentry vehicle.

The new equations of motion become

$$\frac{dX}{dH} = \cot \gamma \tag{8.2}$$

$$mV\frac{dV}{dH} = -\frac{D}{\sin \gamma} - mg \tag{8.3}$$

$$\frac{d\gamma}{dH} = -\frac{g \cot \gamma}{V^2} \tag{8.4}$$

For the case of steep entry angles, the cotangent is small. Also, the velocity is likely to be large during most of the trajectory. So, as a first approximation we will assume that γ changes relatively little, as implied by (8.4). Equation (8.2) then states that the trajectory is a straight line. We will concentrate first on the deceleration of the vehicle as given by (8.3).

8.2.1 Vehicle Deceleration

Assuming an exponential atmosphere and hypersonic flight, the drag force D is given by

$$D = \tfrac{1}{2}C_D A V^2 \rho_0 e^{-H/H_0} \tag{8.5}$$

We will assume that the drag coefficient C_D is constant. Inserting this into (8.3) and rearranging, we find

$$mV\frac{dV}{dH} = -mg + \frac{K_D}{H_0}\left(\tfrac{1}{2}mV^2\right)\rho_0 e^{-H/H_0} \tag{8.6}$$

We have put

$$K_D = -\frac{C_D A H_0}{m \sin \gamma_i} \tag{8.7}$$

which has units of inverse mass density. The entry flight-path angle γ_i is assumed constant throughout the reentry trajectory, so K_D is constant. The quantity K_D is positive, since γ_i is negative. The form of (8.6) suggests changing variables from the velocity V to the kinetic energy $T = \frac{1}{2}mV^2$, since the left side of (8.6) is the derivative of this quantity. The kinetic-energy equation of motion then becomes

$$\frac{dT}{dH} = -mg + \frac{K_D}{H_0}\rho_0 e^{-H/H_0} T \qquad (8.8)$$

We now change independent variables again, replacing the altitude H with the atmospheric density $\rho = \rho_0 e^{-H/H_0}$. Since

$$\frac{d\rho}{dH} = -\frac{\rho_0}{H_0} e^{-H/H_0} = -\frac{\rho}{H_0} \qquad (8.9)$$

the behavior of kinetic energy with air density is given by

$$\frac{dT}{d\rho} + K_D T = \frac{mgH_0}{\rho} \qquad (8.10)$$

This equation is a linear, constant-coefficient differential equation with a gravity-forcing term.

The solution to (8.10) consists of the solution to the homogeneous part plus a particular solution. The homogeneous equation easily separates to give

$$\int_{B_0}^{T} \frac{dT}{T} = -\int_{0}^{\rho} K_D \, d\rho \qquad (8.11)$$

where we have put the air density equal to zero at the initial point of the trajectory. The homogeneous solution is

$$T_h = B_0 e^{-K_D \rho} \qquad (8.12)$$

where B_0 is a constant of integration. The particular solution is given by

$$T_p = +mgH_0 Ei(K_D \rho) e^{-K_D \rho} \qquad (8.13)$$

where the *exponential integral function* is defined as

$$Ei(x) = \int_{-\infty}^{x} \frac{e^x}{x} \, dx \qquad (8.14)$$

Although most pocket calculators do not have a button with this label, this function can be easily handled by numerical techniques. For example, the exponential integral function can be expanded in a Taylor's series as

$$Ei(x) = \gamma_E + \ln x + \sum_{n=1}^{\infty} \frac{x^n}{nn!} \qquad (8.15)$$

where $\gamma_E \approx 0.5772156$ is the Euler constant. The correctness of the particular solution can be verified by direct substitution, using the relation $d/dx \, Ei(x) =$

e^x/x. The complete solution for the energy then becomes

$$T = [B_0 + mgH_0 Ei(K_D\rho)]e^{-K_D\rho} \tag{8.16}$$

For small x, $Ei(x) \approx \gamma_E + \ln x$. So, evaluating (8.16) at high altitude, the constant B_0 is

$$B_0 = \tfrac{1}{2}mV_0^2 - mgH_0[\gamma_E + \ln(K_D\rho)] \tag{8.17}$$

Then, transforming all the way back to our original variables, the behavior of velocity with altitude is given by

$$\tfrac{1}{2}mV^2 = [B_0 + mgH_0 Ei(K_D\rho_0 e^{-H/H_0})] \exp(-K_D\rho_0 e^{-H/H_0}) \tag{8.18}$$

Note that the velocity depends on the exponential of an exponential function. The velocity is thus *very* sensitive to altitude, drag factor, and other parameters within these functions.

We are now prepared to evaluate the deceleration experienced by the reentry body. This is a quantity of great interest to those who must design the structure and contents of such a vehicle. Since the vehicle structure does not sense the acceleration of gravity while in free-fall, it is possible to ignore the gravity terms in (8.3). The aerodynamic deceleration $a = D/m$ can be found from (8.5) and (8.18) without the gravity term as

$$a = -\frac{K_D \sin \gamma_i B_0}{mH_0}\rho e^{-K_D\rho} \tag{8.19}$$

The maximum vehicle deceleration is easily found by setting $da/d\rho = 0$, which gives the result that the maximum deceleration occurs when $\rho = 1/K_D$. Of course, if $\rho_0 < 1/K_D$, the analytic maximum deceleration does not occur above ground level. Then, for reasons not included in the original equations of motion, the actual maximum deceleration will occur precisely at ground level. Inserting this value of ρ into (8.19), the maximum acceleration of the vehicle is given by

$$a_{max} = \frac{V_0^2 \sin \gamma_i}{2H_0 e} \tag{8.20}$$

Note that this does not depend on the drag parameter K_D.

8.2.2 Trajectory Curvature: Small K_D

The case of low-drag reentry bodies is important for the case of the ballistic missile. Low-drag configurations are used since this minimizes the effects of trajectory curvature and thus reduces any possible targeting error caused by local winds and atmospheric conditions. By not slowing down much, this configuration also minimizes atmospheric heating and deceleration structural loading. The reentry vehicle would, however, experience trouble if it hit a raindrop at Mach 15. Also, since the vehicle arrives at the surface at high velocity, this approach is not useful for manned capsules. Inserting our solution (8.18) for the velocity into (8.4), expanding the exponential and *Ei* functions for small K_d, and using the effective gravity, we obtain

$$\frac{d\gamma}{dH} \approx -\frac{mg \cot \gamma}{2}\left[\frac{1}{2}mV_0^2(1 - K_D\rho_0 e^{-H/H_0})\right]^{-1}$$

$$\approx -\frac{\cot \gamma_i}{V_0^2}\left[\left(g_c - \frac{V_0^2 \cos^2 \gamma_i}{R}\right) + K_D g_c \rho_0 e^{-H/H_0}\right] \qquad (8.21)$$

using the binomial theorem to expand the denominator above. Now, make the usual assumption in perturbation theory that the right side of the equation of motion may be evaluated on the approximate solution: constant γ_i with velocity dependence given by (8.18). Integrating (8.21), the trajectory flight-path angle is given as a function of altitude by the approximate result

$$\gamma \approx \gamma_i - \frac{\cot \gamma_i}{V_0^2}\left[\left(g_c - \frac{V_0^2 \cos^2 \gamma_i}{R}\right)(H - H_i) - K_D g_c \rho_0 H_0 e^{-H/H_0}\right] \qquad (8.22)$$

We have assumed that the initial altitude H_i is high enough that the exponential term may be neglected. The first new term in the above is the curvature of the trajectory from gravity, while the second term is the additional curvature due to the deceleration of the vehicle. For steep entry angles, large reentry velocities, and small drag factor, these two additional terms will be small.

Now, expanding the cotangent function in a Taylor's series about γ_i, we have $\cot \gamma \approx \cot \gamma_i - \csc^2 \gamma_i \, \delta\gamma$, where $\delta\gamma$ represents the correction terms in (8.22). Inserting this result with (8.22) into the equation of motion (8.2) for X, there results

$$\frac{dX}{dH} \approx \cot \gamma_i + \frac{\cos \gamma_i}{V_0^2 \sin^2 \gamma_i}\left[\left(g_c - \frac{V_0^2 \cos^2 \gamma_i}{R}\right)(H - H_i) - K_D g_c \rho_0 H_0 e^{-H/H_0}\right] \qquad (8.23)$$

Integrating, the trajectory is given by

$$X = X_i + \cot \gamma_i(H - H_i)$$
$$+ \frac{\cos \gamma_i}{V_0^2 \sin^2 \gamma_i}\left[\frac{1}{2}\left(g_c - \frac{V_0^2 \cos^2 \gamma_i}{R}\right)(H - H_i)^2\right.$$
$$+ \left. K_D g_c \rho_0 H_0^2 e^{-H/H_0}\right] \qquad (8.24)$$

The first two terms above represent the straight-line trajectory we originally assumed. The third term is the parabolic term introduced by gravity, while the last term represents the effect of aerodynamic drag.

Figure 8.2 shows several reentry trajectories for K_D values of 0, 20, and 100 m³/kg. The trajectory with $K_D = 0$ does not suffer from air drag, since the reentry vehicle is infinitely dense. All trajectories began together at an altitude of 70 km, but the figure only shows the lowest 15 km of altitude before impact.

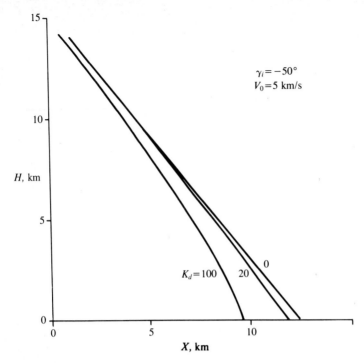

FIGURE 8.2
Low-drag ballistic-reentry trajectories.

8.2.3 Free-Fall: High K_D

Planetary-entry probes are designed to have high values of K_D and usually begin atmospheric entry at a shallow angle. According to (8.20), this produces a smaller value for the maximum deceleration and also ensures that this maximum occurs at high altitude. Many such entry vehicles make observations on the chemistry and structure of the atmosphere during their descent, and this is difficult to do through the ionized sheath produced by hypersonic flight. The entry probe must be decelerated rapidly, so that the atmospheric instruments can observe as much of the atmosphere as possible. The trajectory curves over sharply as the vehicle decelerates, and eventually the vehicle is falling freely in the vertical direction.

A high-drag vehicle will approach a quasi-equilibrium state during this last phase of the entry. In this state of near-equilibrium, air drag will almost balance gravity. Returning to (8.3), this implies that

$$mV\frac{dV}{dH} = \tfrac{1}{2}C_D A V^2 \rho_0 e^{-H/H_0} - mg \approx 0 \qquad (8.25)$$

Solving this expression for the velocity, we find

$$V_{\lim} \approx \left(\frac{2mg}{C_D A \rho_0}\right)^{1/2} e^{H/2H_0} \qquad (8.26)$$

Just as in sport parachuting, this is the "terminal velocity" of the reentry vehicle. Note, however, that it is an exponential function of the altitude above the ground. Most parachute jumps occur within one-half scale height of the earth's surface, so the decrease of the terminal velocity with altitude is not readily noticed. It cannot be ignored for a reentry body.

Since the limiting velocity is a function of altitude, the approximation made in (8.25) is not strictly true. The velocity does indeed change with altitude. The approximation is still valid if the limiting velocity is small, in which case dV/dH will be small. Since parachutes cannot be used at very high speeds, (8.26) gives some indication of the altitude at which parachute deployment will become possible.

8.3 SKIP REENTRY

The earliest treatments of the reentry problem envisioned an airplanelike vehicle performing reentry in a very careful manner. Eugen Sanger, a Swiss engineer, originated a concept for an intercontinental bomber during World War II. This vehicle would be boosted into an ICBM-like trajectory from Germany toward the United States. During its first encounter with the atmosphere, it would drop its bombs on the target and then use its wings to pull up, skipping out of the atmosphere back into another lofted ballistic trajectory, Figure 8.3. With an efficient aerodynamic design, it was hoped that the vehicle would be capable of skipping all the way around the earth, returning to its launch point for another mission.

A similar concept was advanced by Walter Hohmann for return to the earth from the moon or the planets. A returning spaceship would encounter the earth's atmosphere at high altitude in a hyperbolic trajectory. During the first pass through the atmosphere the ship would lose enough velocity to air drag to exit the atmosphere on an elliptical orbit. This would lead to another reentry after one period, and the next elliptical orbit would have about the same perigee but a substantially lower apogee, Figure 8.4. Over several such passes, the ship would lose enough velocity that it would be safe to finally reenter and land. In both these concepts, the vehicle would be moving at a substantial portion of satellite velocity. The apparent gravitational force on the vehicle, given approximately by

$$F_g \approx \frac{GM_e m}{r^2} - \frac{mv^2 \cos^2 \gamma}{r} \qquad (8.27)$$

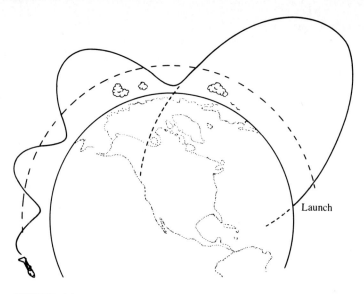

FIGURE 8.3
Sanger antipodal bomber trajectory.

would be small for the Sanger bomber, moving at high suborbital speed, and would be negative (upward) for the Hohmann braking ellipse at perigee. In either case, the apparent force of gravity would be negligible compared to the aerodynamic forces on the vehicle during reentry.

So, we will treat the reentry by assuming a flat earth and zero apparent gravitational force. Describe the velocity vector of the spacecraft by its speed

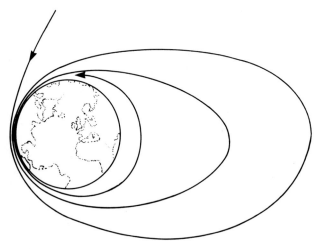

FIGURE 8.4
Hohmann braking ellipses.

V and flight-path angle γ. The equations of motion are

$$\frac{dX}{dt} = V \cos \gamma \qquad (8.28a)$$

$$\frac{dH}{dt} = V \sin \gamma \qquad (8.28b)$$

$$m\frac{dV}{dt} = -D \qquad (8.28c)$$

$$mV\frac{d\gamma}{dt} = L \qquad (8.28d)$$

Here H is the vehicle altitude and X is its downrange distance. The acceleration components are $\dot{V}$ and $V\dot{\gamma}$ resolved along the vehicle axis and transverse to this axis. The drag force D acts along the vehicle axis, and the lift force L on the vehicle acts transverse to the velocity vector. If we assume an exponential atmosphere and hypersonic flight, these forces become

$$D = \tfrac{1}{2}C_D A \rho_0 V^2 e^{-H/H_0} \qquad (8.29)$$
$$L = \tfrac{1}{2}C_L A \rho_0 V^2 e^{-H/H_0} \qquad (8.30)$$

Here C_D and C_L are the drag and lift coefficients, the presented area of the vehicle is A, ρ_0 is the base density of the atmosphere, and H_0 is the atmospheric scale height. Assume that the vehicle is trimmed for its most efficient glide, in which case the presented area A and the lift and drag coefficients C_L and C_D will be constants.

We insert these expressions for the aerodynamic forces into the equations of motion. These can then be integrated if we change to the flight-path angle γ as the new independent variable by multiplying by

$$\frac{dt}{d\gamma} = \frac{2m}{C_L A \rho_0 V e^{-H/H_0}} \qquad (8.31)$$

We expect the flight-path angle γ to be a monotonically increasing function of time during the skip, so it makes an acceptable independent variable. The new equations of motion become

$$\frac{dX}{d\gamma} = \frac{1}{K_L} \cos \gamma \, e^{H/H_0} \qquad (8.32)$$

$$\frac{dH}{d\gamma} = \frac{1}{K_L} \sin \gamma \, e^{H/H_0} \qquad (8.33)$$

$$\frac{dV}{d\gamma} = -\frac{C_D}{C_L} V \qquad (8.34)$$

where the modified lift constant K_L is given by

$$K_L = \frac{C_L A \rho_0}{2m} \qquad (8.35)$$

In this form the equations of motion are easily solved by separation of variables. Equation (8.34) separates to give

$$\int_{V_i}^{V} \frac{dV}{V} = -\frac{C_D}{C_L} \int_{\gamma_i}^{\gamma} d\gamma \qquad (8.36)$$

Here, V_i is the vehicle entry speed and γ_i is the entry angle into the atmosphere, a negative quantity. The solution is simply

$$V = V_i \exp\left[-\frac{C_D}{C_L}(\gamma - \gamma_i) \right] \qquad (8.37)$$

The vehicle will retain more of its initial velocity at exit from the atmosphere if it is an efficient glider (C_L/C_D, called the *lift-to-drag ratio*, is large), or if the skip occurs at a shallow angle.

The altitude equation of motion (8.33) may now be separated to yield

$$K_L \int_{H_i}^{H} e^{-H/H_0} dH = \int_{\gamma_i}^{\gamma} \sin \gamma \, d\gamma \qquad (8.38)$$

Again, the integrals are simple to perform, and the result is easily solved for the altitude dependence

$$H = H_0 \ln \frac{K_L H_0}{\cos \gamma + B} \qquad (8.39)$$

Here,

$$B = K_L H_0 e^{-H_i/H_0} - \cos \gamma_i \qquad (8.40)$$

is the constant of integration evaluated at the atmospheric entry point. It is very desirable that the pull-up be completed before the altitude of the vehicle goes to zero. Of course, the altitude is a minimum when $\gamma = 0$. To keep the argument of the logarithm greater than 1 when $\cos \gamma = 1$, we must require that $1 + B < K_L H_0$. Inserting the value of B from (8.40) and ignoring the usually small exponential term, the lift coefficient must obey the inequality

$$K_L > \frac{1}{H_0}(1 - \cos \gamma_i) \qquad (8.41)$$

Notice that even poor gliders can successfully skip if the entry angle γ_i is very small.

The altitude solution may be inserted into the equation of motion for the horizontal distance flown, (8.32), to find

$$\frac{dX}{d\gamma} = \frac{H_0 \cos \gamma}{\cos \gamma + B} \qquad (8.42)$$

Separating variables,

$$\int_{X_i}^{X} dX = H_0 \int_{\gamma_i}^{\gamma} \frac{\cos \gamma}{\cos \gamma + B} d\gamma$$

$$= H_0 \int_{\gamma_i}^{\gamma} d\gamma - H_0 B \int_{\gamma_i}^{\gamma} \frac{d\gamma}{\cos \gamma + B} \qquad (8.43)$$

when the quantity B is added and subtracted from the numerator of the first integral. The latter integral can take several forms depending on preference and the algebraic sign of B. One form appropriate for $0 \geqslant B > -1$ gives the solution as

$$X = X_i + H_0(\gamma - \gamma_i) - \frac{H_0 B}{\sqrt{1 - B^2}} \ln \frac{\sqrt{1 + B} + \sqrt{1 - B} \tan (\gamma/2)}{\sqrt{1 + B} - \sqrt{1 - B} \tan (\gamma/2)} \Bigg|_{\gamma_i}^{\gamma} \qquad (8.44)$$

At this point we have almost the complete solution to this problem. Equation (8.39) gives the altitude solution for the trajectory, while (8.44) gives the downrange trajectory information, both in terms of the flight-path angle γ as the independent variable. Equation (8.37) specifies the vehicle speed, also as a function of γ. The only missing piece of the solution, then, is the behavior of γ with time, for which we would need to integrate (8.28d). However, we can gain an impression of the behavior of this system by simply plotting the altitude (8.39) versus the downrange distance (8.44). This generates Figure 8.5, showing three skip trajectories for values of $K_L H_0 = 1, 3, 5$. As expected, the trajectories flown by higher-lift vehicles pull up sooner and avoid the densest regions of the earth's atmosphere.

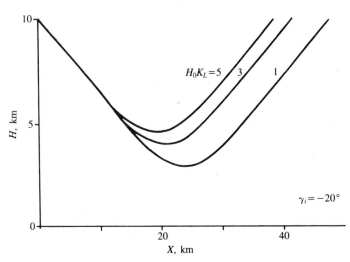

FIGURE 8.5
Skip-reentry trajectories.

Notice from the figure that the exit flight-path angle is the modulus of the entrance flight-path angle. This result can also be shown from (8.39), where the same altitude H can be generated by two values of γ differing only in sign. Thus, in a Sanger bomber trajectory, all the skips begin and end with the same flight-path angle. Once the vehicle has left the atmosphere, it is necessary to return to the two-body problem and calculate the new elliptical orbit for the arc of trajectory outside the atmosphere. Each successive skip will cover a shorter range than its predecessor, since the speed of the vehicle when it leaves the atmosphere drops with each succeeding skip. One practical difficulty with the Sanger skip trajectory is the fact that it is almost impossible to predict more than one or two skips in advance. The final landing point of a vehicle on such a trajectory is thus impossible for the dynamicist to determine, although it would be of great interest to the vehicle's pilot.

8.4 "DOUBLE-DIP" REENTRY

A more sophisticated version of the skip reentry was used by the American Apollo and Soviet Zond capsules during their return from the moon. The Apollo capsule had its center of mass offset from its centerline, causing it to fly with a slight angle of attack, thus producing lift. The first phase of the reentry was flown with the lift vector up and was a skip trajectory as discussed in the last section. However, the Apollo capsule would have left the atmosphere still above escape speed if the skip was flown to completion. With only a few hours of life support available once the service module was discarded, this would have been disastrous. The solution is to roll the vehicle over so the lift vector points *down*, thus using the lift to help hold the vehicle in the atmosphere while it completes its deceleration.

The analysis of the last section is valid until the roll occurs, say, at γ_r, H_r, and X_r, as shown in Figure 8.6. The flip should be on the ascending portion of the skip trajectory, since we have no desire to dive into the ground. After the rollover, the only change is that the lift vector reverses direction. Equations (8.32) to (8.34) of the last section then become

$$\frac{dX}{d\gamma} = -\frac{1}{K_L} \cos \gamma e^{-H/H_0} \qquad (8.45)$$

$$\frac{dH}{d\gamma} = -\frac{1}{K_L} \sin \gamma e^{-H/H_0} \qquad (8.46)$$

$$\frac{dV}{d\gamma} = +\frac{C_D}{C_L} V \qquad (8.47)$$

The last equation still implies that the vehicle is slowing down, since $d\gamma/dt < 0$ on the inverted trajectory.

We must solve these equations of motion and match their initial conditions to the skip trajectory at the rollover point. The altitude equation

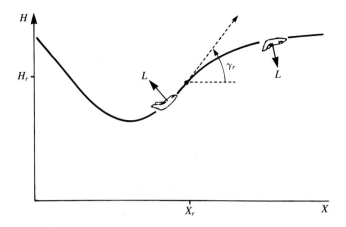

FIGURE 8.6
Rollover maneuver on a skip-reentry trajectory.

(8.46) has solution

$$H_{\text{USD}} = H_0 \ln \frac{-K_L H_0}{\cos \gamma + B_{\text{USD}}} \tag{8.48}$$

The subscript USD will be used for the "upside-down" portion of the trajectory, and the subscript RSU, for "right-side up," will be appended when we need to recall results from the last section. Now, the maximum altitude on the inverted trajectory occurs when $\gamma = 0$. To keep the argument of the natural log positive, we must require that

$$B_{\text{USD}} \leq -1 \tag{8.49}$$

If this condition is not met, then there is no maximum altitude. The vehicle will exit the atmosphere even though it has been rolled inverted.

The integration constant B_{USD} must be evaluated at the rollover point. This point, of course, is the one point in common for both the right-side-up and inverted portions of the trajectory. To evaluate this constant, equate the altitude solution for the skip (8.39) and the inverted (8.48) trajectories at the rollover point γ_r, H_r. This gives

$$H_r = H_0 \ln \frac{K_L H_0}{\cos \gamma_r + B_{\text{RSU}}}$$

$$= H_0 \ln \frac{-K_L H_0}{\cos \gamma_r + B_{\text{USD}}} \tag{8.50}$$

This easily simplifies to

$$B_{\text{USD}} = -2 \cos \gamma_r - B_{\text{RSU}}$$

$$\approx \cos \gamma_i - 2 \cos \gamma_r - K_L H_0 e^{-H_i/H_0} \tag{8.51}$$

Equation (8.51) may be inserted into (8.48) to evaluate the maximum altitude (at $\gamma = 0$) on the inverted trajectory. However, it is more instructive to then solve this expression for the γ_r value in terms of the desired maximum altitude H^+:

$$\cos \gamma_r = \tfrac{1}{2}[1 + \cos \gamma_i + K_L H_0(e^{-H^+/H_0} - e^{-H_i/H_0})] \qquad (8.52)$$

For comparison, (8.49) and (8.51) give the critical rollover point γ_{crit}, the point where the maximum altitude is infinity, as

$$\cos \gamma_{\text{crit}} = \tfrac{1}{2}(1 + \cos \gamma_i) \qquad (8.53)$$

Equation (8.53) can also be found directly from (8.52). They differ only by the last term in (8.52), and since we are likely to want a large maximum altitude H^+, this last term is quite small. The typical rollover maneuver, then, is performed at a point *very* close to the edge of disaster.

The altitude solution can again be inserted into the equation of motion for the horizontal distance and the variables separated, with the result

$$\int_{X_r}^{X} dX = H_0 B_{\text{USD}} \int_{\gamma_r}^{\gamma} d\gamma + H_0 B_{\text{USD}} \int_{\gamma_r}^{\gamma} \frac{d\gamma}{\cos \gamma + B_{\text{USD}}} \qquad (8.54)$$

This has exactly the same form as in the last section. However, the constant B_{USD} must now be less than -1, and this causes a profound change in the correct form of the second integral. The solution is now given by

$$X_{\text{USD}} = X_r + H_0 B_{\text{USD}}(\gamma - \gamma_r) + \frac{H_0 B_{\text{USD}}}{\sqrt{B_{\text{USD}}^2 - 1}} \sin^{-1} \frac{\sqrt{B_{\text{USD}}^2 - 1} \sin \gamma}{B_{\text{USD}} + \cos \gamma} \bigg|_{\gamma_r}^{\gamma} \qquad (8.55)$$

This form of solution for the horizontal distance has the effect of converting the trajectory into a very long inverted glide across the top of the atmosphere.

Figure 8.7 shows several rollover trajectories for an entry angle of $\gamma_i = -10°$ and $K_L H_0 = 2$. The trajectory is generated by plotting (8.55) against (8.48) after the rollover maneuver and using the skip-trajectory equations before the rollover point. The entry altitude was 70 km, and the rollover points are all indistinguishable on this scale. Note carefully the large difference in the X-and H-axis scales. The trajectory with a maximum altitude of about 85 km, in particular, covers almost 10,000 km of distance before it begins its final reentry. The long time spent at high altitude produces a relatively benign aerodynamic heating environment, since the air density is low. This keeps the heating rate within reason while stretching the reentry time to its maximum extent.

Near the end of the inverted trajectory the speed of the vehicle will be low enough that it is no longer possible to neglect gravity in the equations of motion. This too has advantages. The vehicle no longer needs its full lift to stay within the atmosphere, and so it is possible to use lift to produce crossrange by rolling the capsule, or to bring the vehicle down on a precise landing point. Since the Apollo capsule could not change its effective angle of attack, it always produced lift. So, once the landing point was correctly

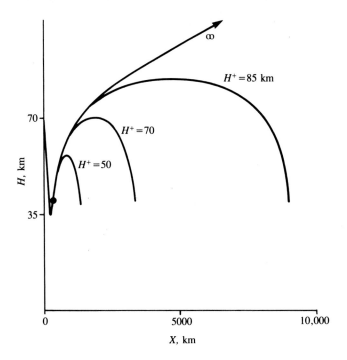

FIGURE 8.7
Rollover-reentry trajectories.

targeted, it was necessary to place the capsule in a slow roll about the velocity vector to average out the now undesired lift.

Of course, it is necessary to remember that we have ignored the gravitational attraction of the earth in this development. At the middle of the reentry, this is an excellent assumption, since the vehicle will be close to circular orbital velocity and the apparent gravitational acceleration will be small. However, near the end of the trajectory gravity will make itself felt in the usual manner. It is less obvious that the reverse happens at the beginning of the reentry: If the vehicle encounters the atmosphere at escape velocity, the apparent gravitational field will be 1 g *upward*. So, the trajectory at both ends of the reentry would need modifications to correctly include gravity and "centrifugal force."

8.5 AEROBRAKING

It seems logical that a lifting body could execute a skip reentry, using its lift force to exit the atmosphere. It is less obvious that lift is not at all necessary to do this, but this is the case. In fact, one of the earliest proposals for a reentry trajectory, the braking ellipses of Hohmann, makes use of this fact. A skip

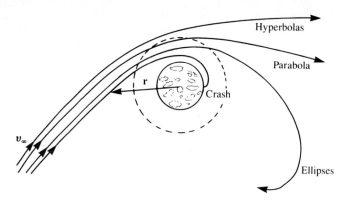

FIGURE 8.8
Aerobraking reentry.

reentry without lift is possible since the earth is not flat but spherical. Until now, we have taken the aerodynamicist's approach of assuming that the earth is "flat" and appending the necessary corrections to obtain valid dynamics. However, in this section we return to the orbital perspective.

Consider the vehicle of Figure 8.8, subject to the gravitational attraction of a planet and to drag from the atmosphere. In vector form, the equations of motion are

$$\ddot{\mathbf{r}} = -\mu \frac{\mathbf{r}}{r^3} - \frac{1}{2} \frac{C_D A}{m} \rho v \mathbf{v} \tag{8.56}$$

where μ is the gravitational parameter, C_D is the drag factor, A and m are the spacecraft area and mass, and ρ is the air density. Assuming an exponential atmosphere and writing R for the radius of the planet, this becomes

$$\ddot{\mathbf{r}} = -\mu \frac{\mathbf{r}}{r^3} - \tfrac{1}{2} B^* \rho_0 e^{-(r-R)/H_0} v \mathbf{v} \tag{8.57}$$

where we have abbreviated $C_D A / m$ as B^* and H_0 is the atmospheric scale height.

Equation (8.57) of course describes the two-body problem with an extra force. Now, approaching a planet from a great distance (or returning to the earth from a great distance), our hyperbolic excess speed v_∞ is essentially fixed. However, with a tiny maneuver long before planetary encounter, we can adjust the periapse radius over a substantial range. Here, a "substantial range" need only run from the planet's surface R to some distance outside the atmosphere. For, if $r - R$ stays large enough in (8.57), the drag term is always negligible, and the spacecraft will fly by the planet on an undisturbed hyperbolic trajectory.

The desired outcome, of course, is to use the air-drag force to retard our speed enough so that we do not leave the vicinity of the planet. Imagine lowering the periapse radius until air drag does slightly retard the speed of the spacecraft. As we drop the point of closest approach deeper and deeper into the atmosphere, the drag force retards our velocity more and more until we reach a trajectory where we exit the atmosphere just at escape velocity for that planet.

This is the first critical trajectory for this problem. If we follow a slightly lower approach path, we will still exit the atmosphere but on an *elliptical* orbit with respect to the planet. Dropping the distance of closest approach further will produce trajectories which exit the atmosphere with lower and lower eccentricities. Finally, the second critical trajectory is encountered when the velocity drops below circular orbital speed as the spacecraft is just exiting the atmosphere. This trajectory terminates on the surface of the planet and does not return to space. Below this trajectory are conventional direct-reentry trajectories. Both critical trajectories must be established by numerical means for a given approach speed v_∞.

If we wish to use aerobraking (or *aerocapture*) to stay in the vicinity of the planet, it is essential that the first approach to the planet be along a trajectory *between* the two critical trajectories. If we exit the atmosphere on an elliptical orbit, our periapse distance will still be near the original value. This means that, unless we maneuver, a second passage through the atmosphere will retard our velocity still further. Small maneuvers may be made at apogee to control the distance of closest approach on the second and subsequent passes. We are thus led to Hohmann's concept of braking ellipses.

One difficulty with aerobraking is that the first passage usually produces the highest heating. During this passage the spacecraft *must* dissipate enough speed to leave the atmosphere below escape speed. The second and subsequent passages may then be made almost at leisure. A second concern is the lack of information on the state of the upper atmospheres of other planets *at the moment the spacecraft will approach*. The outer layers of planetary atmospheres can be quite dynamic, and the two critical trajectories are usually quite close together. An error of only one or two scale heights H_0 (5 to 10 km) could thus produce disaster, as the spacecraft encounters densities lower than planned and leaves the atmosphere at a speed above the speed of escape. The addition of lift to the vehicle for control is thus almost essential.

However, aerobraking produces such large savings in fuel for planetary missions that it is almost certain to be tried within this century. It was considered by NASA for the Venus radar-mapper mission, and is under consideration for returning a reusable space tug to low earth orbit. Such a promising approach will not be ignored forever.

8.6 LIFTING-BODY REENTRY

The reentry of a vehicle capable of lift has become an important practical problem with the advent of the space shuttle. A space shuttle reentry begins

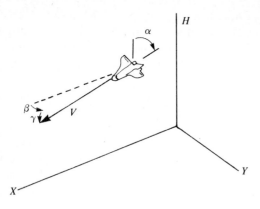

FIGURE 8.9
Vehicle-trajectory reference frame.

with the vehicle executing the first half of a Hohmann transfer from its initial circular orbit outside the atmosphere to a new perigee altitude within the outer fringes of the atmosphere. The reentry begins just before perigee on the transfer ellipse, and the vehicle ceases to follow a keplerian orbit at this point. The reentry is characterized by a very shallow flight-path angle γ, and we will make free use of this fact in what follows. It is also necessary to treat the reentry in three dimensions, for reasons which will shortly become apparent.

Consider Figure 8.9, which shows the vehicle in a rectangular-coordinate reference frame. The velocity vector of the vehicle makes an angle γ with the horizontal, positive upward, and an angle β with the X axis, which we will consider to be the nominal flight direction. The kinematics equations then become

$$\frac{dX}{dt} = V \cos \beta \cos \gamma \approx V \cos \beta$$

$$\frac{dY}{dt} = V \sin \beta \cos \gamma \approx V \sin \beta \qquad (8.58)$$

$$\frac{dH}{dt} = V \sin \gamma \approx V\gamma$$

where we have used the small-γ approximation. Figure 8.10 shows the forces acting on the shuttle, where we have included the possibility that the vehicle may be rolled around the velocity vector, with roll angle α from the vertical. The dynamics equations of motion then are

$$m \frac{dV}{dt} = -D - mg_e \sin \gamma \approx -D - mg_e\gamma$$

$$mV \frac{d\gamma}{dt} = L \cos \alpha - mg_e \cos \gamma$$

$$\approx L \cos \alpha - mg_e \qquad (8.59)$$

$$mV \frac{d\beta}{dt} = L \sin \alpha$$

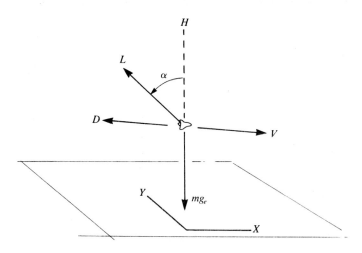

FIGURE 8.10
Vehicle free-body diagram.

The quantity g_e is the effective gravitational force seen by the vehicle, which will consist of the real gravitational force minus the centrifugal "force":

$$g_e = g - \frac{V^2 \cos^2 \gamma}{R + H} \approx g - \frac{V^2}{R} \qquad (8.60)$$

where R is approximately the radius of the earth, ignoring the slight addition due to the vehicle's altitude. We have also used the small-γ assumption in (8.60). We now substitute for the lift and drag forces, assuming our usual exponential atmosphere,

$$L = \tfrac{1}{2} C_L A V^2 \rho_0 e^{-H/H_0} \qquad (8.61)$$

$$D = \tfrac{1}{2} C_D A V^2 \rho_0 e^{-H/H_0} \qquad (8.62)$$

where C_L and C_D are functions of the vehicle pitch angle ψ. When (8.60) to (8.62) are substituted into the dynamics equations of motion, these become

$$\frac{dV}{dt} = -K_D V^2 e^{-H/H_0} - g\gamma + \frac{V^2 \gamma}{R} \qquad (8.63)$$

$$V \frac{d\gamma}{dt} = K_L V^2 e^{-H/H_0} \cos \alpha - g + \frac{V^2}{R} \qquad (8.64)$$

$$V \frac{d\beta}{dt} = K_L V^2 e^{-H/H_0} \sin \alpha \qquad (8.65)$$

The quantities K_L and K_D are given by

$$K_L = \frac{C_L A \rho_0}{2m} \tag{8.66}$$

$$K_D = \frac{C_D A \rho_0}{2m} \tag{8.67}$$

These will not be constants, since we will wish to control both lift and drag during the reentry.

It is advantageous to stay as high as possible as long as possible during this type of reentry. The frictional heating rate is proportional to the work done by air drag, or proportional to $\rho V^2 \times V = \rho V^3$. Wind-tunnel tests have shown that the amount of energy appearing *on the vehicle* is actually proportional to $\sqrt{\rho} V^3$. While the vehicle is at very high velocity, the only way we can reduce this rate is by staying at high altitude. If the vehicle is in a steady glide, then $d\gamma/dt = 0$ and (8.64) becomes

$$K_L = \frac{g - V^2/R}{\cos \alpha V^2 e^{-H/H_0}} \tag{8.68}$$

As the vehicle approaches perigee, $g - V^2/R$ is negative, since while the orbit is "falling" toward perigee the effective gravitational force is upward. Without slowing down, the vehicle will continue to follow a keplerian orbit out of the atmosphere and back to the altitude at which the deorbit burn was made. So, before perigee it is necessary to place the shuttle orbiter in its highest drag configuration. This implies a large angle of attack and ensures that the orbiter decelerates rapidly to below circular orbital speed. In a typical shuttle reentry, the orbiter is only slightly over this value to begin with.

After perigee, the numerator of (8.68) is positive, and lift can be used to force a constant-altitude trajectory. This will ensure that the orbiter will stay as high as possible as long as possible. As shown in Figure 8.11, the lift and drag of the orbiter depend on its angle of attack ψ with respect to the local velocity. In this part of the reentry trajectory, the air density is too low for air to behave as a fluid medium. Individual air molecules simply bounce off the shuttle as if it were a flat plate, and the aerodynamic force $\mathbf{F}$ is simply proportional to the rate of transfer of momentum to the shuttle. Lift and drag are then just the components of $\mathbf{F}$ perpendicular to and parallel to the velocity vector. If the angle of attack is slowly reduced as the vehicle decelerates, lift increases and drag decreases. Equation (8.68) gives the required lift to maintain a constant-altitude glide after perigee. As the speed drops, the amount of lift required increases, until at some point the orbiter is in its maximum lift-to-drag configuration. After this point, a constant-altitude trajectory can no longer be maintained.

At this point, the orbiter will begin to lose altitude, even if it is not banked. Since the orbiter should land at a runway, it is necessary to bring the

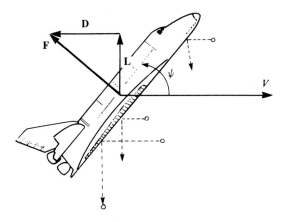

FIGURE 8.11
Aerodynamic forces in free molecular flow.

vehicle down at a precise point on the earth's surface. However, to allow for contingencies, the initial trajectory will purposely bring the vehicle to the landing point at an excessive altitude. It is better to reenter a little long than to be even a very little short of the end of the runway. If the reentry is going well, the extra ground length can be used up by continuing to fly roll-reversal maneuvers. These maneuvers, in which the orbiter banks first one way and then the other, cause the vehicle to fly extra distance laterally and also increase the sink rate. By controlling these two factors, the trajectory can be brought to the proper point above the runway to begin final approach.

8.7 REFERENCES AND FURTHER READING

Since the earliest days of Sanger and Hohmann, it has been realized that the dynamics of reentry and planetary entry cannot really be separated from the heating problem. It is this latter concern that drives the trajectory design. Basic reentry trajectories are treated in Miele. The classic series by Loh covering all phases of the reentry problem has recently been followed by excellent books by Vinh, Busemann, and Culp and by Regan. Research on the heating problem continues, since the "aerospace plane" is also basically a reentry problem, and will require great advances if this vehicle is to become a reality.

Miele, A., *Flight Mechanics,* Addison-Wesley, Reading, Mass., 1962.

Loh, W. H. T., *Reentry and Planetary Entry Physics and Technology,* Springer-Verlag, New York, 1968.

Vinh, N. X., A. Busemann, and R. D. Culp, *Hypersonic and Planetary Entry Flight Mechanics,* University of Michigan Press, Ann Arbor, 1980.

Regan, F. J., *Reentry Vehicle Dynamics,* AIAA Educational Series, AIAA, New York, 1984.

8.8 PROBLEMS

1. On the inverted phase of a rollover reentry, solve the velocity equation of motion (8.47) from the rollover point to the point of maximum altitude ($\gamma = 0$). Match this solution to the skip-phase velocity solution at the rollover point. Show that

$$V_{\text{USD}}(\gamma = 0) = V_i \exp\left[-\frac{C_d}{C_l}(2\gamma_r - \gamma_i)\right]$$

2. A reusable orbit transfer vehicle is returning from geosynchronous orbit and will enter the atmosphere at nearly escape velocity. It will fly a reentry trajectory which includes a skip phase followed by an inverted phase whose maximum altitude H^+ is essentially above the atmosphere. The vehicle must reach this altitude at circular orbital speed for that altitude. That is, the reentry is to end with the orbit transfer vehicle returned to a low-altitude circular orbit. Argue that the rollover flight-path angle is very close to γ_{crit}, and show for small γ's that

$$\gamma_r \approx \frac{|\gamma_i|}{\sqrt{2}}$$

Combine this with the result of the previous problem and provide an expression for the atmospheric entry angle γ_i needed to complete this flight plan.

3. A vehicle performing a skip reentry must be designed to withstand the maximum lift acceleration during the aerodynamic pull-up. Show that the vehicle acceleration due to lift is given by

$$\frac{L}{m} = a_{\text{lift}}$$

$$= \frac{V_i^2}{H_0}(\cos\gamma + B)\exp\left(-2C_d\frac{\gamma - \gamma_i}{C_l}\right)$$

For a very efficient glider, show that the maximum acceleration occurs at $\gamma = 0$ and is given by

$$a_{\text{lift,max}} = \frac{V_i^2}{H_0}(1 - \cos\gamma_i)$$

For $V_i = 20{,}000\,\text{ft/s}$ and $\gamma_i = -20°$, what is the maximum acceleration in feet per second squared and g's? Use $H_0 = 23{,}000\,\text{ft}$ for earth's atmosphere.

CHAPTER
9

THE
SPACE
ENVIRONMENT

9.1 INTRODUCTION

Like any environment, the environment of space dictates the characteristics of
devices intended to operate there and imposes requirements on anyone who
would venture into space. For space is *not* empty. It carries the electromagnetic radiation of the sun: not just visible light, but the entire range from radio
waves to infrared, visible, and ultraviolet light, to x-rays and gamma rays. It is
also filled with the corpuscular radiation of the sun, the solar wind. Some of
this is trapped in the earth's magnetic field, forming intense radiation belts.
And there is matter in space, from microscopic comet dust to macroscopic
meteoroids to asteroids, moons, and other planets. Knowledge of the solar
system is not just a necessary cultural background for astronautical engineering, it is a necessary part of the engineering of any space system.

9.2 THE ATMOSPHERE

Although space is usually thought of as a perfect vacuum, the atmosphere of
the earth (and other planets) often figures in spacecraft operations. Most
satellites occupy low earth orbits within the outermost fringes of the
atmosphere. Space vehicles must penetrate the atmosphere during launch and

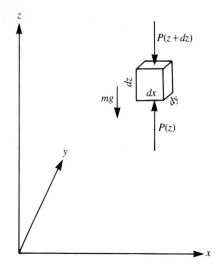

FIGURE 9.1
Hydrostatic equilibrium.

perhaps return through it during reentry. Also, the atmosphere can be used to eliminate the need for rocket propulsion, as in the case of aerobraking. A description of the state of the earth's atmosphere is necessary for both trajectory design and operations.

Consider the small cube of air shown in Figure 9.1. It has density ρ, and therefore mass $m = \rho\, dx\, dy\, dz$. If the atmosphere is in equilibrium, this small parcel of matter is not accelerating and therefore the forces on it must sum to zero. This is a statics problem. The pressure forces on the four vertical faces balance each other, so we are only interested in force equilibrium in the vertical direction. Gravity contributes a force mg downward. To balance this, the pressure on the top of the cube, $P(z + dz) \approx P(z) + dP/dz\, dz$, must be slightly less than the pressure at the bottom of the cube. The pressure force on a face of the cube is $P\, dx\, dy$, so force equilibrium in the vertical direction requires

$$-\rho g\, dx\, dy\, dz + P\, dx\, dy - \left(P + \frac{dP}{dz} dz\right) dx\, dy = 0 \qquad (9.1)$$

After simplification this becomes

$$\frac{dP}{dz} = -g\rho \qquad (9.2)$$

This is called the *hydrostatic equation.*

To make further progress, it is necessary to specify the *equation of state* for the medium. For an atmosphere, this is the perfect gas law

$$P = \rho k T \qquad (9.3)$$

where k is Boltzmann's constant and T is the absolute temperature. However, we cannot simply substitute (9.3) into (9.2) to relate pressure P to density ρ since (9.3) introduces a new quantity, the temperature. Knowledge of the temperature as a function of altitude within the atmosphere would require us to consider the heat balance of the atmosphere: the methods of heat transport within the atmosphere, and sources and sinks for thermal energy. This would take us very far away from our intent. Instead, we will make the assumption that the atmosphere is *isothermal*; that is, assume that the temperature is a constant.

Then differentiating (9.3), we have $dP/dz = kT\,d\rho/dz$, so (9.2) becomes

$$\frac{d\rho}{dz} = -\frac{g}{kT}\rho = -\frac{\rho}{H_0} \tag{9.4}$$

where $H_0 = kT/g$ is termed the *atmospheric scale height*. Separating variables in (9.4), we have

$$\int_{\rho_0}^{\rho} \frac{d\rho}{\rho} = -\frac{1}{H_0} \int_0^H dz \tag{9.5}$$

starting from zero altitude, where the atmosphere has density ρ_0. (For planets without solid surfaces, this level is arbitrarily defined.) Integrating (9.5), we obtain $\ln \rho - \ln \rho_0 = H/H_0$, and taking the exponential of both sides gives

$$\rho = \rho_0 e^{-H/H_0} \tag{9.6}$$

the expression for the "exponential atmosphere" we have already used several times.

Of course, the atmosphere of the earth (and other planets) is not really isothermal. There is a temperature gradient dT/dz, termed the *lapse rate*, in a real atmosphere. Even so, model atmospheres cite the base density ρ_0 and scale height H_0 for different layers in an atmosphere with varying temperature T, using (9.6) as a local description of the behavior. For the earth, the atmosphere can be thought of as having three layers. The lowest 7 to 10 km are the *troposphere*, the layer in which we have weather. During the daytime over much of the earth, this layer is heated from below by the warm ground and conducts heat upward by convection. The lapse rate in this layer during daytime is thus close to the adiabatic value, since convection is a very efficient mode of heat transport. However, after the sun sets, the lower atmosphere usually reverts to radiation and conduction for heat transport, and it becomes more nearly isothermal. Atmospheric conditions in the troposphere are, as we know from personal experience, quite variable.

The *stratosphere* extends upward from the troposphere to about an altitude of 50 to 70 km. This layer is not in convection and is the most nearly isothermal layer of the atmosphere. Above this is the *exosphere*, the natural abode of spacecraft. This layer is directly heated by charged particles from the earth's radiation belts, and the air density may be quite variable here, too. In the case of a solar storm or a disturbance in the earth's magnetic field, the air density in the exosphere may change drastically in a few hours as this layer heats and expands outward.

The individual molecules in the exosphere do not behave as a classical gas. Because the density is so low, the mean free path, the distance a molecule travels before colliding, may be very large. This means that the "newtonian fluid" description of a gas applies in this regime. At the outermost levels, individual molecules can even obtain enough energy from random collisions to escape from the earth entirely. This process is most rapid for light elements like hydrogen and helium, explaining why the earth's atmosphere has so little of these gases, while the atmospheres of the giant planets are so rich in them. The moon, with roughly one-sixth the earth's gravity, is completely unable to retain any permanent atmosphere in spite of the fact that it is at the same distance from the sun. Besides causing satellite-orbit decay, molecules colliding with spacecraft can also cause degradation of the spacecraft materials. Atmospheric oxygen ions are particularly active in this process.

Planetary atmospheres differ greatly. Small Mercury, close to the sun, has no atmosphere, while Venus has an enormous atmosphere. Mars has a surface pressure only 0.6% that of Earth, and the stratosphere may be thought of as beginning at the surface. At the other extreme are the giant planets Jupiter, Saturn, Uranus, and Neptune, whose convective tropospheres may extend downward for a substantial fraction of the radius of the planet itself. Even a moon may have an atmosphere if it is far enough from the sun: Saturn's moon Titan has a thick atmosphere dominated by a natural layer of photochemical smog. Atmospheric data are available for most of these objects, although they have not been studied with the same intensity as the Earth.

9.3 LIGHT AND SPACECRAFT TEMPERATURE

A spacecraft in orbit absorbs heat and light from the sun. Its internal power sources furnish an additional source of energy, most of which will eventually be converted into heat. (Although if the spacecraft is powered by solar cells, then the power source is again absorbed sunlight.) This waste heat must be disposed of or the spacecraft will simply continue to become hotter and hotter. Now, there are three usual mechanisms for heat transport: conduction, convection, and radiation. A spacecraft is denied the use of the first two of these and is forced to radiate away waste heat. Radiating waste heat is simply the inverse process of absorbing the energy in the first place, so we can study both processes at once.

A mythical perfect absorber or emitter of electromagnetic radiation is called a *blackbody*. Perhaps this term is understandable in the case of a perfectly absorbing body, but should a perfectly *emitting* body also be black? The answer is "yes"; a perfect absorber is also a perfect emitter, since the process of absorbing radiation is simply the time reverse of emitting the radiation. (This is why the space shuttle belly tiles are black: Black objects

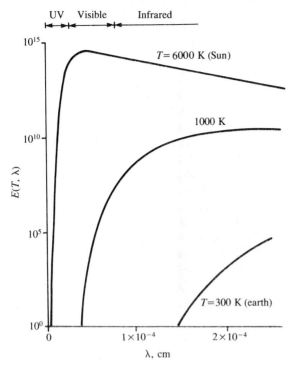

FIGURE 9.2

Blackbody radiation curves.

radiate more heat than any other color.) The way a blackbody radiates energy was discovered by Max Planck near the turn of the century. If the blackbody has temperature T in degrees kelvin, the energy E radiated at wavelength λ cm per unit wavelength interval per square centimeter of emitting surface per second of time per unit solid angle is given by

$$E(T, \lambda) = \frac{2hc^2}{\lambda^5} \frac{1}{e^{hc/\lambda kT} - 1} \tag{9.7}$$

where c = speed of light $\approx 2.998 \times 10^{10}$ cm/s

h = Planck's constant $\approx 6.624 \times 10^{-27}$ erg $\cdot$ s

k = Boltzmann constant $\approx 1.380 \times 10^{-16}$ erg/deg

Figure 9.2 shows several blackbody curves, as they are called, for objects of different temperatures. The 6000 K blackbody curve is a decent approximation to the energy output of the sun, which has an effective blackbody temperature of about 5800 K. As an object becomes hotter, the maximum of the blackbody curve shifts to shorter (more energetic) wavelengths, while the area under the curve (the total energy output per square centimeter per second per solid angle) becomes larger.

However, the blackbody spectra above are less useful to the astronautical engineer than two other results. These were known experimentally before Planck explained the actual mechanism of blackbody emission in detail. Wein's law states that the wavelength at which blackbody emission peaks is given by

$$\lambda_{max} = \frac{0.2897}{T} \tag{9.8}$$

if λ is in centimeters and T is in degrees kelvin. This relation can be derived, of course, by differentiating the blackbody spectrum (9.7) with respect to wavelength and setting the result equal to zero. On the other hand, if the blackbody spectrum is integrated over wavelength (a nontrivial task), the total energy output as a function of temperature alone becomes

$$E(T) = \sigma T^4 \qquad \text{ergs/cm}^2 \cdot \text{s} \tag{9.9}$$

over the entire 4π solid angle. The Stefan–Boltzmann constant σ has a value of 5.672×10^{-5} in cgs units. A spacecraft using a radiator system to radiate heat to space will find that the system disposes of *much* more heat at even a slightly higher temperature.

The Stefan–Boltzmann law can be used to calculate the equilibrium temperature of a spacecraft without internal heat sources. At the earth's distance from the sun, the incoming solar energy amounts to 1.371×10^6 ergs/s $\cdot$ cm^2. This value is called the *solar constant*. If the spacecraft is a perfectly absorbing sphere of radius r, it will intercept energy

$$E_{in} = \pi r^2 \times 1.371 \times 10^6 \text{ ergs/s} \tag{9.10}$$

If the spacecraft is in thermal equilibrium, it must be radiating exactly as much energy as it receives or it would still be heating up or cooling down. So, we can equate (9.10) to the energy output rate over the surface of the spacecraft from (9.9) to find

$$E_{out} = 4\pi r^2 \sigma T^4 = E_{in} \tag{9.11}$$

Solving for the temperature, we find

$$T = \left(\frac{1.371 \times 10^6}{4\sigma}\right)^{1/4} = 278.8 \text{ K} \tag{9.12}$$

This is just 5°C over the temperature at which water freezes. Since the earth itself is such a spherical spacecraft in thermal equilibrium, the average temperature of the earth should be about this value. However, the earth's atmosphere is most definitely not a perfect blackbody emitter. A blackbody at nearly 300 K should radiate most of its energy in the infrared region of the spectrum, as shown in Figure 9.2. However, and luckily for us, at these wavelengths the earth's atmosphere is nearly opaque. Since the earth's atmosphere retains infrared radiation, the earth must warm up further until the leakage of radiation through nearly transparent "windows" in the earth's atmosphere allows balance to be achieved. This is the *greenhouse effect*. Also, any spacecraft which is not a perfect blackbody must also be warmer than

(9.12) predicts. When the spacecraft of the near-disasterous Apollo 13 mission lost electrical power, and therefore its source of internal heat, the internal temperature plummeted to a value near that given in (9.12). The astronauts were *cold*.

The Stefan–Boltzmann law is also of direct utility to the spacecraft designer. The space shuttle, for example, has flat-plate radiators with an area of $111 \, m^2$ or $1.11 \times 10^6 \, cm^2$. Through these radiators flows coolant at a temperature of about 300 K. Now, if these radiators face the blackness of space, they will absorb very little incoming radiation: mainly just the energy of starlight and the universal 3 K microwave background radiation. Under these conditions, the radiator panels should be capable of rejecting energy at a rate of

$$E_{out} = \tfrac{1}{2}\sigma T^4 A = 2.55 \times 10^{11} \text{ ergs/s} \tag{9.13}$$

The factor of $\tfrac{1}{2}$ in (9.13) is necessary to account for the fact that the radiators operate basically only on one side. Their other side is near the payload-bay doors. Now, 10^{10} ergs/s = 1 kW, so the shuttle radiators should carry a heat load of nearly 25 kW when facing off into space.

However, the situation changes if the payload bay faces toward the earth. From low orbit, the earth will cover nearly half the sky, and the radiators will be forced to absorb incident blackbody radiation at the effective temperature of the earth. (The fundamental problem is that a good radiator is also a good absorber!) So, the net energy transfer will now be given by

$$E_{out} = \tfrac{1}{2}\sigma(T^4_{rad} - T^4_{earth})A \tag{9.14}$$

or about 6.7 kW, assuming the earth's effective temperature is 278 K. This is very close to the carrying capacity of 6.3 kW advertised for the shuttle while in orbit. The situation changes again if the payload bay faces squarely into the sun. The sun is nearly a perfect blackbody at an effective temperature of 5800 K, but it does not cover much of the sky seen from the earth. Using the solar constant, the radiator panels will be forced to absorb solar radiation at a rate of 1.52×10^{12} ergs/s, or 152 kW. Since the panels can only radiate at best 25 kW, it is clear that in this orientation the radiators function as solar collectors. Given the specific heat of the coolant and its flow rate through the panels, it is possible to calculate the temperature of the coolant as it returns to the shuttle.

9.4 CHARGED-PARTICLE MOTION

The dilute gas in space is a *plasma*: It is hot enough that it is ionized. Some of the electrons in the outer shells of each atom have been stripped off, leaving the atom with a net positive charge, and the electrons, of course, carry a negative charge. Although plasmas are very hot, at least several thousand degrees kelvin, the interplanetary gas is so dilute that a spacecraft will not notice its temperature. Rather, it sees the gas as a hailstorm of very fast

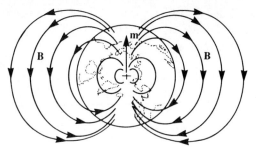

FIGURE 9.3
A dipole magnetic field.

moving individual particles. Near the earth the motion of these particles is governed by the earth's magnetic field. The magnetic field of the earth is nearly a *dipole field*, the simplest magnetic field configuration. For a dipole field, the magnetic field vector **B** is given by

$$\mathbf{B}(\mathbf{r}) = -\mu_0 \nabla \frac{\mathbf{m} \cdot \mathbf{r}}{4\pi r^3} \tag{9.15}$$

where μ_0 is the permeability of free space and **m** is the magnetic dipole moment of the earth. This is a vector whose magnitude sets the intensity of the overall field and whose direction is out of the earth's north magnetic pole.

Now, in any small region of space, a charged particle does not see the overall configuration of the dipole field, Figure 9.3, but only senses the local field at its position. Since the local field is nearly uniform in a small region of space, we will first study the motion of a charged particle in a uniform local magnetic field **B**. The Lorentz force on a particle with electric charge q is given by

$$\mathbf{F} = q\mathbf{v} \times \mathbf{B} \tag{9.16}$$

where **v** is the velocity of the particle, Note that the force on a charged particle in a magnetic field is always perpendicular to the velocity, and hence a magnetic field can do no work on a charged particle. So, the kinetic energy of the charged particle,

$$T = \tfrac{1}{2}mv^2 \tag{9.17}$$

does not change.

As shown in Figure 9.4, let us divide the velocity vector of the particle into v_p, the component parallel to the magnetic field vector **B**, and v_n, the component normal to the field vector. Neither v_p nor v_n can change in magnitude, since the force in (9.16) cannot have a component in either direction. In fact, the only component of the force **F** is directed radially inward, back toward the field vector **B**. So, the motion of a charged particle must consist of (1) a constant "drift velocity" v_p parallel to the field combined with (2) a circular motion centered on the field line. Trajectories of charged particles in a uniform magnetic field are then spirals, centered on the field.

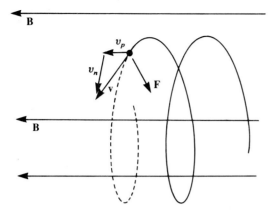

FIGURE 9.4
Particle motion in a uniform magnetic field.

We can obtain the radius of the circular motion by equating the magnetic force in the radial direction to the mass of the particle times its centripetal acceleration:

$$\frac{mv_n^2}{R} = qv_nB \qquad (9.18)$$

This immediately gives

$$R = \frac{mv_n}{qB} \qquad (9.19)$$

which is called the *Larmour radius* of the particle. Electrons, which have a mass 2000 times less than the mass of any positive ion, are bound much more tightly to the field lines than are the atomic nuclei. However, both electrons and nuclei are condemned to spiral around the field lines wherever the latter go. Plasma physicists say that the plasma is "frozen" to the magnetic field. We can thus see how a charged particle in the earth's magnetic field will follow a particular field line to either the north or south magnetic pole and encounter the earth's atmosphere at high latitudes. This is, in fact, the cause of auroras, or northern lights. However, there is another effect which prevents particles in the earth's magnetic field from reaching the atmosphere under most conditions: the magnetic mirror effect. This reflects most particles before they encounter the atmosphere, leading to permanent belts of charged particles around the earth.

9.5 MAGNETIC MIRRORS

In the last section, we considered the motion of a charged particle in a uniform magnetic field. However, the earth's dipole field is only uniform when viewed locally. A charged particle approaching one of the earth's poles sees a slowly converging and intensifying magnetic field. This situation is shown in Figure

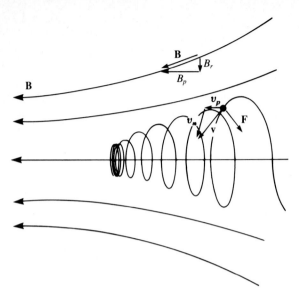

FIGURE 9.5
Particle motion in a converging
magnetic field.

9.5, where, viewed in a cylindrical-coordinate frame, the magnetic field has a
slight radial component inward. As before, we can split the particle's velocity **v**
into components v_p parallel to the coordinate axis and v_n perpendicular to this
axis. Similarly, the **B** field has components B_p along the axis and B_r
perpendicular, positive outward.

In a uniform magnetic field, the velocity component v_p was constant.
However, in a converging field there is a component of the magnetic force
$q\mathbf{v} \times \mathbf{B}$ along the z axis:

$$m\frac{dv_p}{dt} = qv_n B_r \tag{9.20}$$

This force acts to retard the drift velocity v_p, since B_r is negative in a
converging magnetic field. Thus, there is a tendency for a converging field to
slow down, and possibly to reflect a charged particle.

To put this insight on a better analytic footing, note that the magnetic
force is still perpendicular to the velocity vector **v**, and the magnetic field still
does no work on the particle. So, the kinetic energy of the particle,

$$E = \tfrac{1}{2}m(v_p^2 + v_n^2) \tag{9.21}$$

is still a constant of the motion. There is another constant of the motion in
additon to the kinetic energy: The magnetic moment of the particle is also a
constant. This is the product of the effective electric current of the particle i in
its circular path with the area of the path A:

$$M = iA = \frac{qv_n}{2\pi R}\pi R^2 = \tfrac{1}{2}qv_n R \tag{9.22}$$

We now proceed to demonstrate that this is, indeed, a second constant of the motion.

To show (9.22) is a constant of the motion, we will need a few other facts about magnetic fields. First, magnetic fields have zero divergence. This statement, one of Maxwell's four laws of electromagnetic theory, is equivalent to the statement that there is no magnetic "charge." Thus, magnetic field lines always form closed paths. The statement that the field has no divergence

$$\nabla \cdot \mathbf{B} = 0 \tag{9.23}$$

becomes

$$\frac{1}{r}\frac{\partial}{\partial r}(rB_r) + \frac{\partial B_p}{\partial z} = 0 \tag{9.24}$$

in a cylindrical-coordinate frame. This statement can be used to relate the two components of the magnetic field in a slowly converging geometry. Assuming that $\partial B_p/\partial z$ is nearly a constant over the area of the Larmour orbit (this is equivalent to assuming the field is conical on the scale of the orbit), (9.24) becomes

$$\int_0^R \frac{\partial}{\partial r}(rB_r)\,dr = \int_0^R \frac{\partial B_p}{\partial z}r\,dr \tag{9.25}$$

After integrating, this becomes

$$B_{r=R} = -\tfrac{1}{2}R\frac{\partial B_p}{\partial z} \tag{9.26}$$

Now, armed with an approximation to the structure of the magnetic field in the slowly converging case, we can return to the assertion that the magnetic moment of the orbit is constant. Taking the time derivative of (9.22), we obtain

$$\frac{dM}{dt} = \frac{d}{dt}\frac{mv_n^2}{2B_p}$$

$$= \tfrac{1}{2}m\left(2v_n\frac{dv_n}{dt} - \frac{v_n^2}{B_p^2}\frac{dB_p}{dt}\right) \tag{9.27}$$

when the Larmour radius (9.19) is substituted, and the magnetic field changes with time as the particle enters areas of higher field density. Now, energy conservation implies that

$$v_n\dot{v}_n + v_p\dot{v}_p = 0 \tag{9.28}$$

Replacing dB_p/dt with $v_p\,\partial B_p/\partial z$ and substituting (9.28) and (9.26) into (9.27), this latter equation reduces to the statement that $dM/dt = 0$; that is, the magnetic moment of the orbit is constant in a slowly converging field.

We can now detail the motion of the charged particle in a converging magnetic field. Energy requires that the speed of the particle is constant. There is a component of the magnetic force which decelerates the particle's drift velocity v_p. The magnetic moment of the orbit, $M = qv_nR/2$ is a constant. When we substitute the expression for the Larmour radius, this becomes

$$M = \frac{1}{2}\frac{mv_n^2}{B} \qquad (9.29)$$

Now, at some point, the drift velocity v_p will become zero, the entire velocity will now reside in the component v_n, and the particle will be reflected. Depending on the initial value of the magnetic moment M, equation (9.29) allows us to calculate the field intensity B at which reflection occurs. A slowly converging magnetic field can act as a mirror.

9.6 THE VAN ALLEN BELTS

The first satellite launched by the United States, Explorer I, carried charged-particle detectors. These instruments found a belt of high-charged-particle density which had not been expected. Later satellites showed that there were actually two belts, now bearing the name of their discoverer, James Van Allen. The Van Allen belts were not expected for a very simple reason: No source of charged particles to maintain such a belt was known. Without such a source, the natural leakage of the earth's magnetic mirrors would quickly deplete such a belt. To understand the need for a source, first we will consider why any such belt will leak.

A single charged particle in the earth's magnetic field can spiral about a magnetic field line, with reflections near each pole, providing that the magnetic moment M is high enough. The magnetic moment of the particle,

$$M = \frac{mv_n^2}{2B} \qquad (9.30)$$

depends on the velocity of the particle normal to the field direction, v_n. As the particle approaches the poles, the magnetic field intensity B increases, and since the magnetic moment M is a constant, the normal velocity component v_n must also increase to keep M constant. However, the total kinetic energy is also constant, so the velocity component parallel to the field must decrease as v_n increases. When the parallel component v_p becomes zero, the particle is reflected. If this does not happen before the particle hits the earth's atmosphere, the particle will be lost to the belt.

However, there are other particles within the belt, and when two particles approach each other they experience forces from their electrostatic charges. The particles will collide elastically, conserving their total energy. However, the individual velocity components will not be what they were before the collision, and so the magnetic moments of the particles will be different after the collision. If the magnetic moment of one particle is now too low, that particle will encounter the atmosphere on the next approach to the

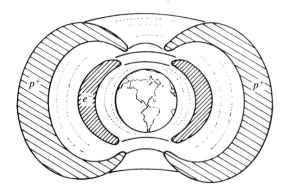

FIGURE 9.6
The earth's radiation belts.

pole and will be lost. Magnetic "bottles" do very well when asked to retain a single particle but become leaky as the particle density increases.

The radiation belts about the earth thus continually lose particles to the earth's atmosphere. Since this loss is most pronounced near the poles of the earth, that is where auroras are most frequent. The charged particles encounter air molecules, and in colliding with them lose energy in exciting the air molecules, which then fluoresce. Weather satellites photographing the earth's poles have shown that there is almost always a continual band of auroral activity around each pole. Auroral activity is visible from lower latitudes only under one of two special conditions. Either the earth's magnetic field can change (it is not constant), causing particles in previously stable orbits to enter the atmosphere, or large numbers of particles can be dumped into the Van Allen belts from the solar wind. This will greatly increase the collision rate in the belts, leading to intense auroral activity, disruption of some radio communications, and greatly increased radiation hazard for spacecraft, even those in low-altitude orbits.

There are two main radiation belts about the earth, as shown in Figure 9.6. Because the reflection points for electrons (light) and ions (principally protons, heavy) are different, the inner belt consists mainly of electrons, while the outer belt consists mainly of protons. At about the same energy per particle, the electrons move quite a bit faster than the ions, since their mass is only $\frac{1}{1800}$ that of a proton. Electrons can only be contained in the stronger magnetic field close to the earth's equator.

To maintain the population of the Van Allen belts against losses to the atmosphere, a source of particles is necessary. This source is the solar wind, a stream of plasma flowing outward from the sun. Its source is the solar corona, the hot $(2 \times 10^9 \, \text{K})$ outer atmosphere of the sun. The speed of the solar wind averages about 400 km/s, although this varies somewhat. Much more variable is the density of particles in the solar wind. This density can increase dramatically after a major solar flare. The solar wind carries along with it the magnetic field of the sun. In the reverse of the process whereby a single

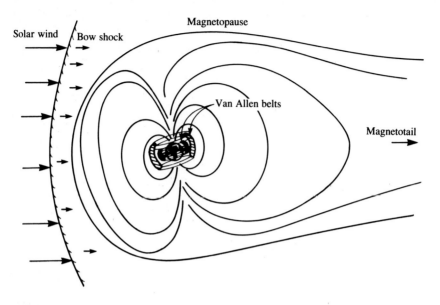

Solar wind Bow shock Magnetopause Van Allen belts Magnetotail

FIGURE 9.7
Structure of the earth's magnetosphere.

particle is trapped by a magnetic field line, large numbers of charged particles can drag the field along with them.

When the solar wind hits the magnetic field of the earth, there occurs a complicated combination of aerodynamic, plasma, and magnetic effects. Because the solar wind is hypersonic with respect to the earth, there is a shock wave upstream of the earth. This is the "bow shock" and is normally located about 10 earth radii out, but this too can vary depending on the pressure of the solar wind on the magnetic field of the earth. Inside the bow shock is the magnetopause, the boundary between the magnetic fields of the earth and the sun. On the leeward side of the earth, the magnetic field of the earth is drawn out into the magnetotail, which is at least several million kilometers long. The large-scale structure of the earth's magnetic fields is shown in Figure 9.7.

At the bow shock, the solar wind is decelerated to subsonic velocities with respect to the earth. Then, at the magnetopause, some solar-wind particles diffuse across the boundary, switching their allegiance from the magnetic field of the sun to that of the earth. Those which continue to diffuse inward in the earth's magnetic field eventually reach a point where they enter stable, trapped trajectories, forming the Van Allen belts. Their residence in the belts is only temporary, however, since they eventually diffuse into trajectories which cause them to enter the earth's atmosphere. Outside the bow shock, a spacecraft is completely naked to the solar wind. Inside the magnetopause, a spacecraft will normally encounter only decelerated solar-wind particles diffusing inward. However, within the belts, the charged-particle density is quite high, sufficiently high to kill an unshielded human being in less than one day's exposure. Only in the narrow region within the inner Van Allen belt and away from the earth's poles is there found a region of low-radiation density. It is within this region that almost all manned spaceflight takes place.

9.7 RADIATION EFFECTS

High-energy charged-particle radiation in space can represent a danger to both manned and unmanned spacecraft. The danger that ionizing radiation poses to biological organisms has been studied for many years. The radiation dosage in living creatures is measured by the rad (*r*adiation *a*bsorbed *d*ose). Radiation damages living creatures by disrupting the structure of the special molecules that form the chemical machine which keeps us running. In small quantities, radiation simply produces small quantities of damaged molecules, most of which the body simply replaces or recycles. The one exception is the DNA molecule, which carries the basic instructions for life. Damage to the genetic material is cumulative, since the DNA molecule is not repaired, and so long-term exposure to low-radiation levels can increase the possibility of undesirable mutations in offspring. At higher levels, radiation produces enough damage that the biochemical machinery is disrupted, and the fragments of damaged molecules can poison the body. Vomiting and nausea are common early symptoms of radiation sickness. Longer-term effects include the disruption of the blood-forming cells in the bone marrow, with the serious consequence of suppressing the body's immune system.

When the radiation dose occurs in a short period of time, so that the body does not have time to regenerate itself, a healthy human can tolerate doses up to about 50 rads. As the dose increases into the hundreds of rads, vomiting and nausea are the first symptoms to appear. Between 300 and 600 rads, many people will survive the immediate effects of the radiation, but an increasing percentage will die weeks later due to the collapse of their immune systems. A dose of 1000 rads will produce immediate incapacitation, and no survivors are expected.

On the earth's surface, we are protected by both the earth's atmosphere and the earth's magnetic field. The magnetic field of the earth is sufficient to deflect charged-particle radiation from the solar wind and most low-energy primary cosmic rays. Only very high energy (relativistic) cosmic rays are able to directly strike the atmosphere, and the air absorbs most of those. However, in space astronauts must depend on the earth's magnetic field alone, assuming their spacecraft is below the Van Allen Belts. In a 300-km-altitude circular orbit, the daily radiation dose to an unshielded human being would be only about 0.1 rad. However, at the maximum intensity point within the Van Allen belts (at about 3000 km over the equator), an astronaut would receive a dose of about 300 rads *per day*. Apollo spacecraft are the only manned vehicles ever to penetrate this region, and they crossed the belts at moderately high latitudes and during the highest speed portion of their flight, so the accumulated dose to their crews was within acceptable limits. Above the Van Allen belts, the dosage rate will depend greatly on the phase of the 11-year sunspot cycle and the occurrence of major solar flares. Even in 300-km low orbits a major solar flare can deliver a dose of 150 rads to an unshielded astronaut. Outside the Van Allen belts, the same flare would rapidly deliver a fatal radiation dose to the same astronaut.

There are two methods available to protect a spacecraft's crew. The first method consists of running for home. In low orbits it should be possible to return to the earth's surface in a short period of time. Even if the space shuttle were only permitted to land within the continental United States, the worst case would involve about an 8-h wait in orbit, still protected from the worst radiation by the earth's magnetic field. A manned vehicle in geosynchronous orbit, however, would also require about 8 h to return to the earth, but most of that time would be spent outside the earth's magnetic field, exposed to the full force of a major solar flare. Similarly, the minimum return time from the moon is about 2 days. In the event of a major solar flare, emergency return to the earth simply cannot be done fast enough to save the lives of the crew in these last two cases.

If it is not possible to run for home, then the crew must be shielded. The earth's atmosphere provides about 10 metric tons per square meter ($10 \, Mg/m^2$) of shielding. Long-duration missions to the planets will have to provide at least half this figure to ensure the safety of the crew. The entire crew area may not have to be covered with a $\frac{1}{2}$ m of lead, however. The major danger comes from solar flares, so provision of a "storm shelter" large enough to contain the crew for several days combined with moderate shielding over the rest of the crew area should suffice. Also, the material of which the shielding is made is much less important than its sheer mass. By proper arrangement of equipment, supplies, and the vehicles' structure and fuel, these can all contribute to shielding the crew.

It is very necessary to not skimp on the amount of radiation shielding, because insufficient shielding may actually *create more radiation than it stops.* High-energy charged particles can trigger "pair production" when they pass near another atom. In this process kinetic energy of the particle is converted into mass according to the famous $E = mc^2$, resulting in the creation of two new particles. These new particles will have lower energy than the original parent, and the parent particle has lost energy too, but they can still contain more than enough energy to cause biological damage. A single relativistic particle can create a shower of thousands of new secondary particles, each with enough energy to cause harm. So, shielding must not only stop most of the original particles but it must also stop most of the secondary shower particles. If this is not true, the crew would be safer outside the spacecraft than they are inside.

Furthermore, low-energy charged particles interact more readily with matter than do high-energy charged particles. This occurs because a subatomic particle also behaves as a quantum wave, whose wavelength λ is given by

$$\lambda = \frac{hc}{E} \tag{9.31}$$

where E is the particle energy, c is the speed of light, and h is Planck's constant. A high-energy particle will have a very small wavelength, so it can easily slip through solid matter, which is mostly empty space. A particle with a wavelength larger than the characteristic size of atoms (about 1 angstrom, or 0.1 nm) can hardly avoid hitting them, however. So, lower-energy particles cause much more damage than very high energy particles. A shield system which simply slows the particles down, creating showers of low-energy secondary particles in the process, is not performing its function.

Charged-particle radiation can also cause difficulties with electronic hardware. This was not true in the early days of the space program, when electronics meant vacuum tubes and relays. However, it has become an increasing problem with the introduction of the transistor, the microcircuit, and VLSI "single-chip" computers. Charged particles propagating through a vacuum tube do not interact with anything. At worst, in *exceedingly* high levels, they constitute a current flowing through the tube that was not anticipated. However, semiconductor materials are another matter entirely. In these materials, a nearly perfect crystal of silicon or germanium is "doped" with impurity atoms to give it the desired electronic properties. High-energy charged-particle radiation can propagate down the channels in a perfect crystal with very little interaction. However, when imperfections are present in a crystal lattice, the radiation will interact mainly with the imperfections. So, semiconductors literally channel the radiation to where it will do the most harm. As digital electronic devices are fabricated at ever smaller sizes, it takes progressively less and less radiation exposure to produce "hard errors": permanent circuit faults. Solar cells are also semiconductor devices and suffer a progressive loss of efficiency when subjected to charged-particle radiation.

Also, as microcircuits have become smaller, it takes less and less energy to produce a "soft error" in digital circuitry. In a soft error, a charged particle creates enough secondary particles within one cell of a microcircuit to change the digital state of that element. Even on the earth's surface, modern microcomputers using 64K or 256K dynamic RAM memory chips commonly incorporate parity-checking circuitry in their memory banks, because a single cosmic-ray primary can easily change a memory element. A soft error does not produce any lasting damage to the spacecraft's electronics. However, if the spacecraft cannot detect that an error has occurred, the error may produce totally unexpected (and possibly catastrophic) results as the spacecraft executes incorrect commands.

Charged-particle radiation can produce one other potentially catastrophic effect on the electronics of a spacecraft. The spacecraft can accumulate a charge of static electricity, which can discharge and destroy electronic circuitry.

Spacecraft accumulate electrons from the neutral plasma surrounding them, since electrons can be held within most materials far better than ions. This can lead to potential differences as high as 10 kV across a spacecraft. The current involved during the discharge is small, but the voltage can be high enough to destroy electronic components. This effect does not occur in the Van Allen belts, where the plasma is sufficiently conductive to short out any static-charge accumulation. It does not occur below the Van Allen belts, where air molecules ionized by collision with the spacecraft serve the same function. Nor does it usually occur if a spacecraft is in sunlight, where solar ultraviolet radiation will knock electrons out of the spacecraft itself (the photoelectric effect) and short out any static charge. However, the effect does occur for satellites in geosynchronous orbit when, twice a year, they are repeatedly eclipsed by the earth. During the time of the eclipse, the spacecraft can accumulate enough charge to destroy itself. The communications satellite DSCS II was lost in this way.

9.8   METEORS, METEORITES, AND IMPACT

The earth is bombarded by a constant rain of small bodies from space. These are called *meteoroids* in deep space, *meteors* when we watch them streak across the night sky, and *meteorites* when they are recovered on the ground. (The separate naming conventions seem to be an example of the old proverb about "a bird in the hand . . ."). Meteors and meteorites are indeed largely different. The random meteors we can see at the rate of a few per hour on any clear night are the size of sand grains or smaller. Rocks large enough to survive entry into the earth's atmosphere and land in museum collections are rather rarer. Meteors have been studied by astronomers for many decades. Camera networks established before the Second World War have obtained simultaneous photographs of many thousands of meteors from two or more observing sites. From this type of data, the path of the meteor through the atmosphere can be calculated, and its velocity found.

Meteors always penetrate the earth's atmosphere at less than 72 km/s. The significance of this velocity can be found from elementary celestial mechanics. It is the velocity with which a meteor in an escape orbit passing through the solar system would encounter the earth in a head-on collision. Since meteor velocities appear to always be less than this value, it appears that meteors are long-term residents of the solar system and that visitors from the outside are quite rare, if they exist at all. Some meteors are members of meteor showers. These occur at the same time each year as the earth passes through the orbit of a swarm of small particles. Several of these swarms have been definitively associated with the orbits of comets. These objects are dirty snowballs several miles in diameter and consist of ices of water, methane, and ammonia combined with small dust particles. When a comet approaches the sun, some of the ices vaporize, forming the gas tail, and the dust particles are swept outward to often form a separate dust tail. Particles which are too large to be ejected from the solar system by the pressure of sunlight continue to follow the orbit of the comet, and eventually the comet's orbit is populated with a diffuse band of small dust particles. So, the first source of meteors is comet dust.

The second source of meteors is more interesting. Three times since the establishment of the meteor camera networks a very bright meteor was observed from at least two sites (enabling its orbit to be calculated), and associated fragments were found. Objects large enough to survive entry have masses of several kilograms before they hit the atmosphere and must be much stronger than flimsy comet dust. They are rocks: the meteorites that can be seen in most major museums. There are many recognizable types of meteorites, but they fall into three major classes: stones, irons, and stony irons. Iron meteorites are the most common type in museums, since they are totally out of place in any terrestrial geological setting and are easily recognized. They are essentially a high grade of stainless steel: iron with a large amount of nickel in alloy form. However, the alloys in iron meteorites cannot be readily produced on the earth. The nickel-iron meteorites have been subjected to a very long period of cooling from the molten state, as evidenced by the large size (several centimeters) of individual crystals of different alloys. To achieve this slow cooling rate, they must have been buried under at least several tens of kilometers of overlying rock. So the second place to look for the source of meteorites is the asteroid belt between Mars and Jupiter. The three orbits of meteorites actually recovered reach from the earth's orbit outward to the area of the asteroid belt.

Although iron meteorites are the most easily recognized in the environment, stony meteorites are the most common type actually seen to fall. These range from highly metamorphosed igneous rocks to breccias (fragments of rocks welded together) to the carbonaceous chondrites. This latter type is the most common type of meteorite in the solar system, but utterly alien to the earth. They can be described as millimeter-sized glass beads held together with road tar. Apparently silicates condensed in the early solar nebula as molten droplets, which then cooled to form silicate beads. When the temperature in the solar nebula dropped still further, long-chain hydrocarbons condensed, coating the silicate spheres and making them sticky. They aggregated to form small objects, which aggregated to form larger objects, which formed asteroid-sized objects as the process continued, finally leading to several major planets in the inner solar system and a small remnant of debris that never came together. These last objects are the present asteroids.

The infall rate for meteorites has been studied from the ground, from satellites, and by locating and surveying large impact sites on the earth's surface. As derived by Vedder (1966) the number of meteorites N per second of mass greater than m impacting on a 1-cm^2 surface is given approximately by

$$\log_{10} N = -17 - \log_{10} m \tag{9.32}$$

at the earth's distance from the sun. This law is apparently valid for meteor masses in the range

$$10^{-13} \, \text{g} < m < 10^{16} \, \text{g}$$

or from microscopic dust to an object which would be several hundred meters across. This ranges from objects which would only produce small pits on a spacecraft to bodies capable of destroying a city. For a large object impacting at many kilometers per second can produce a *very* large crater.

The minimum speed of a meteor near the earth is escape speed from the earth, or 11 km/s. However, typical meteor speeds are likely to be twice or three times this figure, since most objects do not orbit the sun in exactly the same orbit as the earth. These speeds are very much larger than the speed of sound in any physical material, so when objects collide at these speeds it is termed a *hypervelocity impact*. The effects of such an impact are dictated by two facts. First, since the impact occurs at speeds faster than that of sound in both the meteorite and the target, the material which finds itself caught in between cannot get out of the way. Second, the kinetic energy of the meteorite is many times greater than that needed to totally vaporize the meteorite, and a large amount of target as well. Thus, in a hypervelocity impact the meteor and target material is progressively transformed into incandescent vapor which is contained by the meteor "above" and the target "below." This incandescent gas has nowhere to go: It cannot flow out of the way, because this must occur at velocities limited by the speed of sound. The properties of the target have almost nothing to do with the size of crater produced; only the latent heat of vaporization is important. When the meteorite is totally vaporized, the gas is free to expand in all directions, driving shock waves through the material which enlarge the crater to many times the diameter of the original meteorite.

Crater formation can be studied in the laboratory by using light-gas guns to fire projectiles at known speeds at known targets. The size and shape of the crater formed has very little to do with either the type of projectile or the strength of the target. The kinetic energy per unit mass carried by the projectile usually far exceeds the latent heat of vaporization of any known material. Thus, the entire projectile and a large volume of the target are vaporized. Also, the energy in the shock wave propagating outward from the impact site will initially far exceed the mechanical strength of any material. So, large volumes of the target will be liquified, and further outward from the impact target material will be shattered and ejected by the shock. Craters are thus far larger in size than the impacting body. The size of the crater produced thus depends only on the total energy input, or the mass of the object and the impact speed. Figure 9.8 shows a plot of crater size against the mass of the impacting object for several impact speeds. It is clear from this figure that it is not possible to armor a spacecraft against impact with any macroscopic object. Fortunately, (9.32) shows that macroscopic meteors are extremely rare. In the early 1950s, Fred Whipple proposed shielding spacecraft from micrometeorites with a "meteor bumper," a thin sheet of metal suspended several centimeters in front of the main spacecraft hull. Small objects would be broken up and

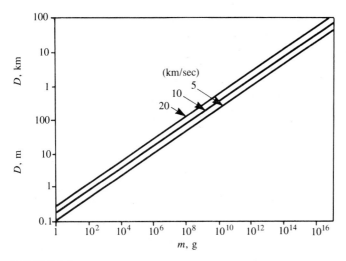

FIGURE 9.8
Crater diameter versus impacting mass at several impact speeds. (*From* Moons and Planets, *by William K. Hartmann,* © *1972 by Wadsworth Publishing Co., Inc. Reprinted by permission of the publisher.*)

considerably slowed by this first impact. Microscopic meteor hits are probably the major cause of infrequent losses of spacecraft-attitude stability.

However, there is an impact danger to spacecraft which is larger than the danger due to natural meteors, and this danger is rapidly increasing. Over 30 years of spaceflight has left many dead spacecraft, empty rocket stages, and random artificial debris in orbit. Of these, the debris is far the most numerous. Explosions in orbit have left thousands of *macroscopic* objects in orbit as spent upper stages detonate, sometimes years after reaching orbit. (This is one major reason payloads are separated from the final booster stage. It is a potential bomb.) These objects will encounter another spacecraft at velocities of several kilometers per second, so their damage potential is just as high as a natural meteor. Also, studies have indicated that the probability of serious damage from artificial debris is already at least an order of magnitude higher than the natural danger. The large Pageos balloon satellite was almost certainly hit and destroyed by a cataloged piece of orbital debris. Also, a space shuttle orbiter has already returned from orbit with a small crater in its windshield caused by an encounter with orbiting debris: The crater was lined with residual artificial materials.

9.9 OUR LOCAL NEIGHBORHOOD

All human activity takes place within our solar system. This system of one star, nine planets, over fifty moons, and assorted thousands of asteroids and comets is, for the foreseeable future, the stage for human activity. Actually, human activity is still limited to the immediate vicinity of the earth. Even most unmanned space activity takes place close to the earth, since interplanetary probes are not common. Furthermore, the scale of the universe is vast beyond our usual imagination.

One of the best ways to visualize the scale of the solar system is to imagine building a scale model. Let us begin with a standard, common 31-cm (1-ft) globe of the earth. On this scale, the earth's atmosphere, less than 81 km (50 mi) thick, is only 1.8 mm (0.07 in) deep on the globe. The earth's breathable atmosphere is probably thinner than the coat of lacquer used to protect the finish of the globe itself. The figure of 81 km, or 50 mi, has another significance: It is the altitude needed to earn Astronaut's Wings from NASA. However, the minimum practical altitude for orbiting the earth is about 93 km (150 mi), or 0.59 cm (0.23 in). All space shuttle missions take place within about this distance from the earth. The shuttle reaches orbit with very little residual maneuvering ability and cannot depart very far from the earth. Most unmanned earth satellites also operate within a few hundred miles from the earth. These satellites are either to study the earth itself, so large separation is not desired, or to study the outer universe, in which case just being above the atmosphere is usually sufficient.

About 1.07 m (3.5 ft) from the earth, in the plane of the equator, is geosynchronous orbit. This orbit provides a seemingly stationary vantage point which is used by communication, weather, and other satellites. It is very useful for providing coverage of large sections of the earth at one time. Still further out, at about 9 m (30 ft) from the earth, is the orbit of the moon. This marks, at least for the moment, the furthest penetration of human beings into the universe. The moon is an airless world, still bearing the scars of its assembly during the last stages of the formation of the solar system. On our scale, the moon would be about 7.7 cm (3 in) in diameter, or about the size of an orange.

When we venture further afield, the sun is the next most obvious object. Although it appears to be about the same size as the moon in our sky, it is located (in our model) 3.5 km, or 2.2 *miles*, from the 31-cm (1-ft) earth, and would be almost 18 m (60 ft) in diameter. This distance is called the *astronomical unit* (AU) and is about 93,000,000 mi, or 150,000,000 km. The sun is a small star, as stars go, and as such burns its nuclear fuel at a pedestrian rate. It is about 4.6 billion years old (a number obtained both from radioisotope dating of solar system rocks and from theoretical models of star evolution) and has about another 5 billion years of life left. It is a large sphere of hydrogen (~80%), helium (~15%), and a slight admixture of everything else. In this it mirrors the average composition of the universe as a whole. Its surface temperature is about 6000 K, but the interior is much hotter.

Between the Earth and the sun are two other planets. Mercury is another cratered, airless world only slightly larger than the moon, about 11.5 cm (4.5 in) across in our model. It is considerably more massive than the moon, however, indicating that it has a larger share of heavier elements than does our moon. Mercury's rotation is locked to its orbital period, performing three rotations on its axis for every two orbits around the sun. Of course, because of its location only 137 m (450 ft) from the sun in our model, Mercury holds the undesirable distinction of being the second hottest planet in the solar system.

The destinction of being the hottest goes to the second planet, Venus. Venus is nearly identical in size to the Earth and is located 2.6 km, or about 1.6 mi, from the sun. It owes its surface temperature of over 800 K to the fact that it is completely covered by clouds. In the atmosphere of Venus the greenhouse effect has run wild. The atmosphere is moderately transparent to normal visible sunlight but is nearly opaque to the infrared-heat radiation the surface tries to radiate back into space. As a result, the temperature of the surface has increased to the point where radiation balance is achieved but at a very uncomfortable temperature. The surface has never been seen through the clouds, which have different compositions at different levels. At least one cloud level appears to be droplets of sulfuric acid. The surface pressure is also high, over 90 atm.

Beyond the orbit of the Earth, the next planet is Mars. Located about 5.3 km, or 3.3 mi, from the sun in our model, it can come as close to the Earth as 1.61 km, or 1 mi, away, again to scale. Mars has the distinction of being the only other planet in the solar system not instantly, totally, and utterly lethal to an unprotected human being. This is not to say that an unprotected human would live long on the surface of Mars, only that death would simply take somewhat longer. Mars is less than half the size of the Earth (12.7 cm, or 5 in, to scale) and has an atmosphere of nearly pure carbon dioxide with a surface pressure of about 0.006 that of Earth. The average temperature is also much lower, so low in fact that the weather patterns on Mars are dominated by the thawing of the atmosphere in the summer hemisphere of the planet, while the atmosphere freezes out at the winter hemisphere polar cap. There are gigantic volcanoes and rift valleys on Mars, indicating that the planet has had an active geology in the not too distant past. Also, there is hard evidence that some time in the past Mars' climate was considerably milder: There are fossil riverbeds on Mars.

Beyond Mars is a long gap, populated only by the asteroids. These have been discussed in the section on meteors. However, at 18.5 km, or 11.5 mi (scale), from the sun, we find Jupiter. Over 3.65 m (12 ft) in diameter in our model, Jupiter is a completely different type of planet from those in the inner solar system. The inner worlds are rocky; Jupiter is a gas planet. Furthermore, it is the only object in the solar system (other than the sun) that is massive enough to retain hydrogen. Jupiter is, in fact, less than a factor of 20 in mass away from being able to sustain fusion reactions in its core; it is a star that didn't quite make it. It controls a miniature solar system in its own right. There are four large moons (one, Ganymede, is larger than Mercury), and at least a dozen smaller moons. These moons are also radically different from moons in the inner solar system: They are largely composed of water ice.

The radical shift in composition in the outer solar system is a matter of the chemical composition of the material in the solar nebula (the stuff from which the solar system formed) and of temperature (distance from the sun). In the inner solar system, only materials able to solidify at high temperatures ever condensed. These include silicon dioxide, aluminium oxide, iron and nickel, and various other substances familiar on earth. The most common substances in the solar nebula, hydrogen and helium, could not condense at all in the inner solar system. The next most common substances, carbon, nitrogen, and oxygen, could only condense if they could be incorporated in minerals stable at moderately high temperatures. However, the situation totally reverses in the outer solar system. There, carbon, nitrogen, and oxygen condensed as their hydrides: methane, ammonia, and water. These three materials form the bodies of the outer planets, and the most stable of these, water, apparently forms the bodies of the moons of the outer planets.

Beyond Jupiter are three other gas giant planets: Saturn at 33.8 km, or 21 mi (scale), from the sun, Uranus at 67.6 km (42 mi), and Neptune at 106.2 km (66 mi). On our scale, Saturn is 2.9 m (9.5 ft) across, while Uranus and Neptune are about 1.1 m (3.5 ft) in diameter. All three are composed largely of methane, ammonia, and water, with at most a small rocky core. They never became large enough to capture much raw hydrogen from the solar nebula. Each one is, of course, colder than the last, although no gas giant planet apparently has a solid surface. Saturn again has a miniature solar system with at least 15 moons, Uranus has five large moons and several small ones, while Neptune has only two known moons. Neptune has yet to be visited by spacecraft, so this last number is subject to revision upward.

At 140 km, or 87 mi, from the sun is Pluto. Pluto is small, probably less than 13 cm, or 5 in, to scale, and has one moon, Charon, which has never been seen separate from the blurry image of Pluto itself. Only four artificial objects to date will penetrate as far as Pluto, and only one of these, Pioneer 11, has already done so. Pluto is not a gas giant planet: It is too cold to support methane, ammonia, or water as a vapor. Beyond Pluto the only other members of the solar system are the comets, orbiting blocks of frozen methane, ammonia, and water ice. Most comets are in very eccentric, long-period orbits around the sun and therefore spend most of their time near the aphelion of their orbits. On our scale, these would be from several thousand to several tens of thousands of miles (scale) from the sun.

Beyond the comets, there is a great open space before the nearest stars are encountered. In this region of the galaxy, stars are separated by about 3 light-years, where the light-year is the distance light travels in a year. To our scale, the nearest stars are about *half a million miles away*. Our galaxy itself is several thousand times larger than this distance and contains at least 10^{10} stars. Beyond our galaxy are others, probably at least another 10^{10} of them. Even the most grandiose human ambitions pale in comparison with the size of the universe. Even though we could conceivably expand throughout the solar system, when faced with the distances between the stars, Robert Browning's couplet is still true:

Ah, but Man's reach should exceed his grasp,
or what's a heaven for?

9.10 REFERENCES AND FURTHER READING

The realization that space is not "empty" was one of the first discoveries of the space program. The study of plasmas in space and their interaction with magnetic fields has grown into a discipline now called *space physics*. The book by White is an excellent introduction to this area, while at a more advanced level Rossi and Olbert discuss the physics of trapped plasmas in the solar system.

It is only within the last 30 years that human beings have attempted to engineer devices to operate within the space environment. The vast body of experience in engineering for the terrestrial environment does not necessarily apply to the new medium, and progress is hampered by the fact that virtually all spacecraft subjected to long exposure to space are not recovered. The two *NASA Technical Memorandums* cited below are valuable compendiums of what is known about the space environment in the vicinity of the earth and most of the other planets. Current information on the space environment, material properties and responses, and design guidelines for space vehicles is also available in a computerized database EnviroNET, managed by the Goddard Spaceflight Center, Greenbelt, Md.

For the student who wishes to pursue the study of the solar system, the works by William Hartmann and by Beatty, O'Leary, and Chaikin are recommended. The latter, especially, contains spacecraft results through the Voyager flybys of Saturn.

White, R. S., *Space Physics,* Gordon and Breach, New York, 1970.

Rossi, B., and S. Olbert, *Introduction to the Physics of Space,* McGraw-Hill, New York, 1970.

Smith, R. E., and G. S. West, "Space and Planetary Environment Criteria Guidelines for Use in Space Vehicle Development," *NASA Tech. Memo.* 82478, vol. I, and *NASA Tech. Memo.* 82501, vol. II.

Hartmann, W., *Moons and Planets,* Wadsworth, Belmont, Calif., 1972.

Beatty, J. K., B. O'Leary, and A. Chaikin, *The New Solar System,* Sky Publishing, Cambridge, Mass., 1982.

Vedder, J. F., "Minor Objects in the Solar System," *Space Sci. Rev.,* vol. 6, p. 365, 1966.

CHAPTER
10

THE
RESTRICTED
THREE-BODY
PROBLEM

10.1 INTRODUCTION

The restricted problem of three bodies holds a special place in the history of celestial mechanics. It is the simplest unsolved and *unsolvable* gravitational problem. In the restricted problem two massive objects, the primaries, execute simple circular orbits about their center of mass. A third body of infinitesimal mass is introduced into this system. The third object is small enough that it does not influence the motion of the primaries, while its motion is determined by the gravitational field of the primary masses. So, although the name of the problem includes the phrase "three-body," the restricted problem is in reality a one-body problem, since the motion of the two primaries is known.

The restricted three-body problem was first defined by Leonard Euler in 1772 in connection with his research on the motion of the moon. He had actually proved the existence of the colinear libration points 7 years earlier, in 1765. Lagrange found the triangular libration points L_4 and L_5 in the same year Euler originated the restricted problem, 1772. The one exact integral of this system was found by Jacobi in 1836. Henri Poincaré proved in 1899 that this is the *only* exact constant of the motion. It was Poincaré who first saw that the restricted problem was not simply unsolved but actually unsolvable in closed form. Although the solution to this problem does exist, it is not an analytic, differentiable function of both the initial conditions and the time. In fact, modern research has shown that the solution space of this problem is *infinitely complex*. It is thus a prototype for many nonlinear dynamical systems which have resisted attempts to find a closed-form solution.

The restricted three-body problem appears often in the actual solar system. It was first formulated by Euler to study the motion of the moon about the earth, perturbed by the sun. If the two primary objects are the sun and Jupiter, then the motion of asteroids and comets can be studied. Taking the primaries to be the earth and the moon, we can study the motion of a spacecraft within the earth-moon system. This study had important conse-quences for the Apollo program, although the influence of the sun cannot be neglected in this case. Finally, most other stars are actually double-star systems. If such systems have planets, the restricted problem can indicate where stable planetary orbits might exist. In this chapter we will discuss the restricted problem assuming the earth-moon-spacecraft problem is the desired application. Other important uses for this dynamical system are possible.

10.2   EQUATIONS OF MOTION

During the 200 years in which the restricted problem has been studied, a convention has emerged on the units and coordinate frames for this problem. In any dynamical problem, we have the choice of three basic physical units: mass, length, and time. The obvious choice for the unit of length is the distance between the two primary objects, so

$$a_{12} = 1 \qquad (10.1)$$

The mass of the third object is insignificant, so it is standard to choose the unit of mass to be the sum of the masses of the two primaries, $m_1 + m_2 = 1$. Then, if we put the mass of the smaller primary $m_2 = \mu$, the mass of the two primary objects is given by

$$m_1 = 1 - \mu \qquad m_2 = \mu \qquad (10.2)$$

Finally, we have the choice of the time unit. Instead of chosing this unit directly, let us pick the time unit such that the gravitational constant $G = 1$. Then, by Kepler's third law, the period of the two primaries in their orbit about each other will be

$$T_{12} = 2\pi \left[\frac{a_{12}^3}{G(m_1 + m_2)} \right]^{1/2} = 2\pi \qquad (10.3)$$

Thus, the choice of the gravitational constant implicitly sets the time unit.

The restricted problem is usually posed in a reference frame rotating with the orbital motion of the primary objects. The origin of this frame, shown in Figure 10.1, is at the center of mass of the two primaries. The frame is arranged such that the two massive objects lie on the s_1 axis. This means that the more massive primary, mass $1 - \mu$, is located a distance μ from the origin

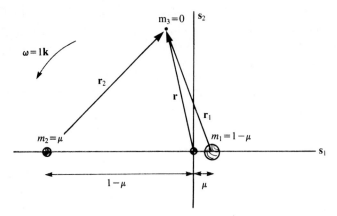

FIGURE 10.1
Geometry of the restricted problem.

of the frame, while the lesser primary, with a mass of μ, is placed a distance $1 - \mu$ on the opposite side of the s_1 axis. Notice that since we have introduced nondimensional units, *anything* can be measured in terms of μ. The quantity μ is called the *mass ratio* of the restricted problem. The use of the symbol μ is standard, but it must not be confused with the μ of the two-body problem, a very different quantity.

The rotating reference frame s has an inertial angular velocity of $\omega = 1\mathbf{k}$, since the rotational period of this frame is 2π. To apply Newton's second law, we must calculate the inertial acceleration of the small mass, but we are free to express this acceleration in the unit vectors of the rotating frame s. The inertial acceleration is given by

$$\frac{{}^i d^2}{dt^2}\mathbf{r} = \frac{{}^s d^2}{dt^2}\mathbf{r} + 2\boldsymbol{\omega} \times \frac{{}^s d}{dt}\mathbf{r} + \boldsymbol{\omega} \times (\boldsymbol{\omega} \times \mathbf{r}) \qquad (10.4)$$

where the position vector of the small mass is

$$\mathbf{r} = x\mathbf{s}_1 + y\mathbf{s}_2 + z\mathbf{s}_3 \qquad (10.5)$$

and all time derivatives on the right of (10.4) are taken with respect to the rotating frame. Thus

$$\frac{{}^s d}{dt}\mathbf{r} = \dot{x}\mathbf{s}_1 + \dot{y}\mathbf{s}_2 + \dot{z}\mathbf{s}_3 \qquad (10.6)$$

$$\frac{{}^s d^2}{dt^2}\mathbf{r} = \ddot{x}\mathbf{s}_1 + \ddot{y}\mathbf{s}_2 + \ddot{z}\mathbf{s}_3 \qquad (10.7)$$

Working out the cross products in (10.4), the inertial acceleration of the small mass, resolved on the rotating frame, becomes

$$\frac{{}^{i}d^2\mathbf{r}}{dt^2} = (\ddot{x} - 2\dot{y} - x)\mathbf{s}_1 + (\ddot{y} + 2\dot{x} - y)\mathbf{s}_2 + \ddot{z}\mathbf{s}_3 \tag{10.8}$$

To complete the equations of motion, we must still calculate the gravitational acceleration on the small mass.

The gravitational force is an inverse-square law, so it is necessary to calculate the radius vectors from each of the primary masses to the third body. The distance between the larger primary and the third object is

$$r_1 = [(x - \mu)^2 + y^2 + z^2]^{1/2} \tag{10.9}$$

since the larger primary is located at $(\mu, 0, 0)$ and the third object is at (x, y, z). Similarly, the distance from the smaller primary (located at $(-1 + \mu, 0, 0)$) to the third object is

$$r_2 = [(x + 1 - \mu)^2 + y^2 + z^2]^{1/2} \tag{10.10}$$

Finally, the gravitational acceleration of the third mass is given by

$$\mathbf{a}_g = -\frac{(1 - \mu)\mathbf{r}_1}{r_1^3} - \frac{\mu\mathbf{r}_2}{r_2^3} \tag{10.11}$$

We are now ready to obtain the equations of motion for this sytem.

The system equations of motion are found by equating the inertial acceleration of the third object to the gravitational acceleration it experiences. Since the third object has "zero" mass, this avoids the necessity to divide the mass of this object from both sides of $\mathbf{F} = m_3\mathbf{a}$. The assumption of zero mass for the third object is more important in the orbit of the two primaries, since we have assumed that these are following a pure keplerian circular orbit. This is true only in the limit that $m_3 \rightarrow 0$. Now, the equations of motion, broken out into the three components in the rotating frame $\mathbf{s}$, are

$$\ddot{x} - 2\dot{y} - x = -\frac{(1 - \mu)(x - \mu)}{r_1^3} - \frac{\mu(x + 1 - \mu)}{r_2^3} \tag{10.12}$$

$$\ddot{y} + 2\dot{x} - y = -\frac{(1 - \mu)y}{r_1^3} - \frac{\mu y}{r_2^3} \tag{10.13}$$

$$\ddot{z} = -\frac{(1 - \mu)z}{r_1^3} - \frac{\mu z}{r_2^3} \tag{10.14}$$

The radii r_1 and r_2 are given by (10.9) and (10.10). These are three second-order, highly nonlinear, coupled ordinary differential equations. The order of this system is no higher than that of the two-body problem we solved in Chapter 2, but this system has resisted attempts to find a global solution for over 200 years. This does not mean, however, that nothing about the solution to this problem has been learned in that time period.

10.3 THE LAGRANGIAN POINTS

One question to ask when faced with a complex dynamical system is: "Does it have equilibrium points?" At an equilibrium point, a particle is stationary with no velocity and no acceleration; the forces in the system balance each other. Equilibrium points cannot exist in a system where the equations of motion depend explicitly on the time. If the system forces are time-dependent, then the points at which they balance must move with time, contradicting the assumption that these points are stationary. The restricted problem does not have time-dependent forces *in the rotating frame.* In the inertial frame the two primary masses move, and the gravitational force expressions contain time explicitly. So, if equilibrium points exist, they can only exist in the rotating frame.

Recall the basic equations of motion for the restricted problem. If we wish to find equilibrium points, we need to set the rotating-frame velocity and acceleration components to zero. This leads to the three equations

$$-x = -\frac{(1-\mu)(x-\mu)}{r_1^3} - \frac{\mu(x+1-\mu)}{r_2^3} \qquad (10.15)$$

$$-y = -\frac{(1-\mu)y}{r_1^3} - \frac{\mu y}{r_2^3} \qquad (10.16)$$

$$0 = -\frac{(1-\mu)z}{r_1^3} - \frac{\mu z}{r_2^3} \qquad (10.17)$$

These three equations are three equations in three unknowns: the three coordinates of any equilibria.

First, notice that (10.17) immediately tells us that $z = 0$. Any equilibrium points must lie within the orbital plane of the two primary masses. This leaves us with (10.15) and (10.16) which, when we set $z = 0$, give us two equations in the two unknowns x and y. The first two equilibrium points are found if we suppose that $r_1 = r_2 = 1$. This reduces both (10.15) and (10.16) to identities, without appearing to specify the values of x and y. Of course, the statement that $r_1 = r_2 = 1$ contains this information implicitly: The two equilibrium points are placed at the vertices of equilateral triangles, with the two primary masses at the other vertices. These are the so-called triangular points. They are shown in Figure 10.2 and are named L_4 and L_5 in honor of their discoverer, Lagrange.

There are three other points, which were discovered by Euler before Lagrange found the triangular points. Equation (10.16) can also be satisfied if $y = 0$. Any remaining equilibrium points must lie on the x axis, and will be roots of (10.15) with $y = z = 0$. This leads to the equation

$$x - \frac{(1-\mu)(x-\mu)}{|x-\mu|^3} - \frac{\mu(x+1-\mu)}{|x+1-\mu|^3} = 0 \qquad (10.18)$$

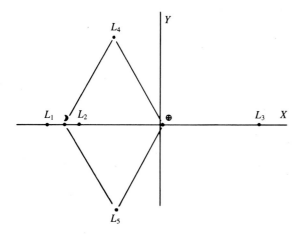

FIGURE 10.2
Equilibrium points for the restricted problem.

When the denominators are cleared, a quintic equation in x results. However, this equation never has more than three real roots for any reasonable value of μ (for example, $0 \leq \mu \leq 1$). Unfortunately, this equation must be solved by numerical techniques to obtain the positions of these three points. For small values of μ, two of these points, L_1 and L_2, bracket the smaller primary, while the third, L_3, is opposite the smaller primary at about unit distance from the larger mass. The remaining problem is to determine the stability of these points.

10.4 STABILITY OF THE LAGRANGIAN POINTS

The five lagrangian points represent equilibrium points in the restricted problem seen from the rotating frame. Seen from inertial space they are circular orbits having the same orbital period as the moon. Having established the existence of these points, the next question to be answered is: "Are they stable?" Of course, a spacecraft placed exactly at one of these points with zero speed will stay there. However, will a spacecraft placed *near, but not exactly at,* one of these points remain in the vicinity of the equilibrium? The existence of equilibria and the stability of equilibria are entirely separate questions. There have been many definitions of what is meant by "stability" in dynamics. Some criteria are so restrictive that few, if any, dynamical systems are stable. Other criteria are so hard to use that the question of stability, by that particular definition, is almost impossible to answer. However, stability problems can always be answered if we use simple *linear* stability theory.

Write the coordinates of the equilibrium point as x_e and y_e. The z_e coordinate is zero at the lagrangian points, and at the equilibria the velocity components must be zero, or it is not an equilibrium point. Now, assume that the spacecraft is placed close to the equilibrium point so that its position can be written

$$x \approx x_e + \delta x \qquad y \approx y_e + \delta y \qquad z \approx \delta z \qquad (10.19)$$

If the equilibrium point is stable, then δx, δy, and δz will all remain small. Equations (10.19) can be substituted in the equations of motion for the restricted problem. The new equations are still coupled and nonlinear equations of motion for the quantities δx, δy, and δz. They are still an exact description of the dynamics of the system and are no more amenable to solution than the standard form of the equations of motion.

However, if we assume that the quantities involving δ are very small, we can use the binomial theorem to expand the nonlinear terms in the equations of motion. For example, the quantity r_1^{-3} appearing in all three of the equations of motion is given by

$$r_1^{-3} = [(x_e + \delta x - \mu)^2 + (y_e + \delta y)^2 + \delta z^2]^{-3/2} \qquad (10.20)$$

Then expansion by the binomial theorem gives

$$r_1^{-3} \approx [(x_e - \mu)^2 + y_e^2]^{-3/2} - \tfrac{3}{2} r_{1e}^{-5}[2(x_e - \mu)\delta x + 2y_e \delta y] \qquad (10.21)$$

restricting ourselves to the xy plane. Terms involving squares or higher powers of δ quantities have been neglected. Here r_{1e} is the radius vector evaluated at the equilibrium point. If these small quantities are really small, then these higher-order terms will be negligible.

When such expansions are carried out on the complete set of equations of motion, we find

$$\delta\ddot{x} - 2\delta\dot{y} - \delta x =$$

$$- \delta x\left\{(1 - \mu)\left[\frac{1}{r_{1e}^3} - 3\frac{(x_e - \mu)^2}{r_{1e}^5}\right] + \mu\left[\frac{1}{r_{2e}^3} - 3\frac{(x_e + 1 - \mu)^2}{r_{2e}^5}\right]\right\}$$

$$+ \delta y\left\{3(1 - \mu)\frac{(x_e - \mu)y_e}{r_{1e}^5} + 3\mu\frac{(x_e + 1 - \mu)y_e}{r_{2e}^5}\right\} \qquad (10.22)$$

$$\delta\ddot{y} + 2\delta\dot{x} - \delta y = + \delta x\left\{3(1 - \mu)\frac{(x_e - \mu)y_e}{r_{1e}^5} + 3\mu\frac{(x_e + 1 - \mu)y_e}{r_{2e}^5}\right\}$$

$$- \delta y\left\{(1 - \mu)\left[\frac{1}{r_{1e}^3} - 3\frac{y_e^2}{r_{1e}^5}\right] + \mu\left[\frac{1}{r_{2e}^3} - 3\frac{y_e^2}{r_{2e}^5}\right]\right\} \qquad (10.23)$$

The z equation is similar but is not particularly instructive here. Several critical things have happened in deriving (10.22) and (10.23). First, all zero-order terms in these equations have vanished. Terms not involving the quantity δ are simply the equations of motion evaluated *at the equilibrium point,* and the equations of equilibrium are automatically satisfied at this point. Second, (10.22) and (10.23) are only a local description of motion near the equilibrium point; they are no longer globally valid equations of motion for the restricted problem. Finally, and most importantly, (10.22) and (10.23) are a set of *constant-coefficient, linear differential equations.* Everything within the curly braces in these equations is evaluated at the equilibrium point. This means that we can actually solve these equations of motion in closed form.

Let us concentrate on the triangular equilibrium points. At the L_4 point, $x_e = -\frac{1}{2} + \mu$ and $y_e = \sqrt{3}/2$. Also, $r_{1e} = r_{2e} = 1$. When these values are inserted into (10.22) and (10.23), they become, after simplification,

$$\delta\ddot{x} - 2\delta\dot{y} - \tfrac{3}{4}\delta x - \frac{3\sqrt{3}}{2}(\mu - \tfrac{1}{2})\delta y = 0 \tag{10.24}$$

$$\delta\ddot{y} + 2\delta\dot{x} - \frac{3\sqrt{3}}{2}(\mu - \tfrac{1}{2})\delta x - \tfrac{9}{4}\delta y = 0 \tag{10.25}$$

If we define the vector $\delta\mathbf{r} = (\delta x, \delta y)$, the position vector relative to the equilibrium point, then (10.24) and (10.25) can be put into matrix form

$$\begin{bmatrix} 1 & 0 \\ 0 & 1 \end{bmatrix}\delta\ddot{\mathbf{r}} + \begin{bmatrix} 0 & -2 \\ 2 & 0 \end{bmatrix}\delta\dot{\mathbf{r}} + \begin{bmatrix} -\dfrac{3}{4} & -\dfrac{3\sqrt{3}}{2}(\mu - \tfrac{1}{2}) \\ -\dfrac{3\sqrt{3}}{2}\left(\mu - \dfrac{1}{2}\right) & -\dfrac{9}{4} \end{bmatrix}\delta\mathbf{r} = 0 \tag{10.26}$$

Now, as is usual with constant-coefficient systems, assume a solution of the form

$$\delta\mathbf{r} = \mathbf{A}e^{\lambda t} \tag{10.27}$$

When (10.27) is substituted into (10.26), we find the characteristic equation

$$\begin{vmatrix} \lambda^2 - \dfrac{3}{4} & -2\lambda - \dfrac{3\sqrt{3}}{2}\left(\mu - \dfrac{1}{2}\right) \\ 2\lambda - \dfrac{3\sqrt{3}}{2}\left(\mu - \dfrac{1}{2}\right) & \lambda^2 - \dfrac{9}{4} \end{vmatrix} \mathbf{A} = 0 \tag{10.28}$$

Equation (10.28) is an eigenvalue-eigenvector problem. Its determinant expands and simplifies to

$$\lambda^4 + \lambda^2 - \tfrac{27}{4}\mu(\mu - 1) = 0 \tag{10.29}$$

This is a quadratic equation in λ^2.

The roots of (10.29) are given by the quadratic formula

$$\lambda^2 = \tfrac{1}{2}[-1 \pm \sqrt{1 - 27\mu(1 - \mu)}] \tag{10.30}$$

For stability, we must require that all four values of λ be pure imaginary numbers. The restricted problem of three bodies has no damping, and it is impossible to get four roots from (10.30) all having negative real parts. If all four λ are to be imaginary, the discriminant of (10.30) must be greater than zero. So, the stability of the triangular points changes when

$$1 - 27\mu(1 - \mu) = 0 \tag{10.31}$$

This gives the two critical values of μ as $\mu_{c1} \approx 0.03852$ and $\mu_{c2} \approx 0.96148$.

Substituting simple trial values of μ into (10.30) will show that the L_4 point is stable whenever $\mu < \mu_{c1}$ or $\mu > \mu_{c2}$. The largest value of μ in the solar system is the value for the earth-moon system, $\mu \approx 0.01213$. Hence, the triangular libration points are stable for any three-body problem occurring within the solar system.

These solutions actually appear in nature. There are two groups of asteroids orbiting the sun in the orbit of Jupiter, spaced roughly 60° ahead and behind that planet. Termed the *Trojan asteroids,* they were not discovered until long after Lagrange had proved the stability of the L_4 and L_5 points. Recently, the Voyager probes discovered several objects at the triangular points of moons within Saturn's system. However, the restricted problem is not a good model for dynamics within the earth-moon system, because the presence of the sun has been ignored. When the sun is included, the triangular libration points are found to be unstable. Our system's L_4 and L_5 points are empty.

To complete the solution, we must resort to numerical methods. For the earth-moon mass ratio, $\mu = 0.01213$, the four eigenvalues λ_i and their normalized eigenvectors $\mathbf{A}_i$ are found to be

$$\lambda_{12} = \pm 0.297931i$$

$$\mathbf{A}_{12} = \begin{bmatrix} 0.77641 \mp 0.3650i \\ 0.5137 \end{bmatrix}$$

for what is termed the *long-period mode,* and

$$\lambda_{34} = \pm 0.954587i$$

$$\mathbf{A}_{34} = \begin{bmatrix} 0.4487 \mp 0.6745i \\ 0.5869 \end{bmatrix}$$

for the *short-period mode.* The period of the short-period mode is nearly that of the lunar orbit itself. The short-period mode can be interpreted as an eccentricity of the underlying circular orbit at L_4. The period of the long-period mode is about 3 months. This mode is an oscillation of the orbit toward and away from the moon. In problems with smaller μ values, the long-period mode has a much longer period.

Figure 10.3 shows the shapes of the short- and long-period modes about L_4. The modes are shown with equal amplitudes, but the axes are unlabeled since in the linear regime the scale of the orbit is unimportant. As the size of the oscillation about L_4 becomes larger, second-order terms in δx and δy become important and the linear theory no longer holds. The general solution for motion near the equilibrium consists of a superposition of both modes:

$$\begin{bmatrix} \delta x \\ \delta y \end{bmatrix} = \sum_{i=1}^{4} \alpha_i \mathbf{A}_i e^{\lambda_i t} \tag{10.32}$$

The constants α are necessary to satisfy any general set of initial conditions.

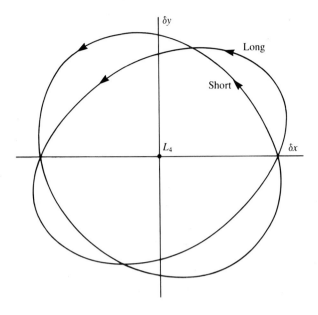

FIGURE 10.3
Short- and long-period modes
at L_4.

Figure 10.4 shows a superposition of both modes, with the long-period mode having twice the amplitude of the short-period mode. Such oscillations are actually observed in the case of the Trojan asteroids in the sun-Jupiter system and must occur for the Lagrange-point satellites in the Saturn-Dione system. However, for the earth-moon case the neglect of the sun is a serious deficiency of the restricted-problem model. Small oscillations about L_4 are not stable when the sun is included; in fact, such a satellite will usually be ejected from the earth-moon system within a month or so.

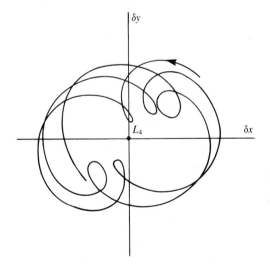

FIGURE 10.4
Superposition of modes at L_4.

10.5 JACOBI'S INTEGRAL

Our study of the two-body problem began with a search for constants of the motion. The two-body problem possesses all the standard conserved quantities: total linear momentum, total energy, and total angular momentum. If we were to follow a similar course with the three-body system of two primary masses and the small third mass, we would find that all these quantities are still conserved. However, because the mass of the third object is zero, these conservation laws would be statements about the two primaries. We would learn nothing about the possible motion of the third mass, since its coordinates and velocity, multiplied by m_3, would disappear from these conservation laws. Now, to completely solve the restricted problem, we would need to have six conservation laws, counted as scalars. The restricted problem, however, possesses *only one* exact conservation law, called *Jacobi's integral*.

To find this conservation law, we begin as if we were looking for an energy-conservation law, but in the rotating frame. Take the three equations of motion from the last section and add them together after multiplying the x equation of motion by dx/dt, the y equation of motion by dy/dt, and the z equation by dz/dt to find

$$\ddot{x}\dot{x} + \ddot{y}\dot{y} + \ddot{z}\dot{z} - x\dot{x} - y\dot{y} = -\frac{1-\mu}{r_1^3}((x-\mu)\dot{x} + y\dot{y} + z\dot{z})$$

$$-\frac{\mu}{r_2^3}[(x+1-\mu)\dot{x} + y\dot{y} + z\dot{z}] \qquad (10.33)$$

Every term in the above equation is the perfect time differential of another quantity. The first three terms are $d/dt\,(\frac{1}{2}v^2)$, where v is the velocity with respect to the rotating frame. The next two terms are the time derivative of $\frac{1}{2}(x^2 + y^2)$, the effective potential energy of the "centrifugal force." Finally, just as in the two-body problem, the gravitational-force terms are the derivatives of the inverse first power gravitational potential.

So, the restricted three-body problem admits of Jacobi's integral that

$$C = \frac{1}{2}(\dot{x}^2 + \dot{y}^2 + \dot{z}^2) - \frac{1}{2}(x^2 + y^2) - \frac{1-\mu}{r_1} - \frac{\mu}{r_2} \qquad (10.34)$$

where C is a constant for a particular orbit of the third mass.

Note that Jacobi's integral is not the total energy of the system. The velocity term does not contain the square of the inertial velocity but the square of the rotating-frame velocity. The rotating frame introduces the "centrifugal potential" terms. If this result is transformed back to its equivalent expression in terms of inertial position and velocity components, the resulting expression resembles a combination of the total energy of the third mass and its total angular momentum. The Jacobi integral is thus peculiar to the restricted problem.

10.6 ACCESSIBLE REGIONS

The first use we made of energy conservation in the two-body problem was to set limits on the maximum distance the satellite could move from the primary. This led directly to the expression for the escape velocity in the two-body problem. A similar use can be made of the Jacobi integral in the restricted problem. Since the velocity appears in the Jacobi integral

$$C = \tfrac{1}{2}v^2 - \tfrac{1}{2}(x^2 + y^2) - \frac{1-\mu}{r_1} - \frac{\mu}{r_2} \tag{10.35}$$

only as v^2, there are boundaries for a given value of C on which the velocity in the rotating frame goes to zero. On these boundaries, the third body must stop and then move back in the general direction from which it came. The expression (10.35) for the Jacobi integral is more complex than the energy expression for the two-body problem, so the mapping of these critical boundaries, called *curves of zero velocity*, is a more complicated matter. To find the curves of zero velocity, we set v^2 to zero in (10.35) to find

$$C = -\tfrac{1}{2}(x^2 + y^2) - \frac{1-\mu}{r_1} - \frac{\mu}{r_2} \tag{10.36}$$

For a given value of C, this function of the position of the third body will yield the critical boundaries which cannot be crossed. Of course, it is necessary to perform this mapping for different values of the constant C.

Let us start with a large, negative value of C. Such a value can be obtained from (10.36) in three ways. First, r_1 could be small, corresponding to the third object close to the more massive primary. Second, r_2 could be small, implying that the small mass is close to the lesser primary. Finally, the quantity $x^2 + y^2$ could be large, implying that the small mass is at a large distance from the z axis. Such a situation is shown in Figure 10.5, where the value of C is -1.7, the value of $\mu = 0.01213$ is appropriate for the earth-moon system, and the curves of zero velocity consist of three nearly circular branches. Motion is physically possible only within the two smaller circles and outside the larger one. The annular region is forbidden to any spacecraft with a value of the Jacobi constant C of -1.7. In fact, (10.36) will yield an imaginary result for the speed v for an object within the annular region. So, for a large, negative value of C, the satellite must be inside a spherelike space about either of the primaries or outside a large cylinder whose axis is the z axis. The first two cases correspond to approximate two-body motion very near a primary; the third corresponds to two-body motion very far removed from either primary. At this value of C, flight between the earth and moon is not possible.

As the value of C is made less negative, the two spherelike shapes will expand and the cylinder will contract. As all the terms in (10.36) become important, the precise shape of the boundaries must be established by numerical computation. As we continue to make C larger, the two spheres distort until they touch each other. At this point, the L_2 lagrangian point, a

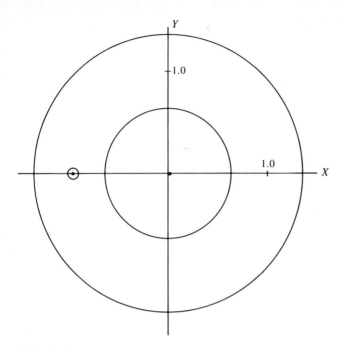

FIGURE 10.5
Curves of zero velocity for $C = -1.7$.

journey from the earth to the moon just becomes possible. This is shown in Figure 10.6, where the value of $C = -1.59411$ is the value at which the curve of zero velocity passes through the L_2 lagrangian point. This is the point between the earth and the moon celebrated in news accounts of lunar voyages. If this point is reached with zero (or very small) velocity, the spacecraft will continue on to the moon. So, to calculate the minimum-energy earth-moon trajectory, we can evaluate the Jacobi constant at L_2 for zero speed and then calculate the speed such a spacecraft must have when it leaves earth parking orbit. The calculation of this speed from (10.35) does not establish either the correct position or the direction of the velocity vector needed to arrive at L_2. For that problem, somewhat more complex mathematics are required.

Continuing to make C less negative, the next contact between the curves of zero velocity is made at the lagrangian point L_1. This is shown in Figure 10.7, where the value of $C = -1.58603$ is appropriate for the earth-moon system. The bottleneck at L_2 has opened, so the contact at L_1 represents the minimum-energy escape trajectory from the earth-moon system. Starting in the vicinity of the earth, the cheapest way to escape to infinity involves a close flyby of the moon. Such a flyby of the moon on leaving the earth-moon system was planned for the Pioneer 10 and 11 probes, but launch delays precluded using this option.

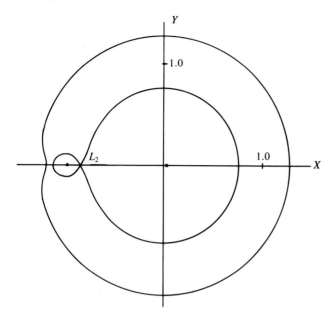

FIGURE 10.6
Curves of zero velocity for $C = C(L_2)$.

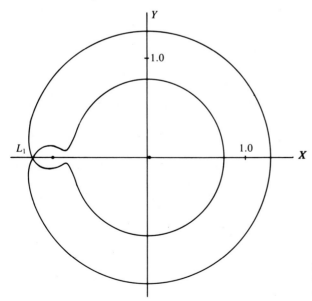

FIGURE 10.7
Curves of zero velocity for $C = C(L_1)$.

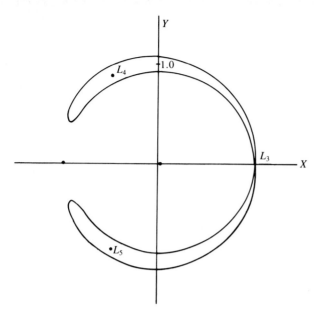

FIGURE 10.8
Curves of zero velocity for $C = C(L_3)$.

As C is made still larger, the bottleneck at L_1 opens still further, and escape is possible through an ever-widening corridor. The last contact point comes for still larger C values, when the inner and outer curves touch at the third colinear point L_3, as shown in Figure 10.8. The forbidden regions split into two teardrop shapes enclosing the triangular libration points L_4 and L_5. Escape from the earth-moon system is now possible by heading away from the moon as well as by heading toward it. However, the triangular points themselves are still within the forbidden regions. As C is made still larger, the forbidden regions shrink until they finally disappear at the equilibrium points L_4 and L_5 themselves. Thus, the triangular points are the most costly points in the entire plane to reach.

Figures 10.9 and 10.10 show a complete set of curves of zero velocity for the restricted problem in the case of the earth-moon mass ratio. Figure 10.9 shows the entire region of the earth-moon system, while Figure 10.10 concentrates on the immediate vicinity of the moon. Such diagrams contain a substantial amount of information about the possible regions accessible to a small object in the restricted problem. It must be emphasized that curves of zero velocity are *not* orbits of the small object. Rather, they are boundaries to the regions accessible to a spacecraft with a given, constant value of the Jacobi constant. Note that the three colinear lagrangian points L_1, L_2, and L_3 are all saddle points on these contour plots of the function C. This implies that these equilibrium points are unstable. It is less intuitively obvious that the closed

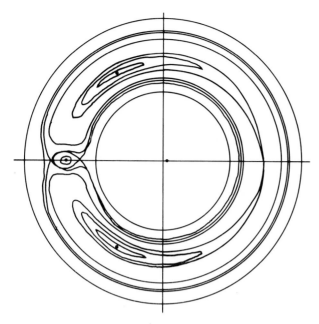

FIGURE 10.9
Curves of zero velocity for the earth-moon system.

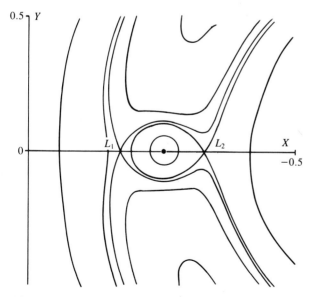

FIGURE 10.10
Zero-velocity curves in the vicinity of the moon.

contours about L_4 and L_5 imply that these points are stable, especially when we realize that these points are the maxima of the function C. However, we have already seen in section 10.4 that this is, indeed, true.

Figure 10.10 shows the region around the moon to larger scale. The dot is drawn at the true diameter of the moon. The region of stable orbits about the moon thus extends out to about 20 lunar radii. If one attempts to orbit the moon at a greater distance, it is possible to transition to an orbit about the earth or to escape from the earth-moon system entirely. However, the perturbations on a lunar satellite induced by the sun are quite large, and often will change the orbit until it contacts the lunar surface. This happened in the lunar orbiter series of spacecraft, when one of the vehicles was crashed on the surface by solar perturbations. However, ground controllers had lost the ability to command this particular probe and were actively trying to crash it so that it would not interfere with the radio transmissions of the next vehicle. Nature performed what human control could not, at least in this instance. Also, the critical curves through L_1 and L_2 are close together, indicating that the energy requirements for a lunar voyage and escape from the earth-moon system do not differ substantially.

10.7 AN UNSOLVABLE PROBLEM

The most startling result in mechanics within the last century is the proof, by Henri Poincaré, that the restricted problem is not solvable in closed form. There is a very real distinction between problems solvable in closed form (i.e., the harmonic oscillator) and those which are not. It is not simply a matter of human cleverness (or lack thereof) which has left the restricted problem of three bodies, and many other problems in nonlinear dynamics, without solutions. In this section we will explore briefly some techniques which have been applied to the restricted problem in an attempt to extract useful information.

Perhaps the technique most widely applied to the restricted problem is the search for periodic solutions. In the problem of two bodies, *every* orbit is periodic, but this is not true in the restricted problem. However, if a special set of initial conditions x_0, y_0, $\dot{x}_0$, $\dot{y}_0$ can be found such that after some time period τ

$$x(\tau) = x_0 \qquad y(\tau) = y_0$$
$$\dot{x}(\tau) = \dot{x}_0 \qquad \dot{y}(\tau) = \dot{y}_0$$

$$(10.37)$$

then the orbit closes upon itself. It returns to the same position in space (in the rotating frame) with the same velocity vector after one period τ. The motion from $\tau < t < 2\tau$ is a repetition of the motion during the first period. Such an orbit may be a special case, but something very valuable has been learned. For the cost of numerically integrating the equations of motion over one period, the motion has been defined for all time. The general approach of directly

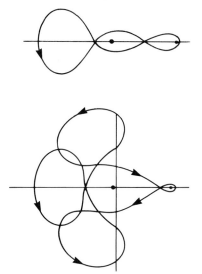

FIGURE 10.11
Periodic orbits in the restricted problem.

integrating the equations of motion for all time quickly runs into problems with both computer round-off error and budget limitations.

The conditions given in equations (10.37) are an example of a boundary value problem in ordinary differential equations. This is a special case, however, since the initial and final points are the same. Once a periodic solution is found, further information can be extracted about the particular orbit. Most important, the stability of the periodic orbit can be studied. This is done by Floquet theory, the theory of time-periodic linear equations. In essence, a time-periodic transformation can reduce a periodic orbit to an equilibrium point in an unusual set of coordinates. Standard techniques can then give information on whether the orbit is stable or not. If it is stable, then a spacecraft could be expected to stay within the vicinity of the periodic orbit, even it if is disturbed by outside perturbations. If the periodic orbit is not stable, then a spacecraft would need stationkeeping capabilities to stay within the vicinity of the periodic orbit. Figure 10.11 shows several examples of periodic orbits in the earth-moon system.

A second major technique used to study the restricted problem is the surface of section. This technique was invented by Poincaré himself, but it only became widely useful after the invention of the digital computer. Consider a dynamical system with two degrees of freedom and one constant of the motion (i.e., the planar restricted problem). The *phase space* of this system is four-dimensional, with coordinate axes x, y, dx/dt, and dy/dt. The Jacobi integral will restrict orbits to a three-dimensional hypersurface within this four-dimensional space. Now, consider all orbits with the same value of the Jacobi integral C. One of the four coordinates x, y, dx/dt, dy/dt is no longer interesting, since we can use the Jacobi integral to solve for the value of this

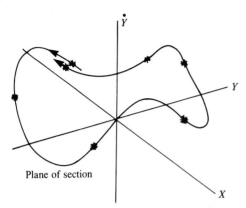

Plane of section

FIGURE 10.12
The surface of section in phase space.

coordinate if we are given the values of the other three. So, this group of orbits fills a three-dimensional space within the four-dimensional phase space. Now, take this three-dimensional space (ignoring one coordinate) and slice it with a plane, the *surface of section*. Every time the orbit crosses this plane, plot its intersection point, as in Figure 10.12. As the orbit is numerically integrated, points build up on the surface of section. Do they form any pattern?

If the restricted problem has a second integral of the motion, then a given orbit must lie on a two-dimensional surface in the three-dimensional subspace, The intersection points on the surface of section will form a group of curves, the cross section of the second integral surface. On the other hand, if there is no second integral of the motion for the restricted problem, the orbit is free to wander throughout the three-dimensional subspace, and points on the surface of section will fill up the entire plane of section. Numerical experiments show that *both behaviors are possible at the same time.* That is, some orbits behave as if there really was a second integral of the motion for the restricted problem, while other orbits with the same Jacobi constant behave as if their motion was random.

Figure 10.13 is a numerically calculated surface of section. There are areas of "good" behavior which form "chain of island" structures. The islands are surrrounded by a "sea" of random orbits, which behave as if there was no second integral of the motion. At the center of each island is an orbit cutting the surface of section repeatedly at exactly the same points. This is a periodic orbit. In fact, it must be a stable periodic orbit, since nearby orbits remain in the vicinity of this orbit, forming the island on the plot. So, stability can be decided by studying a surface-of-section plot.

The "coastline" of the islands would seem to be important, since this is the point where good behavior changes to random behavior, and the orbit becomes chaotic, possibly even escaping from the earth-moon system in this case. However, any attempt to accurately find the coastline produces a final surprise. The coastline is *not* well-defined: It usually consists of a series of ever-finer island-chain structures! The theoretical prediction by Poincaré is that

FIGURE 10.13
A surface of section.

the structure of island chains and chaotic regions repeats *forever* at ever-decreasing scale. Furthermore, there are an infinity of very tiny, stable island chains within the random region. The phase-space structure of the restricted problem is thus *infinitely complex*. The restricted problem of three bodies is not the only dynamical system to show such behavior. In fact, such behavior is the norm, rather than an exception. If the practicing engineer should be required to control such a system, it greatly helps to keep the system within a "good" region!

10.8   REFERENCES AND FURTHER READING

It is not necessary to cite any reference for the restricted problem beyond Szebehely's magnificent book. The study of periodic orbits and surface-of-section techniques are both recent additions to the arsenal of dynamical techniques and can be applied to any other nonlinear system as well. However, the study of chaos, in particular, is too new to cite an introductory book.

Szebehely, V., *Theory of Orbits,* Academic, New York, 1967.

10.9   PROBLEMS

1. Use the Jacobi integral to calculate the speed needed to depart from the vicinity of the earth ($r_1 = \frac{1}{60}$ distance units) to arrive at L_2. The value of C at L_2 is -1.59411. Ignore small terms in the expression for Jacobi's integral.

2. Determine the minimum speed needed departing earth orbit to arrive at L_4.

3. Verify that L_2 is located near $x = -0.83702$. Calculate the λ values at L_2 and the eigenvectors **A** for the case of the earth-moon system, $\mu = 0.01213$. Verify that this equilibrium point is unstable.

4. Show that the inertial-velocity components $\dot{X}$ and $\dot{Y}$ (still resolved along the rotating axes) are given by

$$\dot{X} = \dot{x} - y \qquad \dot{Y} = \dot{y} + x$$

   Show that Jacobi's integral takes the form

$$C = \tfrac{1}{2}(\dot{X}^2 + \dot{Y}^2) + \dot{X}y - \dot{Y}x - \frac{1-\mu}{r_1} - \frac{\mu}{r_2}$$

   When the third object is far from the smaller mass, and ignoring terms of order μ, show that this becomes

$$C \approx E - H \approx -\frac{1}{2a} - \sqrt{a(1 - e^2)}$$

   in terms of the two-body elements. This is *Tisserand's criterion* for the identification of comets after they have undergone a close approach to Jupiter. In consulting Chapter 2, remember that μ has a completely different meaning in the two-body problem.

5. Verify the form of the linearized equations for δx and δy [(10.22) and (10.23)] near an equilibrium point. Also, derive the vertical equation for δz and show that it always has the form of a harmonic oscillator and is therefore always stable.

6. Design a feedback controller for a satellite which is to stationkeep in the vicinity of L_2. (*Hint*: Consider feeding back the amplitudes of the relative position and velocity projected on the "outgoing" unstable eigenvector **A**, the eigenvector associated with a positive real λ value.)

11

INTERPLANETARY
TRAJECTORIES

11.1 INTRODUCTION

The planets were recognized as being different from the stars in antiquity: The Greek word *planet* means "wanderer." Copernicus' rearrangement of the solar system virtually demanded that these objects must be other worlds in their own right, and this is exactly what Galileo found in 1610 when he turned the newly invented telescope to the study of the heavens. With the observation of mountains and "seas" on the moon, and moons revolving around Jupiter, it is only a short leap of the imagination to wonder what these new worlds would be like.

   Indeed, the dream of travel to the planets began quite early. Johannes Kepler wrote to Galileo in April 1610, suggesting that he and Galileo cooperate in the mapping of the moon and planets in anticipation of the celestial voyagers of the future. Kepler went on to write one of the first science fiction stories, the *Somnium,* or *Dream.* With the rise of technology in the industrial revolution, science fiction became more common. Jules Verne and H. G. Wells in the late nineteenth century both dealt with space travel in their writings. By the 1920s science fiction was commonly before the public's eye in the form of pulp magazines, although it was hardly a respected literary genre.

   While these fictional accounts of space travel heightened public awareness, no practical method for accomplishing space voyages seemed possible.

This began to change around the turn of the century with the visionary work of Tsiolkovsky and the practical contributions of Robert Goddard. Both men independently realized that the liquid-fuel rocket was potentially capable of the required performance. Tsiolkovsky anticipated many of the events of the last 30 years, while the more circumspect Goddard set out to actually build the required technology. The founding of the German Society for Space Travel in 1927, the American Interplanetary Society (1930), the British Interplanetary Society (1933), and similar groups in the Soviet Union at about the same time shows that the possibility of spaceflight was recognized by many individuals long before adequate funding was available from national governments. The titles of these groups indicate their real intent, although the intent of governments has not always been so peaceful.

It was Walter Hohmann who first considered the actual problem of finding the best trajectory to another planet. He not only discovered the orbit-transfer strategy which bears his name but also invented the method of "patched conics" and considered the required round-trip time to another planet. Since the development of telemetry, computers, and remote sensing was *not* widely anticipated by science fiction, it was assumed that all planetary missions would be manned. So far this has not proven to be the case. However, the urge to actually stand where our robot explorers have preceded us is very great indeed, and manned missions to the planets will eventually be undertaken.

11.2 THE SPHERE OF ACTIVITY

When an object is "close enough" to the earth, it can be considered to orbit the earth. When it is "far enough" away from the earth, it must be considered to be in orbit about the sun. Since in the patched-conic method we shift point of view when an object is under that planet's control, it is necessary to investigate how large a volume of space is controlled by a particular planet. This is not an exact concept. The gravitational field of the sun influences the motion of an earth satellite no matter how close that satellite is to the earth. The gravity of the earth also influences the motion of an asteroid in deep space. For the case of an asteroid far from the earth, the effect of the earth becomes less as the r^2 denominator in Newton's law of gravity becomes large. For the case of a close earth satellite, the effect of the sun becomes small because the sun produces essentially identical gravitational accelerations on both the earth and the earth satellite.

There is no actual "boundary" between the gravitational fields of the earth and sun. We can, however, formulate approximate definitions. Consider Figure 11.1, where a tiny mass is shown in orbit between the sun, mass M, and a planet of mass m. If we place the small mass at just the correct distance r from the planet, then the gravitational forces essentially balance in this cosmic "tug of war." Of course, it is necessary to include the fact that both the planet and the test mass are themselves in orbit about the sun. Assume that both

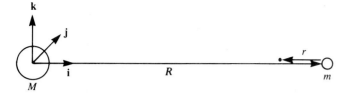

FIGURE 11.1
Activity-sphere calculation.

these orbits are circular and that the test mass stays between the planet and the sun.

The gravitational acceleration produced on the test mass by the sun and the planet is given by

$$\mathbf{a}_g = -\frac{GM\mathbf{i}}{(R-r)^2} + \frac{Gm}{r^2}\mathbf{i} \tag{11.1}$$

where positive acceleration is taken to be to the right in Figure 11.1. Since the planet is in a circular orbit about the sun, it has an angular velocity of

$$\boldsymbol{\omega} = \left(\frac{GM}{R^3}\right)^{1/2}\mathbf{k} \tag{11.2}$$

by Kepler's third law. The test mass then has an inertial acceleration

$$\mathbf{a} = \boldsymbol{\omega} \times [\boldsymbol{\omega} \times (R-r)\mathbf{i}]$$
$$= -\frac{GM}{R^3}(R-r)\mathbf{i} \tag{11.3}$$

directed inward, since this is a centripetal-acceleration term. Equating the gravitational acceleration to the inertial acceleration necessary for the test mass to stay between the sun and the planet, we have

$$-\frac{GM}{(R-r)^2} + \frac{Gm}{r^2} = -\frac{GM}{R^3}(R-r) \tag{11.4}$$

Now, planets in the solar system have masses m very much less than the mass M of the sun. In this case, the distance r controlled by the planet is likely to be small. If we assume that r is small, the denominator $(R-r)^{-2}$ can be expanded with the binomial theorem to give

$$(R-r)^{-2} \approx R^{-2} + 2R^{-3}r + \cdots \tag{11.5}$$

where terms of order r^2 have been dropped. Inserting this in (11.4), we have

$$-2\frac{GM}{R^3}r + \frac{Gm}{r^2} = \frac{GM}{R^3}r \tag{11.6}$$

after canceling the dominant sun terms on both sides. This result can be solved for the ratio r/R as

$$\frac{r}{R} \approx \left(3\frac{m}{M}\right)^{1/3} \tag{11.7}$$

This result goes by several names in the literature.

Comparison with Chapter 10 will show that we have actually calculated the approximate position of the lagrangian point L_2 for the case of small mass ratio m/M. In studying the curves of zero velocity in the last chapter, we did indeed find that the distance from the moon to L_2 approximately bounded the volume of space in which it was possible to orbit the moon itself. Equation (11.7) is also termed the *Roche limit* when r is interpreted as the maximum permitted size of the smaller mass m. If the planet should become as large as radius r, no further mass could be added to it, since small test particles would be torn off by the "tidal" component of the sun's gravity. For example, Mars' moon Phobos comes close to being the same size and shape as its own Roche lobe. This means that the effective gravity on the surface of Phobos, already small for such a small object, is essentially zero in the gravitational field of Mars.

The expression (11.7) is not, however, the usual expression for the radius of the sphere of activity. That radius is given by

$$\frac{r}{R} \approx \left(\frac{m}{M}\right)^{2/5} \tag{11.8}$$

The above equation is derived by a much more complex argument due to Lagrange, in which the errors in considering the test particle as being in a two-body orbit about the planet but perturbed by the sun are equated to the errors in considering the same particle to be in a two-body orbit about the sun and perturbed by the planet. The sphere of activity is defined as the radius r at which both descriptions are equally poor. Lagrange's derivation is not repeated here, since it is much more involved than the chain of logic leading to (11.7). The two expressions are virtually equivalent, since the exponents of $\frac{1}{3}$ and $\frac{2}{5}$ are not very different. However, and most fundamentally, *it does not really matter*, since the concept of the sphere of activity is very nebulous to begin with.

In the method of patched conics, we assume that a spacecraft is influenced only by the gravitational field of the planet when it is within the planet's activity sphere and is influenced only by the gravity of the sun when it is outside any planet's activity sphere. The spacecraft is thus always in a two-body orbit with respect to either the sun or a planet, and we switch points of view when an activity-sphere boundary is crossed. This is at best an approximation, and it is only a decent approximation in the solar system *because the distances involved are very, very large.*

Table 11.1 gives some planetary data and the radii of the activity spheres for the solar system. The table lists the inverse of the planet's mass in solar masses and the orbital semimajor axis and the radius of the activity sphere in astronomical units (AU). The earth's activity sphere is about three times larger

TABLE 11.1
Planetary data

Planet	Reciprocal mass, (solar masses)$^{-1}$	Semimajor axis, (AU)	r, (AU)
Mercury	6,023,600	0.387099	0.00075
Venus	408,520	0.723332	0.00411
Earth	328,900	1.0	0.00621
Mars	3,098,710	1.523691	0.00385
Jupiter	1,047.35	5.2028	0.3222
Saturn	3,498.1	9.53884	0.364
Uranus	22,869	19.1819	0.346
Neptune	19,332	30.0578	0.580
Pluto	130,000,000	39.44	0.00056

than the distance to the moon. A spacecraft departing from the vicinity of the earth on a trip to another planet must leave on a hyperbolic orbit with respect to earth. Long before the spacecraft reaches the orbit of the moon, it will have reached the asymptotic portion of the hyperbola and will essentially be moving along a straight line at constant velocity. On the other hand, the earth's activity sphere is only 0.006 the distance between the earth and the sun. As the spacecraft leaves the control of the earth, it passes into a conic-section orbit about the sun. However, the tiny portion of a solar orbit near the activity sphere, either elliptical or hyperbolic with respect to the sun, will appear to be a straight line at constant velocity. So, because the activity sphere is large compared to the earth but small compared to the solar system, there is a wide area where the two straight lines can be smoothly patched together. It is this fact that makes the method of patched conics a valid approximation in the solar system.

This is not true in the earth-moon system, where the activity sphere of the moon is fully one-sixth the distance from the earth to the moon. The gravity of the earth is not really negligible within the moon's activity sphere, and patched conics is inappropriate for the discussion of earth-moon trajectories. The restricted problem of three bodies is much more valid for this dynamical system.

11.3 LAUNCH WINDOWS AND MISSION DURATION

It should come as no surprise that Hohmann was interested in interplanetary travel when he discovered the orbit-transfer strategy which bears his name. The planetary orbits mostly have very small eccentricities and nearly one common orbital plane. Thus, to a first approximation, the Hohmann transfer is the best (e.g., minimum total Δv) route to follow for interplanetary transfers.

It is certainly not the fastest way to reach the planets, however. The minimum time orbit between two points is a straight line at infinite velocity, a solution which is valid in *any* gravitational field. The Hohmann transfer is actually a local *maximum* in the transfer time.

The time required to go from one planet's orbit at radius R_1 to the orbit of the second planet at R_2 along a Hohmann transfer ellipse with semimajor axis $a_t = (R_1 + R_2)/2$ is given by

$$T_{12} = \pi \left(\frac{a_t^3}{\mu_\odot} \right)^{1/2} \tag{11.9}$$

where $\mu_\odot$ is the gravitational parameter for the sun. A trip from the Earth to Mars, for example, takes about 258 days. A new problem arises in the case of interplanetary transfers, which need not be considered for most low-earth-orbit Hohmann transfers. The departure point on the Hohmann transfer will, of course, be the earth. However, it is desirable to arrive at the orbit of the second planet when that planet also arrives at that point.

Figure 11.2 shows the solution of this problem, also due to Hohmann. Since the departure and arrival points are separated by 180° and since we know the transfer time T_{12}, the angle between the earth at launch and the planet *at launch* can also be calculated. If the planet moves along its orbit with mean motion n_2, then the planet travels $n_2 T_{12}$ rad during the transfer, and the phase angle at launch θ_{12} is just

$$\theta_{12} = \pi - n_2 T_{12} \qquad \text{rad} \tag{11.10}$$

The time when the earth is *nearly* in this angular relationship to another planet is called the *launch window* for that planet. Because of possible difficulties with artificial hardware and the weather, the booster must have some extra Δv

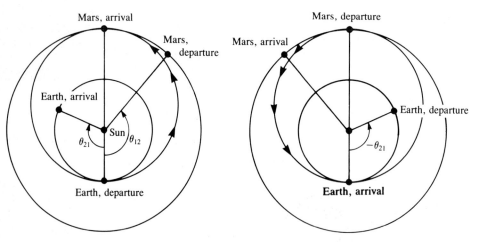

FIGURE 11.2
Outbound (left) and return (right) phase calculations.

TABLE 11.2
Synodic periods and trip times (in days) for planetary missions

Planet	T_{syn}	T_{12}	T_{wait}	T_{trip}
Mercury	115.8	105.4	66.9	277.9
Venus	583.9	146.1	467.0	759.2
Mars	779.9	258.8	454.3	972.1
Jupiter	398.8	997.5	214.6	2209.6
Saturn	378.1	2209.1	363.2	4454.5

margin so that a less-than-optimal trajectory can be flown. This excess booster capability would then determine the duration of the launch window.

The occurrence of a launch window to another planet can be found by calculating θ_{12} for the target object and then searching for dates when the earth and the planet will be in the correct angular relationship. Once one launch window is found, others can be predicted quite easily, since launch windows occur at regular intervals. If the mean motion of the earth is n_1, then the relative angular velocity of the earth to the target planet is $n_1 - n_2$. Another launch window will occur when the earth has had time to "lap" the target planet, or in the case of the faster inner planets, after they have had time to "lap" the earth. The time it takes the earth-to-target angle θ to go through one revolution is

$$T_{syn} = \frac{2\pi}{|n_1 - n_2|} \tag{11.11}$$

and is termed the *synodic period*. Just as it may take a fast race car a long time to lap a slightly slower competitor, the synodic period is long for planets with nearby orbits. It approaches 1 year for outer planets, where n_2 becomes small, and is quite short for Mercury, whose n_2 is large. Table 11.2 gives Hohmann transfer times and intervals between launch windows for some of the other planets. Interestingly, the synodic period is just the period of a planet in ancient ptolemaic astronomy, since it is the period of the planet with respect to the earth.

In the case of a one-way unmanned mission to another planet, our analysis is now complete. However, any manned mission must provide for the return of the crew. This adds another Hohmann transfer for the return journey to the total trip time, but it also introduces another launch-window calculation since the crew must wait at the target planet for the correct phase relation before the return trip can begin. Figure 11.2 also shows a sketch of the return trip, and the fact that it is a mirror image of the outward leg is immediately obvious. However, notice that the return transfer has been rotated to align the positions of the earth at arrival and departure. It is only the phase relationships which are important.

To ensure that the earth will be at the arrival point when the returning ship gets there, the phase angle θ_{21} must be calculated. This is just

$$\theta_{21} = \pi - n_1 T_{12} \tag{11.12}$$

Notice that this is also the phase angle between the earth and the target planet *at arrival at the planet,* but in the opposite sense. For a trip to Mars, the Earth will be θ_{21} rad *ahead* of Mars on arrival at Mars, and the crew must wait for the Earth to be θ_{21} rad *behind* Mars before they can leave. The total shift in the phase angle is thus $2\pi - 2\theta_{21}$ rad, so the wait time is given by

$$T_{\text{wait}} = \frac{2\pi - 2\theta_{21}}{|n_2 - n_1|} \tag{11.13}$$

These figures are also given in Table 11.2, along with the total trip time $T_{\text{trip}} = 2T_{12} + T_{\text{wait}}$. A glance at this table reveals why a manned trip to Mars has not yet been undertaken, since total trip times are far longer than the 10-day "field trips" to the moon of the Apollo program.

11.4 DEPARTURE AND ARRIVAL

In the patched-conic method, the crossing of an activity-sphere boundary requires a change of reference frames. Consider Figure 11.3, which shows the outward crossing of the earth's activity sphere for a typical outer-planet mission. At the crossing point, the spacecraft ceases to be an earth-orbiting object and is now in orbit about the sun. Its position vector with respect to the sun, $\mathbf{R}$, is

$$\mathbf{R} = \mathbf{R}_\oplus + \mathbf{r} \tag{11.14}$$

where $\mathbf{r}$ is its position vector with respect to the earth and $\mathbf{R}_\oplus$ is the position vector of the earth with respect to the sun. (Throughout this chapter,

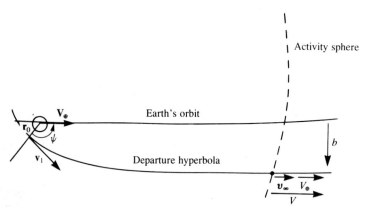

FIGURE 11.3
Departure-hyperbola geometry.

capitalized vectors are with respect to the sun, while lowercase vectors are with respect to a planet.) Since, until the moment of crossing, the earth was considered to be an inertial origin, we must add the earth's velocity vector to obtain the spacecraft's velocity with respect to the sun,

$$\mathbf{V} = \mathbf{V}_\oplus + \mathbf{v} \tag{11.15}$$

where $\mathbf{V}_\oplus$ is the velocity vector of the earth. Notice that (11.15) is simply the derivative of (11.14). With the spacecraft's position and velocity $\mathbf{R}$ and $\mathbf{V}$ available with respect to the sun, the heliocentric orbit of the spacecraft can now be determined. Of course, this requires using the sun's gravitational constant $\mu_\odot$ in the two-body orbit equations.

Equations (11.14) and (11.15) are referred to as the *patch conditions*. They are quite general and serve for any activity-sphere crossing. But if a Hohmann transfer trajectory is used to the other planet, it is possible to be more definite about the departure hyperbola at the earth and the arrival hyperbola at the target planet. Since a Hohmann transfer ellipse will be tangent to both planetary orbits, the velocity vectors will be aligned at the activity-sphere boundary, as shown in Figure 11.3. Equation (11.15) then becomes a scalar relationship.

The earth's activity sphere is very small compared to the size of the solar system. We will commit a very small error, then, if we assume that $\mathbf{V}$ is the speed required at perihelion in the Hohmann transfer ellipse

$$V = \left(\frac{2\mu_\odot}{R} - \frac{1}{a_t} \right)^{1/2} \tag{11.16}$$

where $a_t = (R_\oplus + R_2)/2$ is the semimajor axis of the Hohmann transfer ellipse and R_2 is the orbital radius of the target planet. Then, the speed of the spacecraft relative to the earth at the crossing point is

$$v = V - V_\oplus \approx v_\infty \tag{11.17}$$

Since the size of the activity sphere is very large compared to the size of the earth, it is permissible to assume that the spacecraft has essentially escaped from the earth at the crossing point. Hence the last step in (11.17), where $V - V_\oplus$ has been assumed to equal v_∞, the excess speed at infinity in the escape hyperbola. Comparing (11.17) to (3.3), we notice that v_∞ is just the usual first Hohmann maneuver Δv_1.

Of course, the spacecraft does not maneuver at the crossing point. Only our point of view changes. The actual injection into the departure hyperbola occurred much closer to the earth, at radius r_0. Since energy is conserved along the outgoing hyperbola, the energy relation for hyperbolas

$$E = \tfrac{1}{2} v_\infty^2 = \tfrac{1}{2} v_0^2 - \frac{\mu_\oplus}{r_0} \tag{11.18}$$

can be solved to yield the required injection velocity at perigee in the hyperbola

$$v_0 = \left(v_\infty^2 + \frac{2\mu_\oplus}{r_0} \right)^{1/2} \tag{11.19}$$

If the outgoing hyperbola is entered from a circular parking orbit, then the required injection maneuver can be found by subtracting the circular orbital velocity from (11.19):

$$\Delta v_0 = \left(v_\infty^2 + \frac{2\mu_\oplus}{r_0} \right)^{1/2} - \left(\frac{\mu_\oplus}{r_0} \right)^{1/2} \tag{11.20}$$

As we will shortly see, use of a parking orbit is very desirable.

Still more information can be extracted from the two-body relationships. The semimajor axis of the departure hyperbola is given by

$$a = -\frac{\mu_\oplus}{2E} \tag{11.21}$$

and if the injection maneuver occurs at perigee in the hyperbola, the eccentricity can be found from

$$r_0 = a(1 - e) \tag{11.22}$$

To align the outgoing asymptote of the hyperbola with the earth's velocity vector, it is necessary to perform the injection maneuver at the correct point in the parking orbit. The equation for a conic section

$$r = \frac{a(1 - e^2)}{1 + e \cos v} \tag{11.23}$$

goes to infinity for hyperbolas when the denominator vanishes. As shown in Figure 11.3, we want $v = \psi$ when $r = \infty$. This yields the angle ψ as

$$\psi = \cos^{-1} \frac{-1}{e} \tag{11.24}$$

For an outer-planet mission, as sketched in Figure 11.3, the outgoing asymptote is aligned with the earth's velocity vector. For an inner-planet mission, injection occurs in the opposite direction, and the spacecraft's residual escape speed v_∞ is subtracted from the earth's orbital velocity at the crossing point.

Figure 11.3 is drawn as if the outgoing hyperbola lay in the plane of the earth's orbit. This is not at all necessary. As shown in Figure 11.4, we can sketch a family of outgoing hyperbolas, all of which have asymptotes parallel to the earth's velocity vector and all of which have perigee points an angle ψ back from the direction of this vector. The locus of possible maneuver points thus forms a cone about the direction of the earth's velocity vector. If the spacecraft was directly injected into the departure hyperbola, interplanetary launches could only occur when the rotation of the earth carried the launch site

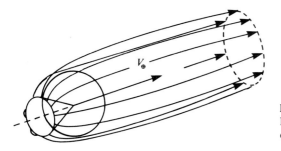

FIGURE 11.4
Departure   hyperbolas   in   three
dimensions.

through this cone. On the other hand, if a parking orbit is used, then the plane
of the parking orbit must pass through the earth's velocity vector.

There is now no requirement that the launch site pass through the
injection point, since the parking orbit itself will arrange for this to occur.
Even if the launch vehicle limits the parking orbit to inclinations near the
latitude of the launch site (which will maximize the payload delivered to the
parking orbit), there will still be a decently long interval each day during which
an acceptable parking orbit can be achieved, and not just one single instant
each day. The errors committed by being slightly out of the earth's orbital
plane or slightly inside or outside the orbit of the earth are assumed to be
small, in keeping with our assumption that the activity sphere of the earth is
very small compared to the size of the solar system. These errors (and the
others we have committed) can be canceled by very small changes in the
injection conditions.

The distance b on Figure 11.3 between the earth's orbit and the outgoing
trajectory is called the *impact parameter*. It can be calculated from Figure 11.3
and the elements of the two-body departure hyperbola. At the arrival planet,
the impact parameter is also needed to be able to calculate the incoming
hyperbolic trajectory. A pure Hohmann transfer to the target planet would
produce $b = 0$, which is not desirable unless an immediate direct landing
attempt is planned. Again, the assumptions of the patched-conic method come
to our aid. Because the activity sphere of the target planet is very small
compared to the size of the Hohmann transfer ellipse, minute changes in the
latter will produce dramatic changes in the impact parameter b at arrival. This
means that the arrival hyperbola can be varied over a broad range for
negligible cost in the initial departure maneuver. Thus, a periapse radius for
the arrival hyperbola can be chosen to suit mission objectives, with the
assurance that this choice will produce nearly negligible changes in the initial
departure maneuver. Also, by arranging to arrive slightly above or below the
plane of the target planet's orbit, the arrival hyperbola can have any desired
orbital inclination.

Arrival at the target planet is just the reverse of departure from the
earth. The incoming Hohmann transfer ellipse will bring the spacecraft to the
vicinity of the target planet with their velocity vectors aligned. At the
activity-sphere crossing, the heliocentric velocity vector of the planet must be

subtracted to find the velocity vector of the spacecraft with respect to the planet, and then the approach hyperbola may be calculated. Of course, unless the spacecraft maneuvers or hits the planet, it will leave the activity sphere, and a new solar orbit can be calculated after the flyby. This new heliocentric orbit can be quite different from the original approach orbit, as we will see in the next section.

11.5 PLANETARY FLYBY

Until now, we have assumed that interplanetary trajectories would be Hohmann transfers. While the Hohmann transfer is the lowest-energy option, it also takes the longest flight time. In particular, the trip times in Table 11.2 are prohibitively long for the outer planets. Under certain conditons it is possible to considerably cut these times while also decreasing the total Δv required. A close flyby of one planet can radically alter the heliocentric trajectory of the spacecraft after the flyby. The spacecraft can either gain or lose energy with respect to the sun so that the new solar orbit can either reach much further out into the outer solar system or drop considerably further inward toward the inner planets. In particular, a flyby of the most massive planet, Jupiter, can provide access to the entire solar system.

Figure 11.5 shows a spacecraft approaching the activity sphere of a planet with velocity vector $\mathbf{V}_1$ with respect to the sun. The planet has an orbital velocity vector $\mathbf{V}_p$, also with respect to the sun. When the spacecraft crosses the activity-sphere boundary, we must change our point of view and calculate the orbit of the spacecraft with respect to the planet. If the position vectors of the spacecraft and planet with respect to the sun are $\mathbf{R}_1$ and $\mathbf{R}_p$ at the crossing

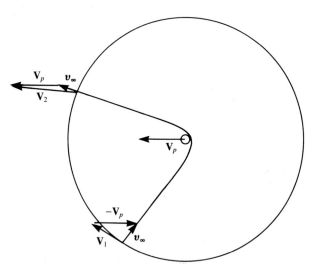

FIGURE 11.5
General activity-sphere crossings.

time, then the spacecraft's position vector **r** with respect to the planet is found by subtraction:

$$\mathbf{r}_1 = \mathbf{R}_1 - \mathbf{R}_p \tag{11.25}$$

Similarly, the velocity vector of the probe $\mathbf{v}_1$ with respect to the planet requires subtracting the planet's orbital velocity:

$$\mathbf{v}_1 = \mathbf{V}_1 - \mathbf{V}_p \tag{11.26}$$

These two equations are just the reverse of (11.14) and (11.15). Since we now have both an initial position and a velocity vector for the spacecraft, the complete two-body orbit of the probe about the planet can be calculated. Of course, the appropriate value of the gravitational parameter μ_p for the planet must be used.

The flyby trajectory will always be a hyperbola with respect to the planet, since the spacecraft approaches the planet from "infinitely" far away. If it does not actually hit the planet, it will recede from the planet and again cross the activity sphere, passing back under the control of the sun. At the outward crossing, the position and velocity vectors of the spacecraft with respect to the sun are obtained by vector addition:

$$\mathbf{R}_2 = \mathbf{r}_2 + \mathbf{R}_p \tag{11.27}$$

$$\mathbf{V}_2 = \mathbf{v}_2 + \mathbf{V}_p \tag{11.28}$$

Of course, (11.25) to (11.28) are vector relations, and both magnitude and direction of the individual quantities affect the outcome. But the speed of the spacecraft with respect to the planet, $|\mathbf{v}_1| = |\mathbf{v}_2| = v_\infty$, will be the same at both the inward and outbound crossings of the activity sphere. The direction of $\mathbf{v}_2$, however, can be greatly different from the direction of $\mathbf{v}_1$. As shown in Figure 11.5, this may lead to a substantially different velocity vector for the spacecraft after the flyby. In fact, for the case sketched in the figure, the new heliocentric speed is more than double the speed at which the spacecraft originally approached the planet. This is characteristic of a *trailing-side* flyby of a planet. A *leading-side* flyby, on the other hand, can considerably reduce the heliocentric speed of the probe.

If the spacecraft's mission is to perform a one-time scientific flyby of the planet, the heliocentric orbit of the spacecraft after the flyby is of little interest. On the other hand, if the planet is being used to perform a "gravitational slingshot" maneuver, then the real intention is for the spacecraft to head somewhere else. The planetary flyby is then used to gain or lose sufficient energy to arrive at the real destination. This virtually requires that such calculations be performed by a computer. The planetary positions and velocities $\mathbf{R}_p$ and $\mathbf{V}_p$ must be obtained from accurate data on their orbits, and the computer must try many different possible flybys to find one leading to a second encounter. When the possibility of still further encounters is included, the required labor quickly passes human abilities.

The usefulness of this technique is greatly enhanced by another important

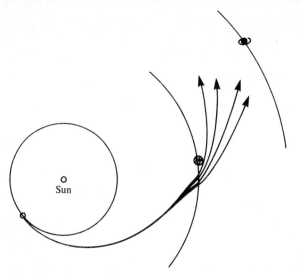

FIGURE 11.6
Some possible trajectories after flyby.

fact: The size of the activity sphere is quite small with respect to the size of the solar system. The spacecraft's heliocentric approach speed $\mathbf{V}_1$ cannot be changed significantly without significant changes in the original heliocentric approach orbit. However, very small changes in this orbit can radically alter the entry point on the planet's activity sphere. This will, in turn, radically alter the flyby hyperbola and the subsequent new heliocentric orbit. As shown in Figure 11.6, this opens up a broad swath of possible trajectories after the flyby, any one of which can be chosen by a very small maneuver performed long before the planetary approach.

Under the right phase conditions between planets, it is possible that one of these trajectories will bring the spacecraft to another planet. If this happens, it may even be possible to repeat this performance a second time. The Voyager II spacecraft has, at the time of this writing, already completed spectacularly successful flybys of Jupiter, Saturn, and Uranus, and is scheduled to arrive in the vicinity of Neptune in August 1989. The Mariner 10 spacecraft used a trailing-side passage of the planet Venus to arrive at Mercury in March 1974. The first flyby of Mercury was then used to further alter the heliocentric orbit of the spacecraft so that its period about the sun was exactly twice that of Mercury itself. In this resonant orbit, two orbital periods of Mercury equal one orbital period of the spacecraft, causing the two objects to meet repeatedly. Mariner 10 performed three successful flybys of Mercury before its attitude-control fuel was exhausted, ending the mission.

This technique also figures prominently in future plans for unmanned solar system exploration. When the earth itself is used as the flyby planet, the technique is called a Delta VEGA trajectory (from Δv, earth-gravity assist). The much-delayed Galileo spacecraft is currently slated to use such a trajectory on its way to Jupiter. At Jupiter, the Galileo probe will perform repeated flybys of the four large moons of Jupiter itself, allowing extended observations of that planet from a range of different orbits. A flyby trajectory is not limited to the plane of the solar system, so the inclination of the post flyby orbit can also be changed. The European Space Agency's Ulysses spacecraft will use a flyby of Jupiter to produce a new heliocentric orbit with perihelion near the sun itself and at an inclination of nearly 90°. This will permit, for the first time, close observation of the polar regions of the sun. Finally, this approach can have utility for manned probes to the planets as well. At least one Mars mission study proposed shortening the waiting time at Mars by returning to Earth via a flyby of Venus.

11.6 OPTIMAL PLANETARY CAPTURE

If a simple flyby of a target planet is intended, the spacecraft does not need a propulsion system, and the trajectory relative to the planet can be optimized to achieve science objectives. But if long-term study of the planet is the goal, then the vehicle must establish a closed orbit about the target planet. Since the spacecraft crosses the planet's activity sphere with a known v_∞, it must be on a hyperbolic trajectory with respect to the planet. Unless the probe hits the planet, it will exit from the sphere of activity of the planet unless it maneuvers.

During the heliocentric phase of the trajectory, small maneuvers on the part of the spacecraft can produce large changes in the position of the entry point of the spacecraft at the activity sphere. This will not produce a noticeable change in the entry velocity v_∞, however. As shown in Figure 11.7, the periapse radius r_p is at our disposal, while the approach speed v_∞ is determined by the heliocentric transfer trajectory. If we elect to enter a circular orbit about the planet by maneuvering at periapse passage, then we need a speed after the maneuver of

$$v_c = \sqrt{\frac{\mu_p}{r_p}} \qquad (11.29)$$

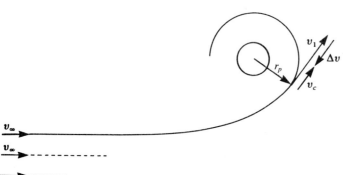

FIGURE 11.7
Optimal capture maneuver.

where μ_p is, of course, the appropriate gravitational parameter for the planet. The energy of the spacecraft with respect to the planet at the crossing of the activity sphere is

$$E = \tfrac{1}{2}v_\infty^2 - \frac{\mu_p}{r_a} \approx \tfrac{1}{2}v_\infty^2 \qquad (11.30)$$

since, in the patched-conic method, we assume that the activity sphere is "large" with respect to the planet. Energy is conserved on the approach hyperbola, so the speed of the spacecraft before the maneuver is given by

$$v_1 = \sqrt{v_\infty^2 + \frac{2\mu_p}{r_p}} \qquad (11.31)$$

Subtracting (11.31) and (11.29) gives the required maneuver as

$$\Delta v = \sqrt{v_\infty^2 + \frac{2\mu_p}{r_p}} - \sqrt{\frac{\mu_p}{r_p}} \qquad (11.32)$$

Now, (11.32) is a function of the radius of the final circular orbit, r_p. Since fuel to perform the maneuver must be transported all the way from earth, it is reasonable to ask if there is a best radius, leading to the cheapest circular orbit. To minimize (11.32) with respect to r_p, we can calculate the derivative and set it equal to zero. This gives

$$\frac{d}{dr_p}\Delta v = -\left(v_\infty^2 + \frac{2\mu_p}{r_p}\right)^{-1/2}\mu_p r_p^{-2} + \tfrac{1}{2}\sqrt{\mu_p}\, r_p^{-3/2} = 0 \qquad (11.33)$$

Multiplying by $r_p^{3/2}$ and rearranging gives

$$2\mu = \sqrt{\mu_p r_p}\left(v_\infty^2 + \frac{2\mu_p}{r_p}\right)^{1/2} \qquad (11.34)$$

Squaring the above, we have

$$4\mu_p^2 = \mu v_\infty^2 r_p + 2\mu_p^2 \qquad (11.35)$$

This gives the minimum Δv circular-orbit radius as

$$r_p = \frac{2\mu_p}{v_\infty^2} \qquad (11.36)$$

and substituting this value of r_p back into the maneuver equation (11.32) gives the minimum maneuver as

$$\Delta v_{\min} = \frac{v_\infty}{\sqrt{2}} \qquad (11.37)$$

Actually, this is not the *absolute* minimum-energy capture maneuver for a given approach speed. By choosing to circularize the final orbit, we have dropped the spacecraft quite deep into the "gravity well" of the planet. The minimum-energy capture maneuver would just prevent the spacecraft from escaping from the planet's gravitational field. This suggests that the final orbit should be elliptical, with an apoapse distance as large as possible. Also, other mission considerations may enter into the choice of the final orbital radius. For example, the orbit of Mariner 9 about Mars was highly elliptical, with a period of nearly 12 h. This choice was made so that the spacecraft could observe the planet from both low and high altitudes, but also so that the period of the spacecraft would be synchronized with the rotation of the earth. Thus the vehicle would be near apoapse when the primary tracking site was communicating with the probe, ensuring that the vehicle would not be hidden behind the planet.

11.7 REFERENCES AND FURTHER READING

Most work in calculating interplanetary trajectories, even using the patched-conic method, requires the use of computers. Both Battin and Bate, Mueller, and White discuss the fundamental two-position vector–time-of-flight orbit determination method required. Battin in particular gives several examples of total Δv plots calculated in this way. Details of modern trajectory work is usually found in the proceedings of the AIAA astrodynamics conferences held each year.

Battin R. H., *Astronautical Guidance,* McGraw-Hill, New York, 1964.

Bate, R. R., D. D. Mueller, and J. E. White, *Fundamentals of Astrodynamics,* Dover, New York, 1971.

11.8 PROBLEMS

1. Our moon has a mass 0.0123 that of the mass of the earth and a radius of 1738 km. How close could the moon orbit the earth without being torn apart by the earth's gravity?

2. Consider placing the small mass outside the planet, as shown in Figure 11.8. Show that in order to stay with the planet, the mass must be located a distance

$$\frac{r}{R} \approx \left(3\frac{m}{M}\right)^{1/3}$$

from the planet. Notice that this result is the same as given in the text for the L_2 point.

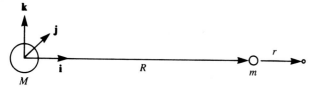

FIGURE 11.8
Outer limit of planetary activity sphere.

3. Consider departing from a high orbit around the earth by the two methods shown in Figure 11.9. In the first method, one maneuver directly injects the spacecraft into the departure hyperbola, with v_∞ specified. If the initial orbit radius is a_0, show that the required maneuver is

$$\Delta v_{\text{direct}} = \sqrt{v_\infty^2 + \frac{2\mu}{a_0}} - \sqrt{\frac{\mu}{a_0}}$$

On the other hand, following Olberth, consider departing from the earth using two maneuvers. The first takes the spacecraft into an elliptical orbit with apogee on the initial circular orbit and perigee at radius r_p. The second maneuver occurs at perigee, and the spacecraft enters a departure hyperbola with the same v_∞ as in the direct method. Show that the two maneuvers are given by

$$\Delta v_1 = \sqrt{\frac{\mu}{a_0}} - \sqrt{\frac{2\mu}{a_0} - \frac{2\mu}{a_0 + r_p}}$$

$$\Delta v_2 = \sqrt{\frac{2\mu}{r_p} + v_\infty^2} - \sqrt{\frac{2\mu}{r_p} - \frac{2\mu}{a_0 + r_p}}$$

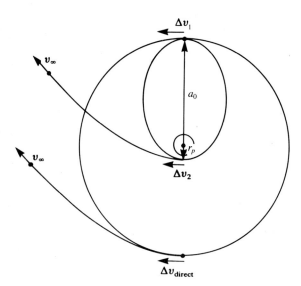

FIGURE 11.9
Olberth's escape maneuver.

4. For a Hohmann transfer to Mars, we need $v_\infty = 2.98$ km/s. Consider departing from a space station at the earth-moon L_4 point, with $a_0 = 384{,}000$ km. Calculate the required Δv for the direct method and for Olberth's method with $r_p = 6578$ km, just outside the earth's atmosphere. Which method is cheaper in total velocity change?

5. Sensitivity analysis is an important part of the planning of any mission, since the flight hardware will never exactly achieve the required burnout velocity. Taking differentials of the energy equation for the departure hyperbola,

$$E = \tfrac{1}{2}v_\infty^2 = \tfrac{1}{2}v_0^2 - \frac{\mu}{r_0}$$

show that the speed error crossing the activity sphere, dv_∞, is related to the error in the burnout speed, dv_0, by

$$dv_\infty = \frac{v_0}{v_\infty} dv_0$$

In the heliocentric portion of the flight, the Hohmann transfer ellipse has semimajor axis

$$a_t = \frac{-\mu_\odot}{2E} = \frac{-\mu_\odot}{V_1^2 - 2\mu_\odot/R_\oplus}$$

where $V_1 = V_\oplus + v_\infty$, $V_\oplus$ is the velocity of the earth, and $\mu_\odot$ is the gravitational parameter of the sun. Write the aphelion distance as $R_2 = 2a_t - R_\oplus$, and by taking differentials show that the aphelion error dR_2 is given by

$$dR_2 = \frac{\mu_\odot V_1}{(V_1^2/2 - \mu_\odot/R_\oplus)^2} dv_\infty$$

6. For a Hohmann transfer to Mars, how large an error in burnout speed, dv_0, will cause an approach error at Mars of 100 km? Modern hardware should be capable of keeping burnout-speed errors to about ± 1 m/s. If this is larger than can be tolerated with a 100-km approach window, how can the trajectory accuracy be brought within tolerable limits?

7. For planetary capture, a high-eccentricity ellipse is cheaper than a circular orbit. Also, a very high eccentricity ellipse is essentially indistinguishable near periapse from a parabolic orbit. If a maneuver is made from the approach hyperbola to a parabola at their joint periapse at distance r_p, show that the required maneuver is

$$\Delta v = \sqrt{v_\infty^2 + \frac{2\mu}{r_p}} - \sqrt{\frac{2\mu}{r_p}}$$

Attempt to minimize this with respect to r_p. Show that it has no true minimum but that the maneuver Δv decreases toward zero as $r_p \rightarrow 0$. So, the practical minimum Δv occurs when r_p is just above the sensible atmosphere of a planet.

APPENDIX
A

VECTORS AND MATRICES

A vector is a geometric object, possessing both a length and a direction. It can be thought of as a directed line segment. Since a vector exists independent of any coordinate frame, vector algebra can be derived without introducing basis vectors and the components of a vector. For example, addition of two vectors consists of placing the "tail" of the second vector at the head of the first and completing the triangle. This is usually referred to as the *parallelogram rule* of vector addition. Dynamicists recognize two different types of vectors: the bound vector and the free vector. A free vector does not have its tail tied to a particular point in space and can be moved around at will, so long as its magnitude and direction are preserved. A bound vector, on the other hand, is tied to a particular point in space as its origin. Examples of bound vectors are position vectors $\mathbf{r}$, which must begin at a specific point in space (the origin), and velocity and acceleration vectors $\mathbf{v}$ and $\mathbf{a}$, which are properties of a particular point in space. A basis vector of a coordinate frame, on the other hand, is a free vector, since it can be translated to any point.

The dot product of two vectors, $\mathbf{A} \cdot \mathbf{B}$, is defined as a scalar equal to the product of the magnitude of the two vectors and the cosine of the angle between them,

$$\mathbf{A} \cdot \mathbf{B} = |\mathbf{A}| \, |\mathbf{B}| \cos \mathbf{A} \cap \mathbf{B} \qquad (A.1)$$

where $A \cap B$ is the angle between A and B. Similarly, the cross product of two vectors is another vector, whose magnitude is the product of the magnitudes of the two vectors times the sine of the angle between them,

$$|A \times B| = |A| \, |B| \sin A \cap B \tag{A.2}$$

and whose direction is determined by the right-hand rule. The product vector is perpendicular to both A and B and in the direction the thumb of the right hand points if the fingers are first aligned with A and then curled in the direction of B. Notice, from the right-hand rule, that the cross product of two vectors in the opposite order will have the same magnitude but points in the opposite direction. Thus exchanging the order of a cross product changes the sign of the result:

$$A \times B = -B \times A \tag{A.3}$$

All the algebraic properties of vectors can be established from these rules. Two vectors are perpendicular if and only if their dot product is zero, $A \cdot B = 0$, while if the cross product of two vectors is zero, then A and B are parallel. Both the dot and cross product are distributive with respect to addition:

$$A \cdot (B + C) = A \cdot B + A \cdot C$$
$$A \times (B + C) = A \times B + A \times C \tag{A.4}$$

although the order of the cross products cannot be changed.

Multiple products obey some very useful identities. Of the four possible multiple products, two $[A \cdot B \cdot C$ and $(A \cdot B) \times C]$ are nonsense. In the scalar double product, $(A \times B) \cdot C$, the order of the dot and cross need not be specified, since one interpretation makes no sense. The cross product must be performed first, otherwise the dot product with a scalar is not defined. The scalar double product allows the dot and cross to be exchanged, however:

$$(A \times B) \cdot C = A \cdot (B \times C) \tag{A.5}$$

The other double product is the vector triple product, $A \times (B \times C)$. In this case the parentheses are necessary, since $(A \times B) \times C$ gives a completely different result. Vector double products can be evaluated with less labor by the $BAC - CAB$ identity

$$A \times (B \times C) = B(A \cdot C) - C(A \cdot B) \tag{A.6}$$

The geometric approach to vectors is very satisfying to the mathematician, but it seems foreign to the engineer accustomed to working with vectors in terms of their components in a given reference frame. Notice, however, that the geometric approach underlies the successful use of vectors at the component level. It is the geometric approach that we use when we make use of (A.1) to calculate the angle between two vectors via the dot product or to test two vectors for parallelism by calculating their cross product (A.2).

A vector is usually represented by its components in an orthonormal set of basis vectors **s**. Although it is not necessary to use orthogonal unit vectors as basis vectors, it is extremely convenient to do so. If the basis vectors are $\mathbf{s}_1$, $\mathbf{s}_2$, and $\mathbf{s}_3$, the vector **A** can be represented in component form as

$$\mathbf{A} = A_1\mathbf{s}_1 + A_2\mathbf{s}_2 + A_3\mathbf{s}_3 \tag{A.7}$$

Since the components of a vector are defined as the dot products of the vector **A** successively with the basis vectors, (A.7) is easily verified. The dot product between two vectors expressed in their components is then found by writing the operation

$$\mathbf{A} \cdot \mathbf{B} = (A_1\mathbf{s}_1 + A_2\mathbf{s}_2 + A_3\mathbf{s}_3) \cdot (B_1\mathbf{s}_1 + B_2\mathbf{s}_2 + B_3\mathbf{s}_3) \tag{A.8}$$

expanding the result, and making use of the simple nature of vector operations between orthogonal unit vectors

$$\mathbf{s}_i \cdot \mathbf{s}_i = 1 \qquad \mathbf{s}_i \cdot \mathbf{s}_j = 0 \qquad i \neq j$$

to find

$$\mathbf{A} \cdot \mathbf{B} = A_1 B_1 + A_2 B_2 + A_3 B_3 \tag{A.9}$$

The cross product of two vectors is similarly found to be

$$\mathbf{A} \times \mathbf{B} = \mathbf{s}_1(A_2 B_3 - A_3 B_2) + \mathbf{s}_2(A_3 B_1 - A_1 B_3) + \mathbf{s}_3(A_1 B_2 - A_2 B_1) \tag{A.10}$$

This is most easily remembered in the form of a determinant

$$\mathbf{A} \times \mathbf{B} = \begin{vmatrix} \mathbf{s}_1 & \mathbf{s}_2 & \mathbf{s}_3 \\ A_1 & A_2 & A_3 \\ B_1 & B_2 & B_3 \end{vmatrix} \tag{A.11}$$

expanded by minors of its first row. The other vector identities above can also be proven by expanding both sides in component form.

When a vector is expanded in its components, as in (A.7), the unit vectors really serve only to remind us which component is which. An alternate method for achieving the same end is to write a vector as a column matrix of its components:

$$\mathbf{A} = \begin{bmatrix} A_1 \\ A_2 \\ A_3 \end{bmatrix} \tag{A.12}$$

The position of the component in the matrix then carries the "which is which" information. The transpose of a matrix is defined as a new matrix with rows and columns interchanged about the principal diagonal (upper left to lower right). The transpose of a column vector **A** is a row vector $\mathbf{A}^T = [A_1, A_2, A_3]$.

The product of two matrices only has meaning if the number of rows of the first equals the number of columns of the second. Then, the element in the *i*th row and *j*th column of the product matrix is generated by multiplying each element of the *i*th row of the first matrix by the corresponding element of the

jth column in the second matrix and adding the results. Matrix multiplication does not commute: Changing the order of a matrix product not only changes the result but may make multiplying the two matrices impossible. It is simple to show, using this rule, that the dot product can be represented in matrix form as

$$\mathbf{A} \cdot \mathbf{B} = \mathbf{A}^T \mathbf{B} = \mathbf{B}^T \mathbf{A} \tag{A.13}$$

while the cross product can be generated as the product of a skew-symmetric matrix involving the components of **A** times a column vector for **B**.

$$\mathbf{A} \times \mathbf{B} = \begin{bmatrix} 0 & -A_3 & A_2 \\ A_3 & 0 & -A_1 \\ -A_2 & A_1 & 0 \end{bmatrix} \begin{bmatrix} B_1 \\ B_2 \\ B_3 \end{bmatrix} \tag{A.14}$$

Matrix products are also used to represent rotations, but this is a topic left for Chapter 1.

Besides using matrices as another way to represent vectors, they also have an algebra in their own right. The transpose of a product obeys the identity

$$(MN)^T = N^T M^T \tag{A.15}$$

The inverse of a matrix is the matrix M^{-1}, which, when multiplied by M, yields the identity matrix **1**, a matrix with 1s on the principal diagonal and 0s in the off-diagonal spaces. Inverse matrices are usually computed by gaussian elimination, a process that is computationally intensive for even moderate-sized matrices. The inverse of a matrix product satisfies

$$(MN)^{-1} = N^{-1} M^{-1} \tag{A.16}$$

which is easily proven by multiplying both sides by MN.

APPENDIX

B

LINEAR SYSTEMS

By careful application of Newton's laws, the equations of any dynamical system can be obtained. Virtually always this leads to a set of coupled, very nonlinear differential equations. There is no standard technique for solving such differential equations. If they are solvable, the solution may only be discovered after much trial and error. And if a solution is not found in closed form, the dynamicists probably are left with the feeling that it is their own lack of skill which is responsible for the failure to find a solution. (This is not always true! There really are unsolvable problems.) There are, however, certain classes of differential equations which can always be solved. By far the most important of these are systems of linear, constant-coefficient differential equations.

Constant-coefficient linear systems are not important for the simple reason that they can always be solved. Rather, they are important because they arise whenever nonlinear equations are expanded about an equilibrium point. Almost any properly engineered system will possess an equilibrium at the desired operating point. The simplest example is the aircraft, a device normally designed to be in equilibrium in straight and level flight. Solution of the linearized equations of motion will then supply stability information about the aircraft in nearly straight and level flight. If the vehicle is not stable the next step is to design a control system, and again constant-coefficient linear systems are the only class of system for which we have a complete control theory.

So, although the universe is not a constant-coefficient linear system, this special case has an importance far greater than would otherwise be expected. Since Newton's second law involves the inertial acceleration, linear dynamical systems are usually encountered in second-order form. That is, assembling the system variables into a vector **X**, termed a *state* vector, the linear system can be put into the matrix-vector form

$$M\ddot{\mathbf{X}} + C\dot{\mathbf{X}} + K\mathbf{X} = 0 \tag{B.1}$$

where the matrices M, C, and K are constants. Such systems can be solved by several techniques. The favorite method in control theory is the use of Laplace transforms. However, in this book we follow the older method of solving systems in the time domain rather than in the frequency domain.

Assume a solution to (B.1) in the form

$$\mathbf{X} = \mathbf{E}e^{\lambda t} \tag{B.2}$$

where both the vector **E** and the frequency λ may be complex. The two derivatives required are then

$$\dot{\mathbf{X}} = \mathbf{E}\lambda e^{\lambda t} \tag{B.3}$$

$$\ddot{\mathbf{X}} = \mathbf{E}\lambda^2 e^{\lambda t} \tag{B.4}$$

Substituting into (B.1) gives

$$(\lambda^2 M + \lambda C + K)\mathbf{E} = 0 \tag{B.5}$$

after canceling the factor of $e^{\lambda t}$, since this can never be zero. If the system (B.1) has N degrees of freedom (the matrices are of order N), then (B.5) represents N equations in $N+1$ unknowns: the N components of **E** and the value λ.

Obviously, the N equations (B.5) are not sufficient to determine $N+1$ unknowns. However, notice that the length of the vector **E** cannot be determined. That is, if the vector **E** is a solution, then the vector $c\mathbf{E}$ is also a solution, where c is a scalar constant. So, we can agree to seek only unit vectors **E** as solutions to (B.5). The fact that the length of **E** cannot be determined arises from the fact that the basic system (B.1) also has this property: If **X** is a solution, $c\mathbf{X}$ is a solution also.

Even when we agree to seek only unit vectors **E** as solutions to (B.5), there is still a difficulty. If the matrix in this equation possesses an inverse, then the only solution to (B.5) is $\mathbf{E} = 0$. While this is indeed a solution to the dynamics (B.1), it merely repeats the fact that $\mathbf{X} = 0$ is an equilibrium point, and it is termed the *trivial solution* for this reason. To force nontrivial solutions to (B.5), the determinant of the matrix must be zero. This leads to

$$|\lambda^2 M + \lambda C + K| = 0 \tag{B.6}$$

a 2Nth-order polynomial in λ. This is called the *characteristic equation* of the system (B.1), or in older works it is called the *secular equation*. There will be $2N$ solutions for λ, since (B.6) is a $2N$th-order polynomial after the determinant is expanded. Since this polynomial has real coefficients, the values of λ are either real or appear in complex conjugate pairs. For the system (B.1) to be stable in the linear sense, the solution (B.2) cannot grow unbounded with time. If we separate a root λ into its real and imaginary parts $\lambda = \lambda_r + i\lambda_i$, where i is $\sqrt{-1}$, (B.2) can be put in the form

$$\mathbf{X} = \mathbf{E}e^{\lambda_r t}(\cos \lambda_i t + i \sin \lambda_i t) \tag{B.7}$$

This shows that we cannot have λ values with positive real parts λ_r if the solution (B.2) is to remain bounded.

Once the $2N$ values λ_j have been obtained, they can be substituted back into (B.5), one at a time, to determine the vectors $\mathbf{E}_j$. Now, we have explicitly arranged for equations (B.5) to be singular by (B.6). Since the N simultaneous equations (B.5) are singular, one of these equations is a linear combination of the others and we cannot solve for all the components of the vector $\mathbf{E}_j$. This is expected, since the magnitude of the vector $\mathbf{E}_j$ cannot be determined. If one component is (temporarily) set equal to unity and one of equations (B.5) is discarded, the remaining equations are $N - 1$ linear equations in the $N - 1$ remaining components of the vector $\mathbf{E}_j$. These can be solved for the remaining components, and then the entire vector can be normalized, if that is desired.

Now, we have sought a solution in the form (B.2) and found $2N$ possible values of λ_j, each of which has an associated vector $\mathbf{E}_j$. Since the original system (B.1) is linear, if two solutions $\mathbf{X}_1$ and $\mathbf{X}_2$ are available, then any linear combination $c_1\mathbf{X}_1 + c_2\mathbf{X}_2$ is also a solution. The most general solution to (B.1) is then a combination of all the individual solutions

$$\mathbf{X}(t) = \sum_{j=1}^{2N} c_j\mathbf{E}_j e^{\lambda_j t} \tag{B.8}$$

where the c_j are complex constants. Evaluating these constants in terms of the initial conditions $\mathbf{X}(t = 0)$ and $\dot{\mathbf{X}}(t = 0)$ requires solving another $2N$ linear equations in $2N$ complex unknowns. If complex λ_j are present, the c_j associated with each conjugate pair of λ_j must also be complex conjugate in order to keep the solution real.

This process is an example of an eigenvalue-eigenvector (λ_j-$\mathbf{E}_j$) problem. For values of N much larger than 1, the number of calculations required is not just large but inhumane. Furthermore, some techniques used in hand calculation are not suited to the case of large N. In particular, it has been found through much experience that expanding the determinant (B.6) to obtain the characteristic polynomial for λ is a very, very bad mistake for values of N much over 3 or 4. Numerical errors in calculating the coefficients of this polynomial can make the characteristic equation totally unreliable. So, instead of performing these computations by hand, standard approaches to this problem have been developed, and an eigenvalue-eigenvector software package is an important part of the software library of any computer center. To use these packages, however, it is necessary to recast (B.1) into standard form.

Define a new state vector $\mathbf{Y}^T = [\dot{\mathbf{X}}, \mathbf{X}]$ of order $2N$. The system (B.1) can then be put into the form

$$\begin{bmatrix} 0 & \vdots & I \\ \cdots & \cdots & \cdots \\ M & \vdots & 0 \end{bmatrix} \dot{\mathbf{Y}} - \begin{bmatrix} I & \vdots & 0 \\ \cdots & \cdots & \cdots \\ -C & \vdots & -K \end{bmatrix} \mathbf{Y} = 0 \tag{B.9}$$

or

$$A\dot{\mathbf{Y}} - B\mathbf{Y} = 0 \tag{B.10}$$

where I is the identity matrix and 0 is the zero matrix. Finally, multiply (B.10) by $-A^{-1}$. The result is the equivalent system

$$\dot{\mathbf{Y}} = (A^{-1}B)\mathbf{Y} \tag{B.11}$$

Now, if we again assume a solution of the form

$$\mathbf{Y} = \mathbf{F}e^{\lambda t} \tag{B.12}$$

and substitute into (B.11), the result is

$$[(A^{-1}B) - \lambda I]\mathbf{F} = 0 \tag{B.13}$$

This is the standard form of the eigenvalue-eigenvector problem. In this form there are software packages available which can reliably and (relatively) quickly calculate the eigenvalues and eigenvectors of even large-order systems. With the eigenvectors $\mathbf{F}$ and eigenvalues λ available, the solution (B.12) can be put into the form

$$\mathbf{Y} = Fe^{\Lambda t}\mathbf{c} \tag{B.14}$$

where $\mathbf{c}$ is a vector of arbitrary constants, Λ is a diagonal matrix of eigenvalues, and F is the matrix of eigenvectors. If the initial conditons $\mathbf{Y}(t = 0)$ are specified, (B.14) becomes

$$\mathbf{Y}(t = 0) = F\mathbf{c} \tag{B.15}$$

since the exponential of a zero matrix ($t = 0$) is the identity matrix. Then, the initial conditions are found to be $\mathbf{c} = F^{-1}\mathbf{Y}(t = 0)$. Substituting this back into (B.14), the solution takes the form

$$\mathbf{Y}(t) = Fe^{\Lambda t}F^{-1}\mathbf{Y}(t = 0) \tag{B.16}$$

Define $\Phi(t, t_0)$ to be

$$\Phi(t, t_0) = Fe^{\Lambda(t - t_0)}F^{-1} \tag{B.17}$$

The matrix function Φ is called the *state transition matrix*. The final form of the solution is then

$$\mathbf{Y}(t) = \Phi(t, t_0)\mathbf{Y}(t_0) \tag{B.18}$$

where t_0 is the start time and t is the final time. The two times specified for the Φ matrix explicitly make this time dependence visible.

The matrix Φ contains all the information needed about the solution to the original system (B.1). It is itself a solution to the differential equations (B.11) and in addition obeys the identities

$$\Phi^{-1}(t, t_0) = \Phi(t_0, t) \tag{B.19}$$

$$\Phi(t_2, t_0) = \Phi(t_2, t_1)\Phi(t_1, t_0) \tag{B.20}$$

which can be proved by direct substitution of (B.18) into the identities or by the definition of the time indices on the Φ matrix itself.

C

ASTRODYNAMIC CONSTANTS

While the acceleration of gravity, $g = 32.16 \text{ f/s}^2 = 9.8 \text{ m/s}^2$, is the way the strength of gravity is cited near the earth's surface, orbital mechanics works with the product of the fundamental gravitational constant G times the mass of the primary body M. This product is usually termed $\mu = GM$ and is very accurately known. Also, orbits near the earth are often specified in terms of altitudes above the surface rather than using the radius vector directly. This forces us to add back in the mean equatorial radius of the earth to obtain radii. For historical reasons (the fact that the actual physical scale of the solar system has only been recently determined to high accuracy), orbits in the solar system are usually specified in terms of the size of the earth's orbit, the astronomical unit (AU). The following includes standard values of the gravitational parameter and length scale for both earth orbits and solar orbits. Values of μ for other objects can be found by knowing their mass in terms of the mass of either the earth or the sun and multiplying by the appropriate factor.

Gravitational parameter μ	
Earth:	$\mu_{\oplus} = 3.98601 \times 10^5 \text{ km}^3/\text{s}^2$
	$= 1.40764 \times 10^{16} \text{ ft}^3/\text{s}^2$
Sun:	$\mu_{\odot} = 1.32715 \times 10^{11} \text{ km}^3/\text{s}^2$
	$= 4.68680 \times 10^{21} \text{ ft}^3/\text{s}^2$

Length units	
Earth radius:	$R_{\oplus} = 6378.135 \text{ km}$
	$= 3963.189 \text{ mi}$
Astronomical unit:	$\text{AU} = 1.49599 \times 10^8 \text{ km}$
	$= 4.90812 \times 10^{11} \text{ ft}$
	$= 9.29568 \times 10^7 \text{ mi}$

INDEX